FORENSIC ARCHAEOLOGY

THE APPLICATION OF COMPARATIVE EXCAVATION METHODS AND RECORDING SYSTEMS

Laura Evis

ARCHAEOPRESS ARCHAEOLOGY

ARCHAEOPRESS PUBLISHING LTD
GORDON HOUSE
276 BANBURY ROAD
OXFORD OX2 7ED

www.archaeopress.com

ISBN 978 1 78491 484 4
ISBN 978 1 78491 485 1 (e-Pdf)

© Archaeopress and L Evis 2016

This book is available direct from Archaeopress or from our website www.archaeopress.com

This book is dedicated to family and friends,
my champions through it all.

Contents

List of Figures

Chapter 1 Introduction

Forensic archaeology is a relatively new sub-discipline of archaeology that emerged out of the potential to apply systematic archaeological excavation and recording to the investigation and research of crime scenes and the recovery of human remains deposited through clandestine activities (Hunter *et al*. 1994: 758). In its developed form it can be defined as a sub-disciple of archaeology that involves the application of archaeological techniques and theories to assist in the process of a forensic investigation by providing evidence for use in legal proceedings (Darvill 2008: 162). Over the past decade this sub-discipline has gained credibility internationally, through the realisation that the utilisation of archaeologists in forensic investigations resulted in an improved rate of evidence recovery and documentation (Crist 2001; Davenport and Harrison 2011; Morse *et al*. 1976; Sigler-Eisenberg 1985; Sonderman 2001). Consequently, forensic archaeologists are increasingly requested to participate in crime scene investigations both nationally and internationally, the majority of which revolve around the recovery of human remains from earth-cut graves.

Existing forensic archaeological literature is dominated by papers and reports that have been written by practitioners both in the academic and commercial sectors of the discipline. These have discussed the sub-field's development and the application of forensic archaeological techniques to various types and stages of forensic investigation, in particular, the excavation and recording of single and mass burials (Blau 2004; Blau 2005; Blau and Skinner 2005; Blau and Ubelaker 2009; Connor 2007; Ferllini 2003; Haglund 2001; Haglund *et al*. 2001; Hunter and Cox 2005; Oakley 2005; Owsley 2001; Schultz and Dupras 2008; Vanezis 1999). Whilst such publications are mostly of the same opinion regarding the value of forensic archaeology in domestic and international contexts, the question of forensic excavation and recording methodology is more divided, with scholars advocating different approaches to the examination of similar types of feature such as pits, ditches, and graves.

The greatest divergence relates to the excavation of single or mass graves. Some practitioners advocate various forms of the Arbitrary Level Excavation method (Bass and Birkby 1978; Brooks and Brooks 1984; Burns 2006; Connor 2007; Haglund *et al*. 2001; Morse *et al*. 1983; Oakley 2005; Pickering and Bachman 1997; Ruwanpura *et al*. 2006; Spennemann and Franke 1995; Stover and Ryan 2001; Ubelaker 1989). Other practitioners suggest that a form of Block Excavation be used (Larson *et al*. 2011). In contrast, some scholars state that graves should be excavated using a form of sectioning, suggesting that either the Demirant or Quadrant Excavation methods be used (Congram 2008; Dupras *et al*. 2006; Hunter 2009; Hunter and Cox 2005; Hunter *et al*. 2013; Ruffell *et al*. 2009). Alternatively, Wolfe Steadman *et al*. (2009) advocate a Vertical Slice Excavation method. Many other academics recommend that graves should be excavated using the Stratigraphic Excavation method (Blau 2005; Blau and Skinner 2005; Cheetham and Hanson 2009; Connor and Scott 2001; Hanson 2004; Hochrein 2002; Hunter *et al*. 2001; Jessee and Skinner 2005; Nuzzolese and Borrini 2010; Powell *et al*. 1997; Schultz and Dupras 2008; Skinner and Sterenberg 2005; Skinner *et al*. 2003).

These divergences emphasise the lack of standardisation in forensic archaeological practice, a problem that can be attributed to the fact that forensic archaeological practitioners have uncritically adopted techniques, principles, and practices from the wider and long-established sub-discipline of field archaeology (Drewett 1999; Hunter *et al*. 1996).

In the field of archaeology, approaches to archaeological excavation and recording vary greatly from country to country, and have evolved to their current state according to the practices advocated by practitioners and professional bodies in their country of origin, and the inherited traditions present in each. Consequently, different excavation methods and recording systems are used by different archaeological practitioners in accordance with their individual preferences. These preferences, however, are largely determined by the site types from which an archaeological practitioner has gained their academic training and experience (Carver 2009; Carver 2011: 107). Thus, if an archaeologist had gained their academic qualifications and field experience in North America, working primarily on prehistoric burial sites lacking stratigraphy, they would be more likely to advocate an Arbitrary Level method of excavation and a Unit Level method of recording (Brooks and Brooks 1984; Drewett 2000a-e; Hester 1997; Hochrein 1997; Joukowsky 1980; Pallis 1956; Pickering and Bachman 1997; Powell *et al.* 1997; Ubelaker 1989; Wheeler 1954; Willey and Sabloff 1980). Whereas, if an archaeologist had gained their academic qualifications and field experience in the United Kingdom since 1980, working primarily on urban cemetery sites with complex stratigraphy, they would be more likely to advocate a Stratigraphic method of excavation and a Single Context method of recording (Balme and Paterson 2006; Barker 1993; Hanson 2004; Harris 1979; Hester 1997; Pallis 1956; Praetzellis 1993; Roskams 2001; Wheeler 1954).

However, the adoption of a variety of different methodological approaches to the excavation and recording of single or mass graves from field archaeology into forensic archaeological practice poses a problem. The primary aim of forensic archaeological investigations is the provision of evidence to legal proceedings. Therefore, when archaeological investigations are conducted within a forensic context the methods utilised, and the evidence retrieved as a consequence of the investigation are held accountable to the admissibility regulations and the legal processes upheld by the courts in the country in which the investigation is being conducted and/or tried.

In general, the legal processes and admissibility regulations state that any techniques used during the course of a forensic investigation must have been subjected to empirical testing, peer review, have known error rates, have standards controlling their operation, and be widely accepted amongst the academic community from which they originate (Edmond 2010; Edwards 2009; Glancy and Bradford 2007; Hanzlick 2007; Klinker 2009; NAS Report 2009; Pepper 2005; Robertson 2009; Robertson 2010; Selby 2010; The Law Commission 2009; The Law Commission 2011). Therefore, if an archaeologist is to be accepted as an expert witness by legal practitioners, and the evidence retrieved as a consequence of an archaeological investigation is to be accepted by a court, the archaeologist must be able to demonstrate that the methods utilised during the course of the forensic archaeological investigation adhered to a widely accepted and tested archaeological investigatory process (Hunter and Knupfer 1996: 37). However, to date, no such forensic archaeological investigatory process has been established. Furthermore, no substantial empirical testing has been undertaken regarding archaeological excavation methods or recording systems, a point which was highlighted in a recent report published by the 'Committee on Identifying the Needs of the Forensic Sciences Community, National Research Council' (NAS Report 2009). As a consequence, much of the work undertaken through excavation by forensic archaeologists does not currently meet the admissibility regulations and legal requirements of the international court systems.

It follows that for the sub-field of forensic archaeology to continue to maintain credibility as a forensic discipline, it is necessary for the various archaeological excavation methods and recording systems advocated by practitioners within the archaeological literature to be empirically tested, error rates to be established, and a peer reviewed protocol to be formulated. This will ensure that evidence gathered as a consequence of a forensic archaeologist's participation within a forensic investigation will not be dismissed from future court proceedings as inadmissible.

Research Question

Against this background, the central question at the heart of this research is: do recognised archaeological excavation methods and recording systems used to recover evidence in forensic cases satisfy the legal tests of admissibility currently applied in the international courts?

Aim of the Research

The aim of this research is to determine which, if any, of the various excavation methods and recording systems currently used in the United Kingdom, Ireland, Australasia and North America fulfil criteria for legal acceptance and best meet the needs of forensic archaeology. Burials and the recovery of human remains are the focus of attention in this book as these represent the majority of work in this sub-field, although the research has wider implications.

Objectives of the Research

Experimental studies conducted by Chilcott and Deetz (1964), Evis (2009), Pelling (2008), Roberts (2009), Scherr (2009) and Tuller and Đurić (2006) compared archaeological excavation methods to determine the impact that different methodological approaches had upon the retrieval of artefacts and the formulation of interpretations regarding an archaeological feature's formation process. In order to expand upon these experimental studies, and to establish the most effective archaeological excavation methods and recording systems to use during forensic archaeological investigations the following objectives were pursued:

To review, analyse and compare published academic literature and published/unpublished archaeological manuals/guidelines.

To identify the origins, development and current use of archaeological excavation methods and recording systems in the United Kingdom, Ireland, Australasia and North America.

To conduct interviews with field and academic archaeologists in order to evaluate how they excavate, and why and when they choose to use particular excavation methods and recording systems.

To create a controlled experiment through which differing archaeological excavation methods, recording systems and the affect of archaeological experience can be directly compared, contrasted and measured.

To examine the affect that factors such as archaeological excavation method, archaeological recording system, and archaeological experience have on archaeological investigations, including: the quality and quantity of evidence recovered, and the consistency of interpretation(s) regarding the formation process of the site.

Chapter 2 Background

The development of archaeological excavation methods and recording systems

In order to examine the separate development and use of different archaeological excavation methods and recording systems, it is necessary to discuss the underlying theoretical principles, the practical application and the origins of the Stratigraphic Excavation method, the Demirant Excavation method, the Quadrant Excavation method and the Arbitrary Level Excavation method and their associated recording systems.

Archaeological excavation and recording

The processes of archaeological excavation and recording provide archaeologists with the methodological tools by which they are able to retrieve and record archaeological data that have been buried as a result of natural and/or man made processes (Barker 1993; Clarke 1978; Pavel 2010). This procedure is often referred to as a process of controlled destruction, which provides archaeologists with a singular opportunity to extract, define and record artefacts, features and stratigraphic units present within an archaeological site (Barker 1993; Carver 2009; Harris 2006; Roskams 2001). Once completed, this process allows archaeologists to formulate interpretations regarding the human activities that occurred at the site during a particular period in the past and changes in those activities through time (Harris 1989; Roskams 2001; Saul and Saul 2002).

Although the processes of archaeological excavation and recording are central to the recovery and documentation of archaeological data in the field of archaeology, methodological approaches vary internationally, with numerous archaeologists advocating differing methodologies for the excavation and recording of the same type of archaeological feature (Barker 1993; Carver 2009; Harris 1979; Joukowsky 1980; Kjolbye-Biddle 1975; Pallis 1956; Pavel 2010; Phillips *et al*. 1951; Roskams 2001; Wheeler 1954). This variation in methodological approaches is primarily due to the differential development of archaeological practice internationally.

The development of Stratigraphic Excavation

The principle of stratigraphy was first conceptualised in the field of geology during the eighteenth century by individuals such as Nicolaus Steno, William Smith and Sir Charles Lyell (Dirkmaat and Adovasio 1997; Gardner 1997; Hochrein 2002; Lyell 1830; Lyell 1832; Lyell 1833; Smith 1815; Smith 1816; Steno 1669). These geologists first recognised the concept of stratification; the process whereby geological layers of strata are continually laid down over time in a series of sequential layers one on top of the other (Barker 1993; Harris 1989; Renfrew 1973). Geological stratigraphy is the overall study and validation of geological stratification with a view to arranging geological strata into a chronological sequence (Barker 1993; Harris 1989; Renfrew 1973).

In archaeology, the process of stratification and the concept of stratigraphy is the same, however, archaeological stratification differs in that strata present within an archaeological site can be the result of both man-made and/or natural processes (Barker 1993; Harris 1989). There are two types of archaeological stratigraphic unit. One is referred to as a layer/deposit/fill and the other is known as an interface.

Layers/deposits/fills have a mass, a physical presence within the site, and have been deposited within the site by either natural processes such as a flood, or anthropogenic processes such as the construction of a floor or the infilling of a pit (Barker 1993; Harris 1989; Harris 2002).

Interfaces are surfaces created as a result of the deposition of a layer/deposit/fill or by the removal of an existing layer/deposit/fill. There are two types of interface – a vertical interface and a horizontal interface (Harris 1989; Harris 2002). Every layer/deposit/fill has two interfaces, one above and one below (Barker 1993: 171). Unlike layers/deposits/fills, such interfaces are not able to be excavated and removed from a site, and must be recorded where they are found (Harris 1989; Harris 2002; Dupras et al. 2006). Although the deposition of a layer/deposit/fill and the creation of an interface can be seen as being contemporaneous they are regarded and recorded as separate stratigraphic units, as there are occasions when the formation of an interface and the deposition of a layer/deposit/fill are not synchronous. A cut into existing layers/deposits/fills is an example of such an occasion where, for example, a ditch has been cut into existing layers of strata and has been left open and at a later point in time been filled in; thus the creation of the ditch and the subsequent infilling are not contemporaneous and are therefore recorded as separate stratigraphic units (Harris 1989; Harris 2002; Roskams 2001).

The process by which archaeological stratification occurs is governed by the following principles:

Principle of Superposition: in a series of layers, as originally created, the upper units of stratification are younger and the lower older, as each must have been deposited on, or created as a result of the removal of, a pre-existing mass of archaeological stratification (Harris 1989: 30).

Principle of Original Horizontality: any layer deposited in an unconsolidated form will tend towards a horizontal position; strata found with tilted surfaces were originally deposited that way, or lie in conformity with the contours of a pre-existing basin of deposition (Harris 1989: 31).

Principle of Original Continuity: an archaeological layer as originally laid down, will be bounded by a basin of deposition or will thin down to a feather edge. If any edge of the layer is exposed in vertical view, part of its original context must have been removed by excavation or erosion and its continuity must be sought or absence explained (Harris 1989: 32).

Principle of Stratigraphical Succession: any unit of archaeological stratification takes its place in the stratigraphic sequence of a site from its position between the lowest of all units which lie above it and the uppermost of all those units which lie below it and with which it has physical contact (Harris 1989: 34; Harris and Reece 1979: 113).

Principle of Intercutting: if a feature or a layer/deposit/fill is found to cut into, or across, another layer/deposit/fill it must be more recent (Darvill 2008: 438).

Principle of Incorporation: all artefactual and ecofactual material found to be contained within a layer/deposit/fill must be the same age or older than the formation of that layer (Darvill 2008: 438).

Principle of Correlation: relationships can be inferred between layers/deposits/fills that exhibit the same characteristics, contain the same range of artefactual and ecofactual material, and occupy comparable stratigraphic positions within related stratigraphic sequences (Darvill 2008: 438).

It is by adherence to these principles that Stratigraphic Excavation is conducted, during which individual stratigraphic units, are excavated in their entirety, in the reverse order in which they were deposited from the latest to the earliest, see Figure 1 (Barker 1993; Darvill 2000; Harris 1989; Harris 2002; Turnbaugh et al. 2002). In addition, any finds or samples retrieved during the excavation of a

STRATIGRAPHIC EXCAVATION METHOD

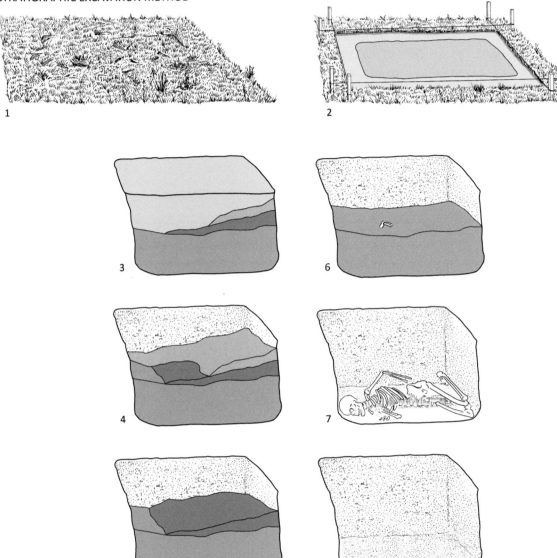

EXCAVATION PROCEDURE

- Locate and define the grave feature.
- Identify and define the first context.
- Allocate the context a unique reference number (context number).
- Plan the context.
- Excavate the context following the principles of archaeological stratigraphy.
- Repeat this procedure until all of the contexts contained within the grave feature have been defined, recorded and excavated.
- Record and lift the human remains.
- Define, inspect and record the grave cut.

FIGURE 1: ILLUSTRATES THE PROCESS OF STRATIGRAPHIC EXCAVATION © EVIS AND GODDARD 2016.

stratigraphic unit are recorded as originating from the stratigraphic unit from which they were obtained (Darvill 2000: 32). When using this excavation technique, an archaeological recording system known as Single Context Recording is used. When using this system of recording, each stratigraphic unit that is identified during the excavation is planned and recorded in three dimensions, as are any artefacts that were found or samples that were taken (Darvill 2000: 32). Consequently, this process allows archaeologists to recover and record the exact physical dimensions of each stratigraphic unit, and the relative chronology of deposition of the individual stratigraphic units present within the archaeological site (Carver 2009; Darvill 2000; Harris 1989; Harris 2002). That, in turn, enables archaeologists to determine the stratigraphic sequence of the archaeological site. If the artefacts recovered during the archaeological excavation can be allocated a specific date of manufacture, then it is possible to allocate a terminus post quem or terminus ante quem for the relative deposition of the stratigraphic units and the creation of archaeological features present at the archaeological site (Barker 1993; Harris 1989; Harris 2002).

The concept of Stratigraphic Excavation has long been recognised by archaeologists. As early as the nineteenth century archaeologists such as J.J.A Worsaae excavating Prehistoric sites in Denmark and Giuseppe Fiorelli excavating in Pompeii applied stratigraphic approaches to their excavations (Gamble 2001: 61), realising as Droop (1915:7-8) so eloquently phrased it – "if objects be taken out in a higgeldy piggeldey manner no subsequent knowledge of the history of the accumulations will be of much avail". However, widespread utilisation of the Stratigraphic method of excavation did not occur until practitioners such as Geoffrey Wainwright excavated rural archaeological sites at Tollard Royal and Gussage All Saints in Dorset, and Philip Barker excavated an urban archaeological site at Wroxeter in Shropshire (Barker *et al.* 1997; Darvill 2000: 32; Wainwright 1968; Wainwright 1979). Ever since, the Stratigraphic method of excavation has been continually advocated by numerous scholars including Barker (1997), Carver (2009), Hanson (2004), Harris (1979; 1989; 2002) and Roskams (2001) to name but a few, and has now become somewhat standardised practice for archaeological investigations conducted in the United Kingdom and Ireland.

The development of Demirant and Quadrant Excavation

The Demirant Excavation method also adheres to the aforementioned principles of archaeological stratigraphy, and is primarily used to excavate circular or negative archaeological features that are too small in size for the Quadrant Excavation method to be used. This excavation method is applied by first dividing the archaeological feature that is to be excavated into two halves. Subsequently, each half is excavated separately, as shown in Figure 2 (Carver 2009; Darvill 2008).

When using the Demirant Excavation system, the recording of stratigraphic units relies on recording the half section face that is exposed once the first half of the archaeological feature has been excavated (Carver 2009; Darvill 2008). This enables the archaeologist to document the sequence of deposition of the stratigraphic units present, at the particular point at which the half section was set up.

Similarly, the Quadrant Excavation method also adheres to the aforementioned principles of archaeological stratigraphy, and is an excavation method that tends to be used to excavate circular or negative archaeological features (Carver 2009; Darvill 2000: 31). This excavation approach is applied by dividing the archaeological feature that is to be excavated into four quarters (Atkinson 1946; Carver 2009; Darvill 2000). Subsequently, each quarter is excavated individually, with two opposing quarters being excavated first, as shown in Figure 3 (Atkinson 1946; Carver 2009; Darvill 2000). Occasionally, when utilising this excavation method, archaeologists leave either a cross-shaped or key-shaped baulk

DEMIRANT EXCAVATION METHOD

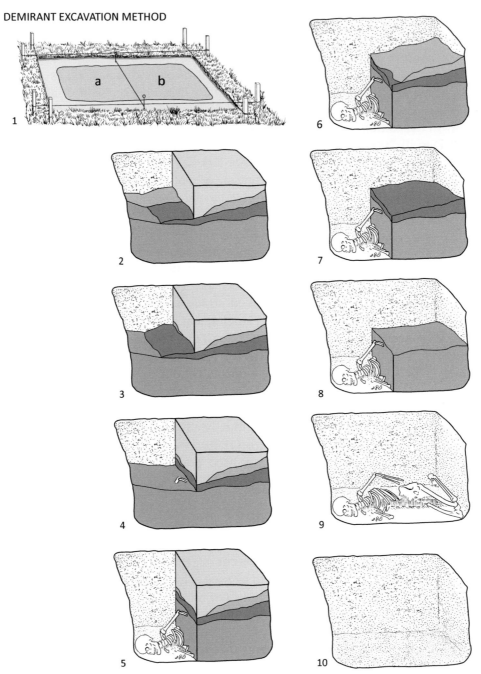

EXCAVATION PROCEDURE

- Locate and define the grave feature.
- Divide the grave feature into two halves.
- Identify and define the first context in the first half.
- Allocate the context a unique reference number (context number).
- Excavate the context up to the dividing section line following the principles of archaeological stratigraphy.
- Repeat this procedure until all of the contexts contained within the first half of the grave feature have been defined and excavated.
- Cover the exposed human remains to protect them from falling material.
- Record the exposed standing section face.
- Excavate the second half of the grave feature following steps described above.
- Record and lift the human remains.
- Define, inspect and record the grave cut.

FIGURE 2: ILLUSTRATES THE PROCESS OF DEMIRANT EXCAVATION © EVIS AND GODDARD 2016.

across the midline of the feature (Atkinson 1946; Carver 2009; Darvill 2000). However, the manner in which this excavation approach is applied is determined by the preferences of the archaeologist responsible for excavating the archaeological feature.

When using the Quadrant Excavation method, the recording of the stratigraphic units present relies not on plans, but on recording the long section and half section profiles that are exposed during the excavation process (Atkinson 1946; Carver 2009; Darvill 2000). This enables the sequence of deposition across the entire archaeological feature to be documented and reconstructed after the excavation has been completed.

The use of sectioning as a method to excavate, record and interpret archaeological features can be dated to the early 1920s and the work of Cyril Fox, who applied this approach to his excavation of Bronze Age round barrows in Wales (Darvill 2000: 31). Subsequently, these methodological approaches have continued to be used regularly in field archaeology to deal with both circular and cut features in the United Kingdom, Ireland and Continental Europe (Cox 2005; Darvill 2000; Darvill 2008; Hunter and Dupras *et al.* 2006; Litherland *et al.* 2011; Ruffell *et al.* 2009).

QUADRANT EXCAVATION METHOD

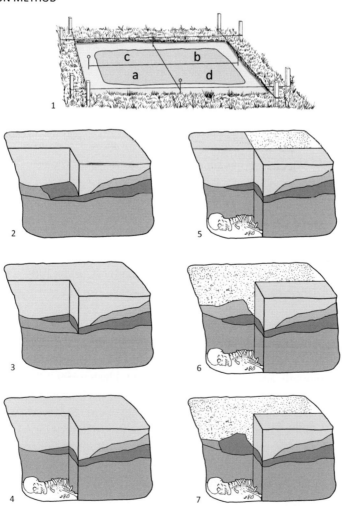

FIGURE 3: ILLUSTRATES THE PROCESS OF QUADRANT EXCAVATION © EVIS AND GODDARD 2016. (1)

QUADRANT EXCAVATION METHOD (continued)

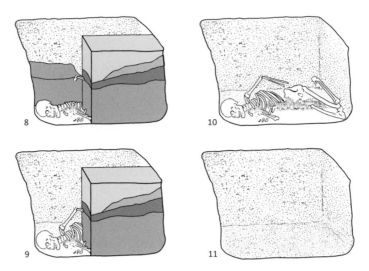

EXCAVATION PROCEDURE

- Locate and define the grave feature.
- Divide the grave feature into four equal quarters.
- Identify and define the first context in the first quarter.
- Allocate the context a unique reference number (context number).
- Excavate the context up to the dividing section lines following the principles of archaeological stratigraphy.
- Repeat this procedure until all of the contexts contained within the first quarter of the grave feature have been defined and excavated.
- Cover the exposed human remains to protect them from falling material.
- Record the two exposed standing section faces.
- Excavate the opposing quarter as outlined above.
- Cover the exposed human remains to protect them from falling material.
- Record the two newly exposed standing section faces.
- Excavate the two remaining quarters following steps described above making sure to cover the human remains after the third quarter has been excavated.
- Record and lift the human remains.
- Define, inspect and record the grave cut.

FIGURE 3: ILLUSTRATES THE PROCESS OF QUADRANT EXCAVATION © EVIS AND GODDARD 2016. (2)

The development of Arbitrary Level Excavation

The excavation method known as Arbitrary Level Excavation is also referred to by other names including – the Unit Level method, the Planum technique and Metrical Stratigraphy (Carver 2009; Darvill 2000; Darvill 2008; Hanson 2004; Hester *et al.* 1997; Lucas 2001).

It is an excavation method whereby an archaeological site is divided into individual excavation units, which are usually 1m x 1m in area (Darvill 2000: 31). The area within these excavation units is then excavated in a succession of separate arbitrary levels, each of which is completed before the next is excavated; such levels are usually 5cm, 10cm or 20cm in depth, see Figure 4 (Darvill 2000; Hester 1997:88). Each arbitrary level is regarded as a separate unit of stratification and artefacts retrieved from each arbitrary level are placed into a depositional sequence in relation to the arbitrary level from which they were recovered, in accordance with the principle of superposition (Darvill 2000; Hanson 2004). When using the Arbitrary Level Excavation method a recording system known as Unit Level Recording is used. When using this system of recording, archaeologists complete plans of each of the arbitrary levels that were excavated in which any relevant soil colours, structures and deposits present are defined and drawn (Darvill 2000: 31; Hester 1997).

The process of Arbitrary Level Excavation originated in the Americas in the early twentieth century, as a result of the increased formalisation of American field archaeology and the rejection of the formally dominant concept

ARBITRARY LEVEL EXCAVATION METHOD

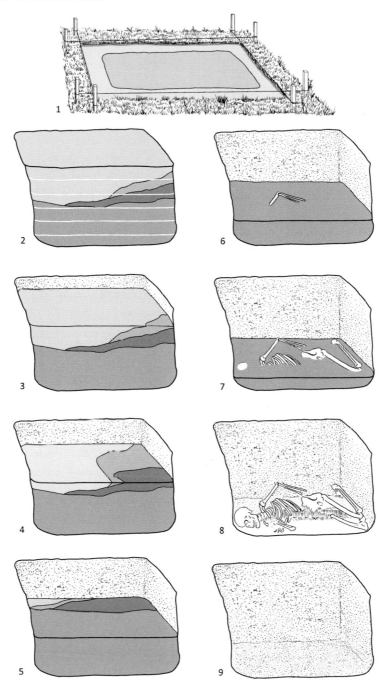

EXCAVATION PROCEDURE

▪ Locate and define the boundaries of the grave feature.
▪ According to professional preference, either create an excavation unit by marking out an area 30cm beyond the boundary of grave feature, or, maintain the boundaries of the grave feature.
▪ Define and record the exposed levelled area.
▪ Excavate the soil within the grave feature or excavation unit to a predetermined levelled spit depth (usually, 5cm or 10cm).
▪ Repeat previous two steps above until the human remains have been exposed.
▪ Circumscribe around the human remains, record and lift them.
▪ Define, inspect and record the grave cut (if it has been left intact).

FIGURE 4: ILLUSTRATES THE PROCESS OF ARBITRARY LEVEL EXCAVATION © EVIS AND GODDARD 2016.

of evolutionism; the belief that non-European peoples lacked any form of identifiable history (Lucas 2001). As a result, increased emphasis was placed upon the retrieval of artefactual evidence from archaeological sites, and the establishment of chronological sequences of material culture assemblages associated with non-European cultures that had existed prior to the European migration to the Americas in the fifteenth century (Browman and Givens 1996; Lucas 2001; Lyman and O'Brien 1999; Willey and Sabloff 1980).

As chronological sequencing of material culture assemblages found at archaeological sites came to dominate the concerns of American archaeologists, existing excavation techniques were adapted in what has come to be referred to as the 'stratigraphic revolution' in which the geological principles of stratification and stratigraphy became widely utilised to interpret archaeological sites, and provided a method by which archaeologists could contrast, compare and prove cross-site material culture sequences (Lucas 2001; Willey and Sabloff 1980). However, unlike in the United Kingdom, Ireland and continental Europe the methodological approaches used in order to understand stratigraphic sequences differed, in that Stratigraphic Excavation usually meant the recognition of evident stratification post-facto or via vertical section as the excavation proceeded in arbitrary levels, rather than excavation by recognition and adherence to identifiable stratigraphic units (Browman and Givens 1996; Lucas 2001).

The adoption of the Arbitrary 'Stratigraphic' Excavation approach has been attributed to the American archaeologists - Nels C Nelson (1916) excavating in South Western America and Manual Gamio (1930) excavating in Mexico. It is widely believed that Nelson adopted this excavation approach after excavating for a season at Castillo in Spain in 1913, during which he witnessed the process of Stratigraphic Excavation (Browman and Givens 1996; Lucas 2001). Subsequently, when he returned to North America it is thought that he was unsure of how to apply the Stratigraphic Excavation method he had utilised in Castillo to his excavations at San Cristóbal (Browman and Givens 1996; Lucas 2001). This resulted in Nelson setting up excavation units within which he excavated using arbitrary one-foot levels and interpreted the stratigraphic sequence via vertical section (Browman and Givens 1996; Lucas 2001).

This Arbitrary Level Excavation method, pioneered by Nelson, was subsequently adopted by both North American and Latin American archaeologists as a standard excavation method and continues to be advocated by archaeologists within American excavation guidance literature (Brooks and Brooks 1984; Browman and Givens 1996; Hester 1997; Spennemann and Franke 1995; Ubelaker 1989).

Despite the widespread adoption of the Arbitrary Level Excavation methodology throughout North and Latin America, not all practitioners utilised this method of excavation. Archaeologists such as George Pepper, Alfred Kidder and Max Uhle preferred to excavate using natural stratigraphy, excavating in accordance to the identifiable stratigraphic boundaries present at archaeological sites (Browman and Givens 1996; Lucas 2001), believing that excavating arbitrarily "would have resulted in the splitting or cross-cutting of strata" (Kidder and Kidder 1917: 340). Although these archaeologists were contemporaries of both Nelson and Gamio, it appears that this excavation method was largely disregarded (Browman and Givens 1996; Lucas 2001). Archaeologists of the period appeared to prefer the Arbitrary Level method of excavation as it was simpler to employ and because the visible layers of stratification evident in North and Latin America tended to be the result of natural geological processes rather than anthropogenic processes (Browman and Givens 1996).

Nevertheless, in more recent American archaeological excavation guidance literature, excavation via natural stratigraphy is increasingly advised whenever possible (Hester et al. 1997). Although, as Lucas (2001) points out, despite the use of natural stratigraphy the layer is still regarded as the primary element of a site; thus features such as pits are often described as disturbances to the stratigraphy rather than stratigraphy in their own right (Joukowsky 1980: 150-7; Lucas 2001: 60), which again differentiates the Natural Stratigraphic Excavation method from the Stratigraphic Excavation method used in the United Kingdom, Ireland and continental Europe.

The adaptation and application of archaeological methods to forensic investigations

In order to effectively discuss the background to the application of forensic archaeology to the investigation of burials and mass graves numerous factors must be considered. Firstly, one must discuss the various roles and responsibilities that may be allocated to a forensic archaeologist during the process of a forensic investigation. One must also evaluate how the differential development of the discipline in the United Kingdom, Ireland, Australasia and North America has resulted in discrepancies in how the discipline's practitioners have come to define the discipline, and the impact that this has had upon the archaeological excavation methods that they have chosen to adopt. Finally, one must consider what qualifications and what level of experience a forensic archaeologist is expected to have gained in order to be qualified to participate in a forensic investigation.

Crime scene to court

The acceptance and utilisation of forensic archaeologists in criminal investigations of burials and mass graves has been a relatively recent development (Hunter *et al.* 1994: 758). Occasions during which forensic archaeologists are requested to participate in a forensic investigation in the United Kingdom, Ireland and Australasia are relatively rare, compared with their North American counterparts (Hunter and Cox 2005). This is largely due to the fact that these regions have lower homicide rates (Blau 2004; Donlon 2009; Skinner and Bowie 2009).

In recent years forensic archaeologists have been increasingly integrated into crime scene operations. This is due to greater public awareness of the discipline, and the fact that if a forensic archaeologist is not used, a defence barrister may call upon a forensic archaeologist to testify that the methodologies utilised were unsuitable, with the aim of having the evidence that was collected by unspecialised investigators dismissed from the court proceedings (Hunter and Dockrill 1996: 43).

However, the decision on whether to employ a forensic archaeologist in a crime scene investigation is left to the discretion of the case's Senior Investigator (Hunter and Cox 2005; Menez 2005). Nonetheless, if a forensic archaeologist is brought in to assist with a forensic investigation, there are several requirements that the archaeologist must fulfil.

Firstly, upon arrival at the crime scene the forensic archaeologist must make sure that their entrance is recorded in the crime scene log (Pepper 2005: 109). Furthermore, due to the forensic nature of the investigation, the archaeologist must ensure that they are wearing a disposable crime scene suit and have either brought, or been provided with, sterile or new equipment in order to prevent contamination (Hunter and Knupfer 1996: 31; Pepper 2005: 177). It is advisable that the forensic archaeologist also has their sterile equipment photographed prior to starting any investigative work, to satisfy the court that no cross contamination may have occurred. It is then that the forensic archaeologist will be briefed by the Senior Investigator. Such briefings usually outline the background of the case and involve a discussion of what evidence the investigators believe may be present at the crime scene (Pepper 2005).

If the investigators require the archaeologist to identify the location of a clandestine grave, the archaeologist will then establish a grid, and then proceed to inspect the site in order to identify and record the presence of surface indicators associated with the creation of a grave (Dupras *et al.* 2006: 38; Killam 2004; Rodriguez and Bass 1985).

There are four types of surface indicators that may be present at an interment site. The first indicator is changes to vegetation (Hunter and Cox 2005: 30; Hunter and Martin 1996; Killam 2004). These changes are induced by the process of digging the grave and placing human remains within it. Depending on the species of plants present at the gravesite, the increase in nutrients, provided by the decomposition of the body, will either promote plant growth or retard plant growth on top of and around the grave (Hunter and Cox 2005; Killam 2004; Morse *et al*. 1983).

The second indicator is the presence of disturbed soil. Soil disturbances occur due to the fact that "one can never dig in the ground and put the dirt back exactly as nature had put it there originally" (Bass and Birkby 1978: 6). Such disturbances take the form of colour or textural differences between the disturbed soil contained within the grave and the undisturbed soil surrounding it (Dupras *et al*. 2006; Hunter and Cox 2005: 32; Killam 2004; Morse *et al*. 1984). Additionally, due to the fact that space is taken up within the grave by the placement of human remains, perpetrators are often left with an excess of soil after backfilling. Consequently, this soil is often piled on top of the grave resulting in the creation of a soil mound (Hunter and Cox 2005: 31; Killam 2004).

The third indicator is the presence of depressions on the ground surface. Primary depressions are triggered by the settlement of soil contained within the grave over the course of time (Dupras *et al*. 2006; Killam 2004). This results in the grave's surface sinking beneath that of the surrounding undisturbed ground. Secondary depressions are caused by the collapse of the victim's abdominal cavity during decomposition (Morse *et al*. 1983). When this occurs, the soil that had been resting on top of this cavity falls into it, resulting in the creation of a small secondary depression on the grave's surface, over the area where the victim's abdomen once was (Killam 2004: 35).

The fourth indicator is the presence of animal activity (Dupras *et al*. 2006; Rodriguez and Bass 1985). When human remains are contained within a grave it often attracts scavengers. These scavengers attempt to gain access to the remains by burrowing into the grave, resulting in signs of digging around the area in which the grave is located (Dupras *et al*. 2006; Killam 2004: 36).

In addition to surface indicators, the forensic archaeologist may also employ geophysical surveying techniques such as: earth resistivity, magnetometry, electromagnetic systems, or ground penetrating radar (Bevan 1991; Cheetham 2005; Davenport 2001a; Davenport 2001b; Dupras *et al*. 2006; Killam 2004). These techniques allow the archaeologist to measure and map the properties of the investigative area's sub-surface. Such properties include its "acoustic, electrical, magnetic and electromagnetic" characteristics (Cheetham 2005: 65). It is by identifying regions within the survey area that have measurably different properties from those surrounding it that the archaeologist is able to detect possible grave locations. These regions are commonly referred to as 'anomalies' or 'hot-spots' (Cheetham 2005: 65).

If any surface indicators or sub-surface anomalies are identified during the search phase of the investigation they are drawn, photographed and/or videoed, and digitally recorded using a Total Station, or its technological equivalent (Hunter and Dockrill 1996: 52; Menez 2005). Moreover, if potential evidentiary artefacts are located they are also recorded using the same techniques. Once recorded, these artefacts are then signed over to the investigator who is in charge of exhibits, so that they can be packaged and labelled in the appropriate manner. This ensures that the chain of custody is maintained, and that the artefactual evidence is admissible in court (Hunter and Cox 2005).

Once potential grave locations have been identified, the forensic archaeologist must then conduct an invasive evaluation in order to confirm or discount the presence of a grave. The invasive evaluation involves two steps. The first step is to excavate a test pit, away from the potential grave locations, in order to evaluate the natural geological composition of the site (Congram 2008: 794; Wright *et al*. 2005:

147). The findings from this test pit will then be used to identify areas in which the natural geology of the site has been disturbed, potentially due to the excavation and backfilling of a grave.

The second step is to excavate rectangular sondages across the width of each of the potential graves in order to identify disturbances to the natural geology of the site, and to locate the presence of human remains (Hunter and Cox 2005). If human remains are found, the forensic archaeologist can then confirm to the Senior Investigator that they have identified the grave. It is at this point that the Coroner or Medical Examiner is notified that human remains have been found, and their permission is requested to continue the investigation (Hunter and Cox 2005; Hunter and Knupfer 1996).

Having obtained permission to continue, the forensic archaeologist then proceeds to excavate the grave. There are a variety of approaches to excavating such features. Practitioners such as Brooks and Brooks (1984), Burns (2006), Haglund *et al.* (2001), Oakley (2005), Pickering and Bachman (1997), Ruwanpura *et al.* (2006), Spennemann and Franke (1995), Stover and Ryan (2001), and Ubelaker (1989) state that access trenches should be excavated around the outline of the identified grave cut, and that excavation should proceed in a series of arbitrarily defined spits, removing both the material contained within the grave feature and the access trenches surrounding it. Any human remains or artefacts found during the excavation process should be left upon a soil pedestal until they have been recorded. Such items can then be removed and transferred to the relevant processing facility.

Similarly, Bass and Birkby (1978), Connor (2007), and Morse *et al.* (1983) recommend that the grave should be excavated in a series of arbitrarily defined spits, and that any human remains or artefacts uncovered during the excavation should be left upon a soil pedestal. However, unlike the former set of academics they suggest that the grave cut should be maintained throughout the excavation process.

In contrast, Larson *et al.* (2011) suggest that a trench should be excavated around the outline of the grave cut until the bottom of the grave has been reached. The grave should then be undercut and a forensic platform slid underneath the human remains and the in situ grave fill. This 'grave block' is then transferred into a specially designed evidence bag and shock-resistant container. The 'grave block' is then transported back to the laboratory and excavated in controlled laboratory conditions.

Unlike the previous academics Dupras *et al.* (2006), Hunter and Cox (2005), Hunter *et al.* (2013), Litherland *et al.* (2012), and Ruffell *et al.* (2009) state that the grave should be excavated using the Demirant Excavation method. When using this method, the excavation process is contained to within the boundaries of the grave cut. Once the outline of the grave has been defined, the grave is then divided into two halves, each of which are excavated separately. Likewise, Hunter (2009) suggests that the grave cut should be maintained, and that the feature should be divided and excavated following the Demirant Excavation method. However, he recommends that a baulk, along the mid-line of the grave, should be left in place until the end of the excavation process.

Alternatively, Congram (2008) recommends that the Quadrant Excavation method be used. When using this method the grave cut is maintained, but the grave feature is divided into four different quarters, each of which are excavated separately.

In contrast, Wolfe Steadman *et al.* (2009) advocate a method in which the grave's boundary is maintained, but the grave's fill is excavated by vertically slicing through the fill from one side of the grave to the other. The size of the individual slices is dependent upon the size of the grave that is being excavated, and is left to the discretion of the archaeologist.

Conversely, Blau (2005), Blau and Skinner (2005), Cheetham and Hanson (2009), Connor and Scott (2001), Hanson (2004), Hochrein (2002), Hunter *et al.* (2001), Jessee and Skinner (2005), Nuzzolese and Borrini (2010), Powell *et al.* (1997), Schultz and Dupras (2008), Skinner and Sterenberg (2005), and Skinner *et al.* (2003) argue that the grave should be excavated using the Stratigraphic Excavation method. When using this method, the grave cut is maintained, and each of the fills contained within the grave are defined and excavated in isolation, in the reverse order of their deposition – "last in = first out" (Congram 2008: 794).

During the excavation, it is crucial that the forensic archaeologist identifies and records any geotaphonomic features that may be present. These features are categorised into six groups: stratification, tool marks, bioturbation, sedimentation, compression/depression, and internal compaction (Hochrein 1997; Hochrein 2002). Such features can indicate: the method by which the grave was constructed, whether the grave had been left open prior to the placement of human remains within it, whether the perpetrator(s) stood in the grave when constructing it, whether the victim was alive when placed into the grave, and whether the grave had been disturbed by scavengers during the time period between the human remains interment and subsequent excavation (Hochrein 1997; Hochrein 2002; Menez 2005; Wright *et al.* 2005). Such features can therefore provide evidence to determine whether the crime was pre-meditated, which will have significance when the case goes to court, and assist the forensic archaeologist in reconstructing the sequence of events that occurred from the point at which the grave was created up until its excavation (Cox *et al.* 2008; Hunter and Cox 2005).

In addition to geotaphonomic evidence, the grave may also contain entomological, toxicological, and botanical evidence. If such evidence is identified, it is best practice to request assistance from the relevant specialist to ensure that the correct sampling and collection procedures are adhered to (Cox *et al.* 2008; Dirkmaat and Adovasio 1997; Hunter *et al.* 2001).

In regards to the human remains, once they have been uncovered it is a legal requirement for a medical professional to declare that "life is extinct" (Pepper 2005: 111). It is also the point at which a forensic pathologist is required to attend the scene in order to supervise the removal of the human remains and any sampling (Hunter and Cox 2005: 6; Pepper 2005: 111). If the remains are skeletonised, a forensic anthropologist may also be requested to assist with the removal and recording of the skeletal remains (Hunter and Cox 2005: 7; Hunter *et al.* 1996).

It is essential that the entire excavation process is documented either in written form – site journals or pre-prepared pro-formas, or in a digital format – digital photographs or digital videos, preferably both (Cheetham *et al.* 2008; Dirkmaat and Adovasio 1997; Hunter and Cox 2005; Pepper 2005). Such records should contain: descriptions of the stratigraphic contexts associated with the grave, a list of the evidence that was found in the grave as well as the exact spatial location in which each piece was found, and a plan of the grave structure (Dupras *et al.* 2006; Hunter and Cox 2005; Hunter *et al.* 1996; Menez 2005). This recording procedure is important, as the process of excavating a grave is destructive, and the records made during the excavation are the only remaining evidence of what was contained within the grave, and, provide the means by which the forensic archaeologist can construct and support their narrative of the sequence of events that occurred from the point at which the grave was dug to the time at which it was excavated (Adams and Valdez 1997; Crist 2001; Dupras *et al.* 2006: 107; Saul and Saul 2002).

Having completed the excavation the forensic archaeologist will then be required to write a report. This report will be submitted to the court as evidence. Such reports are required to:

Contain the expert's qualifications, relevant experience and accreditation.

Provide references of any literature that the expert has relied upon whilst making the report.

Contain a statement summarising all of the facts that were given to the expert which were influential in forming their opinion.

Outline which of the facts stated in the report are within the expert's expertise.

State who carried out any examination, measurement, test, or experiment, which the expert has used in the report, and provide the qualifications, relevant experience and accreditation of that person.

Provide a summary of the expert's findings.

If there is a range of opinion regarding matters considered in the report, the expert should summarise these opinions, and give the reasons for their own.

Contain a summary of the conclusions reached.

Include a statement which explains that the expert has understood their duty to the court, and that they have complied and will continue to comply with this duty throughout the proceedings.

Include the same declaration of truth as is found in a witness statement.

Having provided the court with the report the forensic archaeologist may then be required to appear in court as an expert witness. This, however, is dependent upon the defendant's plea. If they plead 'guilty' the forensic archaeologist's report will not be contested in the courtroom (Dilley 2005: 196). Whereas, if the defendant pleads 'not guilty', on the basis that the expert evidence is inaccurate, or the investigation was mishandled in some manner, then the forensic archaeologist may be required to give evidence to the court as an expert witness (Dilley 2005: 196). In this role, the forensic archaeologist must assist the court, including barristers, judges, and juries, to understand the technical or scientific data contained within their report that the court deems to be beyond the understanding of the average lay person (Dilley 2005; Hampton 2004). In order to be effective during this process, the forensic archaeologist must ensure that they are "fully cognisant with the contents of their report" (Dilley 2005: 197). This will ensure that the forensic archaeologist is able to present and defend their findings in a convincing and credible manner, during both the examination-in-chief and cross-examination stages of the court proceedings.

International perspectives

Forensic archaeology can be defined as the practical application of archaeological methods and theories to aid in the identification, investigation and interpretation of sites of legal interest (Cox *et al.* 2008: 216; Dupras *et al.* 2006: 5; Killam 2004; Oakley 2005). Most often, forensic archaeological work is focused upon the recovery of human remains (Borić *et al.* 2011; Gojanović and Sutlović 2007; Jankauskas *et al.* 2005; Komar and Buikstra 2008; Rainio *et al.* 2001). However, in recent years, the discipline's practitioners have become involved in criminal investigations that have sought to recover buried stolen goods, weaponry, and ransom money (Cheetham and Hanson 2009; Oakley 2005). Forensic archaeologists have also assisted in civil investigations in which land boundaries have been under dispute. It is through involvement in such cases that the discipline of forensic archaeology has become widely accepted as a specialist forensic discipline, both domestically and internationally.

However, despite the disciplines wide scale acceptance within the forensic community, differences in terms of the origins and subsequent development of the discipline internationally, have resulted in discrepancies in how its practitioners have come to define the sub-field of forensic archaeology and the manner in which they have utilised archaeological methodologies during the process of a forensic investigation.

It was in North America where the beneficial application of archaeological techniques and theories to forensic investigations was first recognised by anthropologists such as: Krogman (1943), Morse *et al.* (1976), and Snow (1982) who highlighted how the application of archaeological methodologies to the excavation and recording of graves would result in a better understanding of the grave's formation process and a greater rate of evidence recovery. As with traditional archaeology, universities providing training facilities in forensic archaeology in North America are subsumed under the auspices of anthropology departments, resulting in individuals training in the field of forensic anthropology to regard themselves as "cross-trained between physical anthropology and archaeology" (Scott and Connor 2001: 101).

The extent to which the discipline has developed in North America can be gauged by the fact that North American forensic anthropologists have become increasingly involved in homicide investigations, particularly in cases where investigators are attempting to locate and recover buried human remains (Connor 2007; Dupras *et al.* 2006; Larson *et al.* 2011; Schultz and Dupras 2008; Skinner and Bowie 2009). Moreover, such practitioners are now included in the majority of State and Federal mass disaster teams, and have assisted in recovery operations such as: the World Trade Centre investigation in New York in 2001, and the Rhode Island night club fire investigation in 2003 (Blau 2005).

Due to the increase in interest in forensic anthropology and archaeology universities such as: the University of Montana, Texas State University, Boston University, the University of Tennessee, the University of Florida, the University of Indianapolis, Adelphi University, Mercyhurst Archaeological Institute, St. Mary's University, the University of Toronto, the University of Windsor, and Simon Fraser University have developed courses specialising in forensic archaeology and anthropology. Furthermore, forensic archaeology field schools have been developed by establishments such as Florida State University and the American Academy of Forensic Sciences (AAFS) in conjunction with local law enforcement departments in order to train police investigators in forensic archaeological techniques (Crist 2001; Hunter *et al.* 1996: 21; Schultz and Dupras 2008).

Notably, North American forensic anthropologists are not only being used domestically. Due to the recognition of the disciplines capabilities in regards to the recovery of human remains, numerous organisations have been founded including: Physicians for Human Rights (PHR) in 1986, Necrosearch International in 1991, the International Commission on Missing Persons (ICMP) in 1996, and the POW/MIA Accounting Command (JPAC) in 2002, all of whom employ forensic anthropologists to recover the remains of missing persons, whether it be working in collaboration with organisations such as the United Nations, as was seen with the use of North American forensic anthropologists in the missing persons recovery operations during the International Criminal Tribunal for the Former Yugoslavia (ICTY) and the International Criminal Tribunal for Rwanda (ICTR), or recovering the remains of American soldiers who have been killed in action (Dupras *et al.* 2006; Hunter and Cox 2005; Juhl 2005; Skinner and Bowie 2009: 94).

Similarly, in the United Kingdom and Ireland, forensic archaeological expertise has been "increasingly utilised in the search, location and recovery of materials associated with buried crime scenes" (Davenport and Harrison 2011; Oakley 2005: 170; Ruffell *et al.* 2009). Unlike North America, forensic archaeology in the United Kingdom and Ireland is not regarded as a sub-discipline of forensic anthropology, but as a discipline in its own right, concerned primarily with excavation and field skills (Blau and Ubelaker 2009; Hunter and Cox 2005; Hunter *et al.* 1996). Thus, when forensic archaeologists are invited to partake in forensic investigations, they are first and foremost archaeologists, rather than physical anthropologists (Scott and Connor 2001). The widespread acceptance of forensic archaeology in the United Kingdom owes much to its formal accreditation by the Council for the Registration of Forensic Practitioners (CRFP), allowing for its acceptance amongst other forensic fields that have long been utilised in forensic investigations (Ebsworth 2000; Kershaw 2001). However, since the disbandment of the CRFP in 2009, formal accreditation of the discipline and its practitioners has been transferred to the Chartered Institute for Archaeologists Forensic Archaeology Special Interest Group, which was formed in 2011, which accredits its members according to their experience, publications and qualifications.

The increase in awareness of the discipline of forensic archaeology also resulted in the formation of university post-graduate courses at: Bournemouth University, Cranfield University, the University of Bradford, and the University of Dundee, each of which focus on training individuals in the principles and practices of forensic archaeology (Menez 2005). In addition, increasing involvement of forensic archaeologists in criminal investigations dealing with the recovery of human remains has resulted in the integration of forensic archaeological methods into the basic training programmes of Police Search Advisors (POLSA) and Crime Scene Investigators (CSI) (Hunter and Cox 2005: 6).

The presence of forensic archaeological expertise in the United Kingdom and Ireland has also led to the formation of organisations such as: the Centre for International Forensic Assistance (CIFA) in 2001, and the International Centre of Excellence for the Investigation of Genocide (INFORCE) in 2001. Both of these organisations employ forensic archaeologists to assist with the recovery of human remains in investigations of human rights abuses, war crimes, genocide, and mass disasters (Cox 2009: 32; Hunter and Cox 2005: 23).

In Australasia, the situation is somewhat different. Despite Australian forensic archaeologists becoming involved in the international investigations of mass graves at Serniki (1990) and Ustinovka (1991) in the Ukraine (Wright 1995), and in domestic contexts such as the search for Samantha Knight (McDonald 1999) and the exhumation of Sally Anne Huckstepp (Donlon 2009), and universities such as: James Cook University, the University of Adelaide, Flinders University, the University of Queensland, the University of New England, and the Australian National University offering degree programmes in forensic archaeology (Colley 2004; Donlon 2009), the expertise of forensic archaeologists in domestic cases is relatively underutilised as an investigative tool in Australia, New Zealand, New Guinea and the surrounding Pacific Islands (Bedford *et al.* 2011; Blau 2004; Blau 2005; Donlon and Littleton 2011; Oakley 2005; Tayles and Halcrow 2011). Moreover, the discipline of forensic archaeology is also yet to be recognised by the Senior Managers of Australian and New Zealand Forensic Laboratories (SMANZFL) and its associated Scientific Advisory Group (SAG) (Donlon 2009). However, in the relatively few cases in which Australasian practitioners have been invited to assist in forensic investigations, forensic archaeologists and forensic anthropologists have been utilised separately.

Despite the underutilisation of forensic archaeology in Australasia, the discipline's practitioners have established an organisation known as Australian – Forensic Archaeology Recovery (Aus-FAR) in 2003, whose members provide forensic archaeological expertise when required (Blau 2005: 20). Additionally, Australian forensic archaeologists have become involved in the creation of Australian Disaster Victim Identification (DVI) programmes and have been added to their list of mass disaster responders. This has resulted in Australian forensic archaeologists becoming involved in disaster recovery operations such as: the Bali Bombings in 2002, and the Thailand Tsunami in 2004 (Blau 2005).

Due to the differential development and categorisation of forensic archaeology internationally, it is evident that practitioners have come to define their roles within forensic investigations differently, resulting in a variety of approaches being utilised during the recovery of human remains in forensic contexts. Congram (2008) and Tuller and Đurić (2006) have argued that those individuals that classify themselves as forensic anthropologists and archaeologists, such as in North America, tend to be body-centric in their approach to the excavation of human remains, as their primary goal is to retrieve the remains in order to establish an accurate biological profile and determine the manner and/or the cause of death.

In contrast, in the United Kingdom, Ireland and Australasia, where practitioners define the sub-field of forensic archaeology as a separate speciality, and are employed solely as forensic archaeologists, their approach to the excavation of a forensic grave differs (Menez 2005; Thompson 2003). Scholars such as Congram (2008), Hanson (2004) and Tuller and Đurić (2006) state that such practitioner's primary focus during the excavation is to understand the grave's formation process in order to interpret and reconstruct the sequence of human activity surrounding the construction of the grave, during which human remains form a part of the material evidence collected, and are viewed as a depositional event within the overall sequence of the grave's formation process (Connor and Scott 2001; Saul and Saul 2002; Tuller and Đurić 2006).

Qualifications and experience

As a consequence of participating in a forensic investigation a forensic archaeologist may be required to appear in a court as an expert witness (Dilley 2005; Pepper 2005). In such circumstances, the forensic archaeologist will have to demonstrate to the court that they have gained enough experience and relevant qualifications in the sub-field of forensic archaeology to qualify to be considered as an expert by the court (Dilley 2005; Hunter *et al.* 1996).

In terms of the appropriate level of experience and qualifications that a forensic archaeologist should have in order to participate within a forensic investigation and qualify as an expert witness in court, scholars are unanimous in their claims that a forensic archaeologist should have gained at least an undergraduate degree in archaeology or anthropology, and be an experienced field archaeologist (Cheetham and Hanson 2009; Cox *et al.* 2008; Crist 2001; France *et al.* 1992; Hunter *et al.* 2001; Sigler-Eisenberg 1985; Spennemann and Franke 1995; United Nations 1991; Wright *et al.* 2005). Such qualifications and experience, they argue, will ensure that when a forensic archaeologist is called upon to participate within a forensic investigation they will be able to successfully evaluate, adapt and implement excavations in a variety of contexts rapidly, and record as second nature (Connor and Scott 2001; Hunter and Cox 2005).

However, despite statements claiming that archaeological and anthropological qualifications and experience are vital components in defining a forensic archaeologist's competence, and determining if they are suitably qualified to be deemed as an expert by the court, little research has been conducted to gauge how much experience a forensic archaeologist must have in order to become sufficiently experienced to ensure an effective and complete recovery of all of the forensic evidence present at a crime scene (Cheetham and Hanson 2009; Cox 2009; Haglund 2001; Hunter and Cox 2005; Kershaw 2001). There are a few references within the literature that refer to the need for forensic archaeologists to have 'many years experience' (Haglund 2001; Hunter and Cox 2005; Sigler-Eisenberg 1985) but only Scott and Connor (2001) quantify the amount, stating that in order to be deemed as a capable practitioner the forensic archaeologist should have at least "3 years full-time fieldwork experience" (Scott and Connor 2001: 104). Within organisations such as the American Academy of Forensic Sciences (AAFS), other forensic disciplines such as forensic engineering and criminalistics have set levels of experience required for a candidate to be accepted as an associate member, ranging from 4-5 years with a baccalaureate degree, 3-4 years with a masters degree, and 2-3 years with a doctorate (AAFS 2012). As yet, no such standards exist for the forensic anthropology section (AAFS 2012). However, currently, in the United Kingdom, the Chartered Institute for Archaeologists Forensic Archaeology Special Interest Group is in the process of developing a rank system for forensic archaeological practitioners, based upon their level of experience and qualifications.

In a legal context, as experience is regarded as one of the key factors in establishing oneself as a credible expert witness, the current lack of a quantified and internationally agreed upon level of experience necessary for a forensic archaeologist to be recognised as a capable practitioner, may have serious ramifications in terms of the admissibility of a forensic archaeologist as an expert witness in a court of law (Cheetham and Hanson 2009; NAS Report 2009).

Legal concerns: How international legislation and admissibility regulations impact forensic archaeological investigations

In order to appreciate the impact that international admissibility regulations and legislative requirements have on forensic archaeological investigations, it is necessary to examine a number of factors. Firstly, one must examine the domestic legislative acts and the admissibility regulations that are currently in use in the countries involved in the study, including: the United Kingdom, Ireland, Australia, New Zealand, Canada and the United States. In addition, as forensic archaeologists are also employed in international contexts, to aid in the investigation of genocide, crimes against humanity, war crimes, and crimes of aggression, it is necessary to consider the legislative acts and the admissibility regulations upheld by the International Criminal Court (ICC), and the impact that these have on the work undertaken by forensic archaeologists. Finally, one must evaluate whether current forensic archaeological practice meets the admissibility and legislative requirements of these different legal systems, and if not, how this impacts the status of forensic archaeology as an accepted forensic discipline.

Domestic contexts

When forensic archaeologists are called upon to participate in forensic investigations within domestic contexts, there are a number of legislative acts and admissibility regulations in place which influence how forensic archaeological investigations are conducted, and outline what is expected of forensic archaeologists when acting in the capacity of an expert witness during court proceedings.

The major legislative acts and admissibility regulations that are currently in place within the various domestic courts of the countries involved in the study are as follows:

The Police and Criminal Evidence Act (1984), the Criminal Justice Act (2003), the Criminal Evidence (Experts) Act (2011), and the Criminal Procedure Rules (2011) in England and Wales.

The Criminal Procedure (Scotland) Act (1995), and the Criminal Justice (Scotland) Act (2003) in Scotland.

The Criminal Evidence Act (Northern Ireland) (1923), the Police and Criminal Evidence Act (1984), and the Criminal Justice (Evidence) (Northern Ireland) Order (2004) in Northern Ireland.

The Criminal Justice (Forensic Evidence) Act (1990), and the Rules of the Superior Courts (Evidence) (2007) in Ireland.

The Criminal Procedure Ordinance (1993), the Evidence Act (1995), the Evidence Amendment Act (2008), and the Practice Note CM7 Expert Witnesses in Proceedings in the Federal Court of Australia (2009) in Australia.

The Evidence Act (2006), the Practice Notes – Expert Witness – Code of Conduct (2011), and the Criminal Procedure Act (2011) in New Zealand.

The Canada Evidence Act (1985), the Criminal Code (1985), and the Federal Court Rules – SOR/98-106 in Canada.

The Frye Standard (1923), the Daubert Standard (1993) and the Federal Rules of Evidence (2011) in the United States.

These domestic admissibility regulations and legislative acts establish five key requirements that must be considered by the court to determine if the expert testimony and the evidence retrieved by the forensic archaeologist during the course of the forensic archaeological investigation can be deemed as reliable, and therefore admissible in the court.

Empirical testing

The first requirement is whether the techniques that were used by the forensic archaeologist during the course of the forensic investigation have been empirically tested. Currently, little testing has been done in the field of archaeology, or forensic archaeology, to comparatively assess the various excavation methods and recording systems utilised by forensic archaeologists to excavate burials. The only significant experimental research that has been conducted in this area thus far has been completed by Evis (2009), Pelling (2008), and Tuller and Đurić (2006) each of whom comparatively assessed two different excavation methods – Arbitrary Level and Stratigraphic excavation using mass graves or grave simulations. However, these studies failed to use large enough sample sizes to thoroughly compare these two excavation methodologies, and failed to compare archaeological recording systems (Evis 2009; Pelling 2008; Tuller and Đurić 2006). Another weakness of these studies is that they failed to test other excavation methodologies, such as the Demirant Excavation method and the Quadrant Excavation method, each of which have been used by practitioners in forensic casework (Congram 2008; Dupras *et al.* 2006; Hunter 2009; Hunter and Cox 2005; Ruffell *et al.* 2009). As a result, these three studies should be regarded as pilot studies that need to be expanded upon by further experimental research. Therefore, currently, forensic archaeologists should not state to the court that the methodological approaches that they used to excavate and record a forensic burial have been thoroughly tested, and as a result, the court could deem these methodological approaches as potentially unreliable, and consequently inadmissible.

Peer review

The second requirement is whether the techniques that were used by the forensic archaeologist have been subjected to peer review and have been published. Within the discipline of forensic archaeology and archaeology, numerous scholars have published articles, books, and archaeological manuals discussing excavation methods and recording systems, including: Balme and Paterson (2006), Barker (1993), Blau and Ubelaker (2009), Burns (2006), Carver (2009), Hanson (2004), Harris (1979), Hester (1997), Joukowsky (1980), Kjolbye-Biddle (1975), Pallis (1956), Phillips *et al.* (1951), Roskams (2001), and Wheeler (1954) to name but a few. However, depending on their own personal preferences, training and experience different scholars advocate and publish different methodological approaches. Therefore, if a forensic archaeologist was to justify their methodological approach to the court by referring to publications, they would be able to do so, as a number of excavation methods and recording systems have been published and peer reviewed. However, the fact that there is no standardised and widely accepted methodological approach for the excavation and recording of burials, could arguably weaken forensic archaeologists attempts to demonstrate to the court that the techniques that they used during the forensic investigation were reliable, and thus admissible.

Error rates

The third requirement is whether the techniques that were used by the forensic archaeologist have known or potential rates of error. As stated earlier, due to the fact that there has been very little testing of archaeological excavation methods and recording systems, there are no published error rates for individual excavation methods and their associated recording systems. This means that this requirement has not yet been met by the discipline of forensic archaeology. Consequently, forensic archaeological techniques could be deemed by the court to be potentially unreliable, as there is no data to suggest otherwise, and lead to the forensic archaeologist's testimony and evidence collected as a result of their participation in the investigation being dismissed from the court proceedings.

Professional standards

The fourth requirement is whether the techniques that were used by the forensic archaeologist have standards controlling their operation, and whether these standards were maintained during the course of the forensic investigation. In terms of an accepted standardised protocol for domestic forensic archaeological investigations, the only organisation that has created one is the Chartered Institute for Archaeologists Forensic Archaeology Special Interest Group in the United Kingdom, who published 'Standards and Guidance for Forensic Archaeologists' in October of 2011 (Powers and Sibun 2011).

This document outlines a protocol for forensic archaeologists to follow when conducting forensic archaeological investigations in the United Kingdom. However, in terms of providing a standardised methodological approach for the excavation and recording of clandestine burials, this guidance document is vague, stating that "features are investigated and excavated using the archaeological technique most suited to the specific circumstances of the case" (Powers and Sibun 2011: 16) and that the "forensic archaeologist must undertake all excavation with the aim of maintaining stratigraphic integrity, maximise the recovery of evidence, carrying out in situ evidence recording, and ensure that the details of the scene are reproducible through the production of written, illustrative and photographic records" (Powers and Sibun 2011: 16). This guidance document, therefore, does not provide any specific procedures for the excavation and recording of clandestine graves, and merely reiterates statements outlined in text books dedicated to the discipline of forensic archaeology (Blau and Ubelaker 2009; Connor 2007; Dupras *et al.* 2006; Hunter and Cox 2005).

As this guidance document fails to outline a specific excavation and recording protocol for forensic archaeological investigations, it provides forensic archaeologists in the United Kingdom with the freedom to excavate using any archaeological approach that has been used previously, or that has been published, and does not provide specific standards for forensic archaeologists to follow when conducting a forensic archaeological investigation. Furthermore, as not one of the archaeological excavation methods and recording systems that are published and used by forensic archaeologists to excavate and record clandestine burials have been thoroughly tested, these methodological approaches and this guidance document do not satisfy the requirements of the British or international courts.

The fact that no other domestic forensic archaeological organisations, working in the other geographical regions covered by this study, have attempted to create standard protocols for forensic archaeological investigations also means that, in international contexts, the discipline of forensic archaeology is failing to meet the fourth requirement for admissibility (Law Commission 2011). This could result in the findings and testimonies of forensic archaeologists being deemed as inadmissible and dismissed from court proceedings.

Widespread acceptance

The fifth requirement is whether the techniques that were used by the forensic archaeologist have gained widespread acceptance within the particular field to which they belong. As with the second requirement, several scholars have published and advocated a variety of different archaeological approaches to the excavation and recording of clandestine burials (Bass and Birkby 1978; Burns 2006; Cheetham and Hanson 2009; Congram 2008; Connor 2007; Dupras *et al.* 2006; Haglund *et al.* 2001; Hunter 2009; Hunter and Cox 2005; Hunter *et al.* 2001; Larson *et al.* 2011; Pickering and Bachman 1997; Skinner *et al.* 2003; Wolfe Steadman *et al.* 2009). The approaches that they recommend vary according to when and where they received their training and experience, and their own personal preferences. Furthermore, many of the techniques that these scholars advocate originate from field archaeology and have been used for a long period of time, and continue to be used on a regular basis within the commercial archaeology sector. Thus, if a forensic archaeologist was required to prove that the techniques that they used during the forensic investigation were widely accepted within the field of archaeology, they would be able to provide many examples to attest to the fact that they were.

However, as with the second requirement, the fact that all of the methodological approaches that are currently used within the field of archaeology and forensic archaeology are generally accepted, as long as the selection of each approach is justified, may in turn, weaken a forensic archaeologist's attempts to highlight to the court that the technique that they chose was the most reliable and suitable to use. This is because there is no data, as yet, from which a forensic archaeologist can reliably state that one methodological approach is more reliable and consistent than another, and therefore, the forensic archaeologist may find that their testimony is regarded with caution, due to the potential for the technique that they chose to be unreliable.

The primary reason why forensic archaeology has, to date, remained unaffected by the legislative requirements and admissibility regulations set by domestic courts was recently highlighted in a review of expert evidence conducted by the Law Commission which stated that "expert evidence is often trusted like no other category of evidence" (The Law Commission 2011: 12) and "cross-examining advocates tend not to probe, test or challenge the underlying basis of an expert's opinion evidence" (The Law Commission 2011: 13) as they "do not feel confident or equipped to challenge the material underpinning the expert opinion" (The Law Commission 2011: 13), leading to forensic archaeologists and the methods they utilise remaining unquestioned in a court environment.

However, in light of recent reviews of expert evidence conducted in the United States, Canada, Australia and the United Kingdom (Australian Law Reform Commission 2005; Edmond 2010; Edwards 2009; Glancy and Bradford 2007; NAS Report 2009; Robertson 2009; Robertson 2010; Selby 2010; The Law Commission 2009, The Law Commission 2011), admissibility regulations are now being increasingly enforced and expert witnesses and the legal practitioners with whom they are working are now required to prove, prior to appearing in court, that the methods used by the forensic archaeologist during the course of a forensic investigation meet the admissibility requirements of the court. If the forensic archaeologist or the legal practitioner fails to do so, the evidence gathered by the forensic archaeologist in support of the case will be deemed unreliable and dismissed from the court proceedings (NAS Report 2009; The Law Commission 2009, 2011). This is illustrated by the case of Hunter v. United States, 48F. Supp. 2d 1283, 1288 (D. Utah 1998) during which the government sought to prosecute Hunter for defacing the Santa Clara River Gorge Shelter site in the Dixie National Forest and looting other archaeological sites (Hutt 2006). Two archaeologists were commissioned to assess the damage to the site and estimate how much it would cost to repair and restore the site back to its former state, and in turn, how much Hunter would be required to pay in damages (Hutt 2006). However, the two reports contradicted each other, and the Judge presiding over the case stated that the methods used by each of the archaeologists in order to complete their archaeological site damage assessment reports "lack[ed] sufficient reliability to be used for sentencing purposes" (ID. At 1291.) (Hutt 2006). Consequently, both of the reports that had been submitted as evidence by the two archaeologists were dismissed from the court proceedings, and were not taken into consideration by the judge when he calculated the amount of damages that the defendant would have to pay towards the restoration of the archaeological site (Hutt 2006).

International contexts

In addition to domestic investigations, forensic archaeologists are also requested to assist in investigations in international contexts. Such investigations require forensic archaeologists to assist in the search, location, recovery and recording of evidence from scenes of crime relating to genocide, crimes against humanity, war crimes, and crimes of aggression. Currently, investigations into these four types of crime are conducted under the auspices of the International Criminal Court (ICC). Prior to the establishment of the ICC in 2002, after the ratification of the Rome Statute, investigations of these crimes were conducted and prosecuted by International Criminal Tribunals, such as the International Criminal Tribunal for the Former Yugoslavia (ICTY) and the International Criminal Tribunal for Rwanda (ICTR) (Cordner and McKelvie 2002; Cox *et al.* 2008; Kittichaisaree 2001).

Unlike the domestic court systems discussed previously, the ICC does not follow an adversarial system of justice. Rather, it uses a hybrid system, of adversarial and inquisitorial systems, utilising the strengths of both approaches to investigate crimes under its jurisdiction. Under this hybrid system, it is important to note that there is no jury; instead, professional judges are elected to preside over the trial, determining the weight of the evidence brought before them and the sentence that ought to be given to the accused. As a result, the judges who preside over such trials are not required to act as "gate keepers against bad science" on behalf of juries, as their professional training and experience ought to provide them with the ability to consider each piece of evidence that has been presented to them, and ascertain the weight that it ought to be given (Klinkner 2008; Klinkner 2009; Roberts and Willmore 1993; Schabas 2006). Therefore, the ICC does not have strict procedures or rules governing the admissibility of evidence or expert witness testimonies, and tends to follow an approach of "admit everything, determine weight later" (Klinkner 2009; Zahar and Sluiter 2008: 384).

There are however, 'Rules of Procedure and Evidence' that are followed by the ICC, which state in Rule 64 that "evidence that is ruled irrelevant or inadmissible shall not be considered by the Chamber" (The International Criminal Court 2013: 21). Therefore, if the methodological approaches utilised by forensic archaeologists during the course of the investigation are proved to be unreliable, then the evidence retrieved by the forensic archaeologists and the testimony provided by the forensic archaeologist in charge of the investigation, would then lack relevance to the case and be deemed inadmissible, and ought not to be considered by the trial chamber when determining the sentence of the accused. However, as there have been relatively few studies focused on evaluating the reliability of forensic archaeological excavation methods and recording systems, there is no data for the defence or prosecution to use to critique or defend the methodological approaches used by forensic archaeologists during such investigations (Klinkner 2009).

This issue is exacerbated by the fact that when forensic archaeologists are called to participate in such investigations they are employed by the prosecution, and are directed by the prosecution's senior investigator in all operational matters (Klinkner 2009). At no stage during this investigative process are the defence's forensic archaeological experts given the opportunity to observe or conduct independent investigations at these scenes of crime. Given that excavation is a destructive process, and once a grave, be it a mass or single grave, has been excavated it cannot be re-excavated, any operational errors in terms of methodological or recording processes cannot be observed at the scene, and any evidence that may have contradicted the prosecution's report's findings will have been lost during the investigative process (Barker 1993). Therefore, the current system of providing the defence with the prosecution's forensic archaeological report post-facto to review and critique clearly has a flaw, as issues that may have gone unnoticed or were overlooked by the prosecution's team during the investigation will not be present in the report. Thus, the defence's forensic archaeological experts will not have the entire picture of the investigative process, and will not be able to evaluate it fully.

This issue also applies to the process of employing independent experts to evaluate reports provided by the prosecution in cases where the defence and prosecution disagree over evidential reliability matters, as they too will only be able to review the 'polished' end of investigation report. Given that such investigations are planned over a long period of time, the fact that courts are aware of who they wish to prosecute, and the vast amount of money that is invested in investigating these crime scenes, it would be beneficial and possible for the court to inform the defence of when and where forensic archaeological investigations are to take place, and to invite them to send defence forensic archaeological experts to observe these investigations as they take place. This will ensure that they, the defence, are satisfied that the evidence that is collected and recorded is being done so in a reliable manner.

Furthermore, as with the domestic legal systems, the reliance on the cross-examining advocates' and the professional judges' experience to highlight and evaluate the reliability of the methods used by experts during the course of an investigation poses a problem. Given the vast array of different forensic disciplines that are required to assist with such investigations and present evidence in ICC trials, these legal practitioners cannot be expected to have gained sufficient experience and knowledge in each of these disciplines to be able to effectively evaluate each discipline's methodological approaches to determine if they are reliable enough to be admitted (Boas 2001; Klinkner 2009; Roberts and Willmore 1993; The Law Commission 2011: 13). Moreover, as there are no admissibility requirements in place, current procedure does not demand that methods be shown to be reliable before they are admitted. Thus, each judge may use different evaluative criteria when assessing the admissibility of particular methods, which may result in judges having contradicting opinions as to whether a particular methodological approach was appropriate (Boas 2001: 59). This may explain why, to date, the "reliability of particular scientific methods and techniques appears to be more fully canvassed as a question of weight, rather than admissibility" and highlights why the methodological approaches of disciplines such as forensic archaeology, have remained unquestioned during trials at the ICC and International Criminal Tribunals for the Former Yugoslavia and Rwanda (Boas 2001: 59).

It is evident, through reviewing the admissibility regulations and the legislative requirements of courts in both domestic and international contexts that forensic archaeology has, to date, failed to satisfy the requirements of the international court systems. This is primarily due to the fact that practitioners in the sub-field of forensic archaeology have yet to test the methodological approaches that they use and advocate. This has resulted in a multitude of excavation methods and recording systems continuing to be used without any proof that they produce consistent and accurate results. Furthermore, the fact that there is no internationally agreed upon, standardised, peer reviewed protocol for forensic archaeological investigations, weakens the credibility of the sub-field, as it implies that there is a lack of professionalism within the discipline, and that its practitioners use ad hoc approaches during forensic investigations. Consequently, the court systems discussed in this research should not allow evidence retrieved through excavation by forensic archaeologists into their courts. Therefore, if the discipline of forensic archaeology is to continue to be accepted as a forensic discipline, these weaknesses must be addressed, otherwise, forensic archaeologists' testimonies and the evidence that they collect during the course of a forensic investigation may be dismissed from the court proceedings, as the methodological approaches underpinning the investigation were deemed unsafe, potentially unreliable, and therefore inadmissible.

The search for standardisation in forensic archaeological investigations

In order to evaluate how the discipline of forensic archaeology has attempted to address the issue of the lack of empirical testing of archaeological excavation methods and recording systems. One must first evaluate existing research that has been conducted into archaeological excavation methods and recording systems. After which, one needs to discuss how this research can be improved upon and expanded in order to ensure that the discipline of forensic archaeology meets the admissibility regulations and the legislative requirements of the international court systems.

Testing of archaeological techniques

The process of controlled excavation is perhaps the most common aspect of archaeology with which archaeologists are associated. The process of "excavation is the prime method available to archaeologists for the recovery of new data" (Clarke 1978: 63), whereas the process of recording provides the only current means of preserving the new data obtained through the excavation activity. It is universally accepted that the excavation and recording of sites, be it in archaeological or forensic contexts, by competent personnel with adequate facilities represents both the most important and best means of acquiring additional information (Clarke 1978; Haglund 2001; Hunter and Cox 2005; Scott and Connor 2001; Sigler-Eisenberg 1985).

However, excavation methods and recording systems have, to a large extent, developed independently from one another, each evolving to their current state according to the intellectual, technical, and theoretical developments within the discipline, and the professional and ethical codes of practice advocated by professional bodies such as: the Chartered Institute for Archaeologists (CIfA) in the United Kingdom, the Institute of Archaeologists of Ireland (IAI) in the Republic of Ireland, the Society for American Archaeology (SAA) in the United States, the Canadian Archaeological Association (CAA) in Canada, the Australian Archaeological Association (AAA) in Australia, and the New Zealand Archaeological Association (NZAA) in New Zealand. Consequently, different combinations of excavation methods and recording systems are utilised by different archaeological practitioners, in accordance with their individual preferences, experience, academic training, and the legislative requirements present within the geographical area in which they are working.

Therefore, numerous practitioners advocate different excavation methods and recording systems as the works of Barker (1993), Carver (2009), Harris (1979), Harris (2006), Joukowsky (1980), Kjolbye-Biddle (1975), Pallis (1956), Phillips *et al.* (1951), Roskams (2001) and Wheeler (1954) demonstrate. However, as different excavation methods such as: the Stratigraphic method, the Schnitt method, the Gezer method, the Quadrant method, the Arbitrary Level method, and the Box method, and different recording systems such as: Single Context Recording, Multi-Context Recording, Unit Level Recording, and Pre-determined Strata Recording continue to be utilised and advocated (along with their theoretical justifications) within academic literature these different techniques have come to be attributed the status of 'standardised methods' within the countries, traditions, schools and institutes in which they are practiced (Bar-Yosef and Mazar 1982; Browman and Givens 1996; Collis 2002; Daniel 1978; Darvill 2000; Desert Archaeology 2008; Drewett 2000; Fowler 1977; Greene and Moore 2010; Hester 1997; Joukowsky 1980; Kidder and Kidder 1917; Lucas 2001; Lymen and O'Brien 1999; Willey and Sabloff 1980).

Although a variety of excavation techniques and recording systems have been allocated the status of a "standardised method" the suitability of certain excavation methods and recording systems have been called into question and sparked debate within the academic community (see Harris 1989; Harris *et al.* 1993; Phillips *et al.* 1951 for an overview). However, little research has been conducted to directly measure, contrast and compare differing excavation methods and recording systems, or to understand the impact that variations in approaches to excavation and recording may have on data recovery and interpretation(s). This is largely because no two archaeological sites are exactly alike and the excavation of an archaeological site cannot be replicated, as the process of excavation is one of destruction. Methods cannot, normally, be tested and compared on the same set of deposits, although there are examples of archaeological sites that have been re-excavated and the findings of previous archaeological investigations re-evaluated – Ian Hodder's work in Çatalhöyük and the re-excavation of the 'Lazete 2' Srebrenica mass grave site being two prominent examples (Çatalhöyük 2011; ICTY 2007).

Additionally, in recent experimental research conducted by Evis (2009), Pelling (2008), and Tuller and Đurić (2006) it is apparent that the assumption that excavation methods are suitable for use on the basis of their widespread adoption and 'standardised' status should be called into question and re-evaluated.

Each of these experimental studies directly compared two standardised excavation methods – the Arbitrary Level Excavation method and the Stratigraphic Excavation method. In Tuller and Đurić's (2006) study they compared these two excavation methods in the excavation and recovery of skeletal remains within two mass graves in an attempt to identify which method was the most effective. Their results established that the Stratigraphic method of excavation was the most effective at maintaining the provenience and articulation of the remains. However, each grave was excavated by differing teams of experts, one consisting of archaeologists and the other of a mix of pathologists and biological anthropologists, and therefore was not an objective or direct comparison of each methodological approach (Tuller and Đurić 2006).

In Pelling's (2008) study, the primary aim was to define which of these two methods was the most effective at recovering 'evidence' from simulated graves which included: artefacts, geotaphonomic features and stratigraphic deposits and interfaces. Pelling's study found the Stratigraphic method of excavation to be the most effective and resulted in 94.4% of material evidence placed within the grave being recovered in situ, whereas 50% of material evidence was found in situ using the Arbitrary Level Excavation method (Pelling 2008). In addition, the Stratigraphic approach recovered 100% of geotaphonomic features introduced into the grave whereas the Arbitrary Level approach failed to recover any (Pelling 2008). However, Pelling's (2008) study only simulated one grave for each method and therefore was not a large enough sample size to effectively compare these methodological approaches.

In light of this, Evis (2009) conducted a similar experiment to Pelling and created eight mock graves in order to assess which of these two excavation methods was the most effective at correctly interpreting the grave's formation sequence and recovering 'evidence' which, again, included: artefacts, geotaphonomic features, and stratigraphic deposits and interfaces. Evis enlisted four archaeologists to excavate two graves each (Evis 2009). The results of the study indicated that the Stratigraphic Excavation method was the most effective, recovering an average of 71% of all defined evidence types, whereas the Arbitrary Level Excavation method recovered an average of 56% (Evis 2009). In terms of consistency of interpretation of the grave's formation sequence, again the Stratigraphic method of excavation proved to have a higher rate of consistency, correctly identifying an average of 71% of the deposits and interfaces present and an average of 62.5% of the geotaphonomic features introduced into the graves, whereas the Arbitrary Level Excavation method correctly identified an average of 51% of the deposits and interfaces present and an average of 12.5% of the geotaphonomic features (Evis 2009). In addition, through conducting the experiment with multiple archaeologists it became apparent that additional variables may have affected the percentage of evidence recovered and the consistency of subsequent interpretations. Factors such as experience and time caused variation in the results gained, findings that were supported by experiments conducted by Roberts (2009), Scherr (2009) and Tuller and Đurić (2006). However, this study only compared four graves for each methodological approach and therefore was not a large enough sample size to effectively compare these two excavation methods. Furthermore, the experiment only compared two excavation methods and therefore did not take into account or test all of the published methodological approaches used in the field of archaeology. Also, the experiment did not assess the affect of variables such as sight, sound, touch and smell which could have had an influence on the overall findings.

Improvement and expansion of existing research

Because of the limitations of the aforementioned studies they should be regarded as pilot studies that need to be expanded upon through further research and experimentation. Moreover, not one of the discussed studies comparatively assessed archaeological recording systems. This is an important aspect of the archaeological process not to have investigated, as the records produced during the course of an excavation will provide the evidence from which the archaeologist will construct and justify their narrative of how the archaeological feature or site under investigation was formed. Therefore, the impact that different recording systems may have on the overall representation and interpretation of an archaeologist's findings is as yet unknown, and must also be addressed through further research and experimentation.

This research aimed to complete such tasks by increasing the sample size and increasing the number of methodological approaches assessed. It also comparatively assessed recording systems to determine whether different approaches had any impact on the representation and subsequent interpretation of archaeological data. It is also recognised that archaeological excavation and recording is a highly perceptual skill and can be affected by a number of extrinsic factors. However, an evaluation of these factors was outside the remit of this research project.

Despite the limitations of the previous studies their findings have significant implications for the field of archaeology, as the processes of excavation and recording provide the raw data to which theories are applied and interpretations are formed. If certain excavation methods, recording systems and additional influential variables are proven to result in the loss of contextual information, artefacts, features or deposits, then any interpretations based on assemblages excavated and recorded via these methods may be incomplete, or at least subject to different interpretation. Subsequently, any individuals reading the interpretations may be mislead and the results potentially misused, one needs only to refer to the 'Ahnenerbe' and the 'Amt Rosenberg' organisations in Nazi Germany to highlight this point (Arnold 1990; Arnold 1996; Pringle 2006). Members of these archaeological organisations focused their investigations on proving Gustaf Kossinna's 'Kulturkreis Theory' which stated that geographical regions could be associated with specific ethnic groups based on the presence of material culture, and that cultural diffusion was a process whereby influences and ideas were passed on by more advanced societies to the less advanced societies with whom they came into contact (Arnold 1996: 550). As a result, "wherever a single find of a type designated as Germanic was found the land was declared ancient German territory" (Sklenář 1983: 151), thus providing the theoretical justification for the Nazi's expansion into Czechoslovakia and Poland, as it was deemed that the Nazis were claiming back their ancient ancestral heritage (Arnold 1990; Arnold 1996; Pringle 2006). Moreover, in a forensic context, it could result in a perpetrator not being convicted or the wrongful conviction of an innocent individual (Cheetham and Hanson 2009; Gould 2007).

It was therefore important for this research to be undertaken as it empirically tested the excavation and recording techniques currently used in the field of archaeology and resulted in the formulation of bespoke recommendations for archaeological investigations conducted in forensic contexts. The development of this research project was timely, as it was directly responding to the latest legislative requirements of the international courts, and will help to ensure that archaeologists and the evidence retrieved as a consequence of their involvement in a forensic investigation will meet the legislative requirements and the admissibility regulations of the international courts, and will continue to be accepted in international court proceedings in the future.

Chapter 3 Methodology

The first two objectives of the research were to review, analyse and compare published academic literature and published/unpublished archaeological manuals/guidelines, and to identify the origins, development and current use of excavation methods and recording systems in the United Kingdom, Ireland, Australasia and North America. To achieve these two objectives the author took the following approach:

In order to collect and review published academic literature, a thorough literature search was performed. This literature search resulted in the collection of 400 data sources, forming a substantial literature pool that was used to inform the research and underpin the subsequent discussion.

In order to obtain unpublished/published archaeological manuals/guidelines, 499 archaeologists, archaeological companies, organisations, institutions, museums and libraries in the United Kingdom, Ireland, Australasia and North America were contacted (Appendix A). Through contacting these establishments 153 archaeological manuals and guidelines were obtained. The data relating to excavation methods and recording procedures in these manuals/guidelines were analysed against a set of analytical criteria (Appendix B). These criteria were developed by reading through each of the manuals/guidelines and identifying points of difference in excavation approaches and recording methods. Consequently, any variation in approach to excavation or recording prompted the creation of a new criterion, against which all of the other manuals/guidelines would be evaluated. Each criterion was specifically developed to initiate a 'yes' or 'no' response when a manual/guideline was being assessed against it. This ensured that all data entered into the system was in a binary format (1 = Yes, 0 = No). The data entry process is explained in Figure 5.

It was necessary to use a binary data entry format as this enabled each of the manuals/guidelines to be directly compared against one another, and analysed to determine the extent of similarities and differences in the

Organisation	Manual usage				Manual creation year					
	The organisation has got its own manual	The organisation has not got its own manual	The organisation uses another organisation's manual	The organisation is in the process of updating their manual	1960-1969	1970-1979	1980-1989	1990-1999	2000-2009	2010-2019
University College London	1	0	0	0	0	0	0	0	1	0
SHARP Archaeology	1	0	0	0	0	0	0	0	1	0
Archaeology South East	0	1	1	0	0	0	0	1	0	0
Museum of London	1	0	0	0	0	0	0	1	0	0
Archaeology Project Services	0	1	1	0	0	0	0	1	0	0
CFA Archaeology Ltd	0	1	1	0	0	0	0	1	0	0

FIGURE 5: AN EXAMPLE OF DATA ENTRY INTO THE SPREADSHEET SYSTEM © EVIS 2016.

excavation and recording approaches presented in the manuals/guidelines. Once the data was processed in this manner, it was possible to evaluate the extent and causes of variation in the approaches to excavation and recording in the geographical areas being studied. This data also assisted in the development of the experimental phase of the research as it directly dictated which excavation methods and recording systems were to be experimentally compared.

The third objective of the research was to conduct interviews with field and academic archaeologists in order to evaluate how they excavate, and why and when they choose to use particular excavation methods and recording systems.

In order to achieve this objective, and to assist with the evaluation of variation, and reasons for variation within individual excavation methods and recording systems, and between systems identified during the data gathering phase of the research, the author arranged and chaired conferences at the Theoretical Archaeology Group (TAG), the Society for American Archaeology (SAA), and the Chartered Institute for Archaeologists (CIfA) annual conferences (Appendix A). This enabled the author to obtain the conference's participant's perspectives on how, why and when they choose to use a particular excavation method or recording technique. Additionally, during the experimental phase of the research project, the author was able to conduct interviews with experimental participants 'live' in the field, to ascertain how, why and when different archaeologists choose to use particular excavation methods and recording systems, and the extent to which a certain approach to excavation or recording can be justified. These interviews were based around a consistent series of open and closed questions that enabled the answers to be directly compared and analysed (Appendix C).

Excavation method and recording system selection

The results from the manual/guideline analysis revealed that four distinct excavation methods are used in field archaeology to excavate negative features (Figure 6).

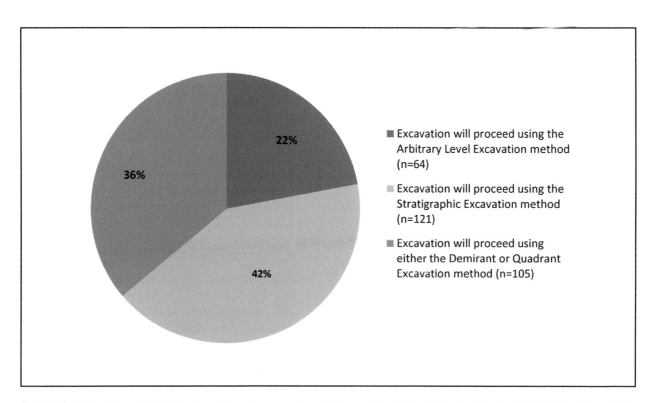

22%

36%

42%

■ Excavation will proceed using the Arbitrary Level Excavation method (n=64)

■ Excavation will proceed using the Stratigraphic Excavation method (n=121)

■ Excavation will proceed using either the Demirant or Quadrant Excavation method (n=105)

FIGURE 6: ILLUSTRATES THE EXCAVATION METHODS THAT ARE CURRENTLY BEING USED IN FIELD ARCHAEOLOGY TO DEAL WITH NEGATIVE FEATURES © EVIS 2016.

These are the Stratigraphic Excavation method, the Demirant Excavation method, the Quadrant Excavation method, and the Arbitrary Level Excavation method. Additionally, each of these excavation methods has a particular recording system associated with its use. These include the Single Context Recording system, the Standard Context Recording system and the Unit Level Recording system. The application of these methods in terms of excavating and recording a negative feature, such as a grave, is understood as follows:

The Stratigraphic Excavation method and Single Context Recording system

The Stratigraphic Excavation method relies on the recognition that any single action, whether it results in the creation of a positive record, such as a deposit, or a negative record, such as a cut, is identified and recorded as an individual stratigraphic unit or context during the excavation process. This method relies on the understanding that within any stratigraphic sequence, a chronologically earlier context will be found to be sealed or cut by a chronologically later context, and that by excavating each of these individual stratigraphic units in their entirety, in the reverse order in which they were created, it is possible to accurately reconstruct the formation sequence of an archaeological site or an individual feature.

Whilst excavating using the Stratigraphic Excavation method, a recording system known as the Single Context Recording system is used. When using this system, the uppermost, chronologically latest context is defined, allocated a unique identifying number often referred to as a context number, and is then planned and excavated. This procedure is then repeated until all of the contexts contained within the archaeological site, or individual feature, have been removed. Both during and after an individual context's excavation the archaeologist proceeds to complete a pro-forma known as a context recording form. This sheet contains a variety of prompts which allow an archaeologist to provide a description of the context, discuss what the context contained, note down any sampling that was undertaken, write down what photographs were taken, and provide an interpretation of what they believe the context may be. This pro-forma also contains a space in which to record the context's stratigraphic relationship with those contexts that were identified and excavated before and after it. In order to establish this relationship, each of the individual context plans are laid over one another in chronological order. If a relationship is found to exist between contexts, it is recorded on the pro-forma by writing the current context's unique identifying number in the central box; those contexts that have been proven to stratigraphically precede the current context's formation are then written in the box below, and those that stratigraphically succeed the current context's formation are written in the box above.

The Demirant Excavation method and Standard Context Recording system

The Demirant Excavation method is also known as Half-sectioning. When using this methodology, the feature that is to be excavated is divided into two halves. The archaeologist then proceeds to excavate each half separately. Once the first half has been removed, the archaeologist inspects the exposed section of the remaining half of the feature. It is by examining and recording this section that the archaeologist is then able to identify and verify the presence of, and the stratigraphic relationships between, individual contexts contained within the feature. Such information is then used by the archaeologist to guide their excavation of the remaining half of the feature.

When using the Demirant Excavation method a recording system known as the Standard Context Recording system is used. When using this system, as each context is found, a pro-forma known as a context recording form is filled out. This form contains the same prompts as the context recording form used in the Single Context Recording system. However, unlike the Single Context Recording system, this system does not rely on overlaying plans of individual contexts to determine stratigraphic relationships. Instead, it relies on using the section that was exposed and recorded during the excavation process. Nevertheless, once such relationships are established, they are recorded on the pro-forma in the same way as described for the Single Context Recording system.

The Quadrant Excavation method and Standard Context Recording system

The Quadrant Excavation method involves dividing the feature to be excavated into four equal sectors. The archaeologist then proceeds to excavate each sector individually. Usually, when using this excavation method, the archaeologist leaves a staggered, cross-shaped baulk along the axes of the feature during the excavation process. This ensures that any artefactual or ecofactual material that may be present in the central area of the feature is recovered, and also enables the archaeologist to inspect and record sections along the entire length and width of the feature. As with the Demirant Excavation method, it is by examining and recording these sections that the archaeologist is then able to identify and verify the presence of, and the stratigraphic relationships between, individual contexts. Once these sections have been drawn, the archaeologist then removes the standing baulks in order to ensure that any artefactual and/or ecofactual material contained within the baulks is recovered.

The Quadrant Excavation method uses the same recording system as the Demirant Excavation method; this system is known as the Standard Context Recording system. However, unlike the Demirant Excavation method, sections from both the length and width of the feature are used to evaluate the stratigraphic relationships between individual contexts.

The Arbitrary Level Excavation method and Unit Level Recording system

The Arbitrary Level Excavation method is also known as Spit Excavation, Planum Excavation or Metrical Excavation. The first step of this method is to define the boundaries of the feature that is to be excavated. Once this has been completed an area, usually 30cm beyond the boundaries of the feature, is demarcated to form an excavation unit. After the excavation unit has been established, soil is removed from within the boundaries of the excavation unit in a succession of pre-determined spits, usually ranging from between 5cm and 10cm in depth. If artefacts are located whilst excavating an individual spit they are left upon a soil pedestal until the excavation of the spit has been completed and their horizontal and vertical locations have been recorded, after which the artefacts and pedestals are removed. This process continues until the feature contained within the excavation unit has been completely excavated.

Whilst using the Arbitrary Level Excavation method, a recording system known as the Unit Level Recording system is used. This recording system relies on the use of a pro-forma, known as the unit level record. This pro-forma is filled out once an individual spit has been excavated. It contains a series of prompts which require the archaeologist to describe and discuss the presence of any features, deposits, fills, artefacts, or disturbances within the excavation unit. It also contains a planning grid in which the archaeologist should draw a plan of the excavation unit, in which any features, artefacts, fills, or deposits that are discernable are drawn and annotated. The pro-forma also has a space in which to record the surface and ending elevations of the individual spit, a column in which to list the type and number of samples taken, tables in which to record the artefactual/ecofactual material that has been recovered, and the number and type of photographs taken.

As the four aforementioned excavation methods and their associated recording systems are the most commonly used approaches identified during the manual/guideline analysis for excavating negative features, they were the techniques that were experimentally tested.

Experimental design

In terms of experimental design, the most effective method by which an experiment could be created that would allow for a direct comparison of different excavation methods and recording systems was through creating an archaeological site simulation. This approach is one that has been used in several experimental archaeology projects. For example, Crabtree (1990), Fowler (1980), Nash and Petraglia (1987), and

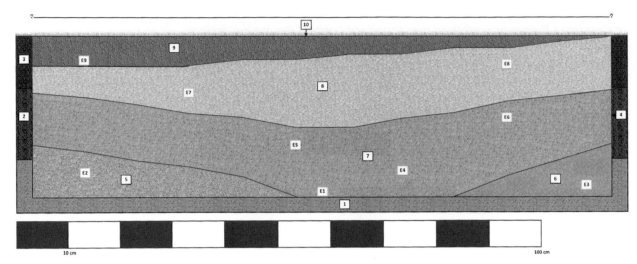

Context number	Description
1	Natural
2	Subsoil
3	Topsoil and turf
4	Feature cut
5	Fill 1
6	Fill 2
7	Fill 3
8	Fill 4
9	Fill 5
10	Replaced turf

Evidence number	Description	Location (length, width, depth)
E1	Dress	55 cm, 20 cm, 0 cm
E2	Two pence coin	10 cm, 35 cm, 25 cm
E3	Lighter	105 cm, 2 cm, 28 cm
E4	Fake nail	70 cm, 10 cm, 25 cm
E5	ID card	50 cm, 20 cm, 20 cm
E6	Earring 2	90 cm, 20 cm, 15 cm
E7	Kirby grip	30 cm, 15 cm, 10 cm
E8	Earring 1	90 cm, 35 cm, 5 cm
E9	Cigarette papers	10 cm, 10 cm, 4 cm

FIGURE 7: ILLUSTRATES THE DESIGN OF THE GRAVE SIMULATION © EVIS 2016.

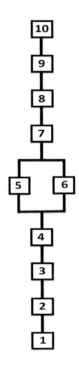

Context number	Description
1	Natural
2	Subsoil
3	Topsoil and turf
4	Feature cut
5	Fill 1
6	Fill 2
7	Fill 3
8	Fill 4
9	Fill 5
10	Replaced turf

FIGURE 8: THE HARRIS MATRIX OF THE GRAVE SIMULATION © EVIS 2016.

Riley and Freimuth used an archaeological site simulation to study site formation processes. Davenport *et al*. (1988), France *et al*. (1992), and Isaacson *et al*. (1999) used an archaeological site simulation to test geophysical equipment, and Chilcott and Deetz (1964) used an archaeological site simulation to evaluate excavation methods and excavator proficiency.

In the case of this experiment, as it ultimately aimed to determine which excavation method and recording technique best meets the needs of forensic archaeology, the experiment used a grave simulation. By using a simulation such as this it enabled evidence such as artefacts, geotaphonomic features, stratigraphic contexts (deposits, fills, cuts, interfaces) and stratigraphic relationships to be created and placed into a stratigraphic sequence at measured points with known and defined properties. In order to ensure that the grave simulation was able to be replicated easily, and to better control the experimental process the grave simulation was simple in design.

The design of the grave simulation is shown in Figure 7, Figure 8 and Figure 9.

In terms of artefact selection, the dress (E1) located at the bottom of the grave replicated the placement of a body. Unfortunately, due to resource constraints, it was not possible to use a plastic skeleton. Furthermore, as the experiment was repeated at various sites around the United Kingdom, it was not possible to transport

E1 Dress

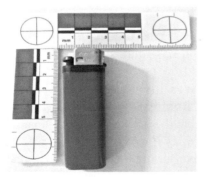

E2 Two pence coin

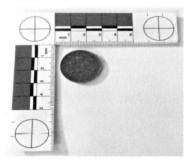

E3 Lighter

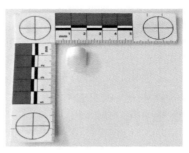

E4 Fake nail

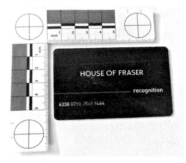

E5 ID card

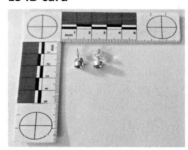

E6 and E8 Earrings

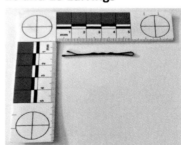

E7 Kirby grip

E9 Cigarette papers

FIGURE 9: ILLUSTRATES THE MATERIAL EVIDENCE ITEMS E1-E9 © EVIS 2016.

a plastic skeleton to these locations (Appendix D). The other artefacts included in the grave (E2-E9) were chosen to represent items that a perpetrator might lose out of their pocket whilst creating a grave, or a victim may have on their person whilst being buried. All of these artefacts were selected to ensure that participating archaeologists would recognise them if they came across them during the excavation process, and are types of artefacts that, according to the literature, are often found by forensic archaeologists during the course of excavating a clandestine burial. In addition, by including artefacts of various sizes in the experiment it made it possible to determine if certain excavation approaches have a greater tendency to recover smaller artefacts than others. Another variable taken into consideration when selecting these artefacts was whether they would preserve during the short time period between the grave's creation and excavation, in variable soil conditions. By reviewing the results of existing experimental studies conducted by Janaway (1996: 58-81; 2002: 380-399) it was evident that these artefacts would preserve during the short time frame between the grave's creation and subsequent excavation.

The stratigraphic contexts contained within the grave were designed to overcome the weaknesses of previous grave simulation studies conducted by Evis (2009) and Pelling (2008). One of the weaknesses of Evis's (2009) and Pelling's (2008) studies was that the stratigraphic contexts included in their graves contained a number of fills that were flat bottomed and topped, and were either 5cm or 10cm in depth. Although such fills are easy to replicate, they are unusual in archaeological contexts, and potentially favour an arbitrary level form of excavation, providing an unfair advantage to this technique during experimental testing. Furthermore, one of the criticisms of Arbitrary Level Excavation is that it is unable to recover angled deposits/fills, as the method relies on the use of levelled spits, usually 5cm or 10cm in thickness, to remove the fills contained within the feature, thereby introducing artificial divisions of space and time (Hanson 2004; Harris 1979; Harris 1989; Harris 2002; Harris 2006; Komar and Buikstra 2008). That, in turn, could result in the creation of a biased and potentially incomplete interpretation of the stratigraphic sequence, as the original physical dimensions of the deposits/fills are destroyed during the excavation process (Hanson 2004; Harris 1979; Harris 1989; Harris 2002). Therefore, this grave was designed to include multiple angled fills, enabling the author to evaluate whether the criticisms noted above had any foundation during experimental testing. The straight edged and flat bottomed design of the grave cut was chosen because it made replication of the grave easier and less time consuming, as attempting to accurately replicate a curved edged and bottomed grave could increase the potential for inaccuracies during repetitive experimental testing.

The soils chosen to represent the various fills contained within the grave were selected for three reasons. Firstly, the play sand (Context 5/fill 1 and Context 6/fill 2) and the sterilised topsoil (Context 8/fill 4) could be purchased from a mainstream supplier which has retail outlets across the United Kingdom meaning that the same brand of soil products could be purchased at each experimental location, ensuring that the soils contained within the feature were consistent. Secondly, Context 7/fill 3 and Context 9/fill 5 were constructed from the spoil created from the excavation of the grave. The justification for using the spoil to construct these contexts was that the archaeologists who participated in the experiment were familiar with excavating this type of soil as it was the soil that was present in the regions in which they conducted their archaeological fieldwork, and therefore, the archaeologists would be able to differentiate these fills from the others contained within the grave. Thirdly, by using the play sand to represent context 5/fill 1 and context 6/fill 2 and the spoil to represent context 7/fill 3 and context 9/fill 5 it was possible to detect whether an excavation technique was able to distinguish and define individual contexts, or whether the excavation technique resulted in these contexts being mistakenly joined, as context 5/fill1 and context 6/fill 2 were at the same depth in the grave but were not connected, and context 7/fill 3 and context 9/fill 5 were separated in their entirety by context 8/fill 4.

Another weakness of Evis's (2009) experimental research was that the graves were too large. This resulted in archaeologists taking up to 31 hours to excavate an individual grave. Due to the fact that the majority of the participants in the experiment were commercial archaeologists, who were only able to participate in the experiment for a maximum of one day, it was decided to decrease the size of the overall grave from 120cm x 75cm x 85cm to 110cm x 40cm x 30cm. This reduced the length of time that it took for archaeologists to excavate the grave, and in turn, allowed for more experimental graves to be produced and excavated than in Evis's (2009) and Pelling's (2008) studies.

Participant selection

Participants were gained by inviting the various organisations that donated their archaeological manuals/ guidelines to the research to participate in the experiment. Each of the organisations were provided with an outline of the excavation methods and the recording systems that were going to be tested, and were asked if they could provide archaeologists that had experience in using one or all of these techniques. Additionally, it was requested that any individual who volunteered to participate in the experiment did not suffer from colour blindness as this could have biased a participant's ability to distinguish and define contexts contained within the grave simulation.

This approach resulted in a grab sample of participants. In total, 50 individuals participated in the experiment, 40 had archaeological training, and ten acted as controls and had never received any archaeological training. Each archaeologist was allowed to choose which method they wished to use, and were able to choose which tools they would like to use to excavate and record the grave. In regards to the controls, they too were allowed to choose which tools they wished to use and were given the freedom to excavate and record the grave using whatever approach they deemed fit. In total, ten different archaeologists tested each of the four excavation methods and recording systems, producing data against which each of the excavation methods and recording systems could be compared.

Due to the nature of the field of archaeology, the participating archaeologists had varying levels of experience. For this reason, this variable was taken into consideration during the experiment, to determine whether it affected the quality and quantity of evidence recovered, and the consistency of interpretation(s) regarding the site's formation process. In order to collect data regarding the experience level that the participating archaeologists held, the author interviewed the participants before the experiment began, and asked questions relating to their archaeological experience (the length of time that they had worked as an archaeologist).

The final objective of the research project was to examine the affect that factors such as archaeological excavation method, archaeological recording system, and experience have on archaeological investigations including, the quality and quantity of evidence recovered and the consistency of interpretation(s) regarding the site's formation process. The data that has allowed for this objective to be completed was collected during the experimental phase of the research and has been used to determine:

Which excavation method was the most productive and consistent in terms of evidential recovery (evidence includes: artefacts, geotaphonomic features and stratigraphic contexts – deposits/fills/ interfaces/stratigraphic relationships).

Which recording system provided the most consistent and informative record of the evidence and depositional sequence present in the grave simulation.

Which excavation and recording method provided the most consistent interpretation-based narrative of the simulated grave's formation process.

To assist with the assessment of which excavation and recording technique provides the most consistent interpretation-based narrative of the simulated grave's formation process, the author went into the field, prior to the experimental phase of the research, and created the grave simulation. This procedure was recorded using a digital camera and a video camera. During filming, the author narrated the exact steps taken to form the grave. These steps are summarised in Figure 10.

This digital narrative along with the steps outlined in Figure 10 was then compared to the narratives produced by each of the participating archaeologists to evaluate which excavation and recording technique provided the most consistent and accurate interpretation of the simulated grave's formation process.

Stage	Description
Stage 1	The feature was cut (C4) through the top soil and turf (C3), subsoil (C2) and natural (C1)
Stage 2	The dress (E1) was placed along the base of the cut feature (C4)
Stage 3	Fill 1 (C5) started to be added to the feature overlaying the dress (E1)
Stage 4	A two pence coin (E2) was added to fill 1 (C5) at 10 cm L, 35 cm W, 25 cm D
Stage 5	The rest of fill 1 (C5) was added to the feature, covering the two pence coin (E2)
Stage 6	Fill 2 (C6) started to be added to the feature
Stage 7	A lighter (E3) was added to fill 2 (C6) at 105 cm L, 2 cm W, 28 cm D
Stage 8	The rest of fill 2 (C6) was added to the feature, covering the lighter (E3)
Stage 9	Fill 3 (C7) started to be added to the feature overlaying fill 1 (C5) and fill 2 (C6)
Stage 10	A fake nail (E4) was added to fill 3 (C7) at 70 cm L, 10 cm W, 25 cm D
Stage 11	More of fill 3 (C7) was added to the feature, covering the fake nail (E4)
Stage 12	An ID card (E5) was added to fill 3 (C7) at 50 cm L, 20 cm W, 20 cm D
Stage 13	More of fill 3 (C7) was added to the feature, covering the ID card (E5)
Stage 14	Earring 2 (E6) was added to fill 3 (C7) at 90 cm L, 20 cm W, 15 cm D
Stage 15	The rest of fill 3 (C7) was added to the feature, covering earring 2 (E6)
Stage 16	Fill 4 (C8) started to be added to the feature overlaying fill 3 (C7)
Stage 17	A kirby grip (E7) was added to fill 4 (C8) at 30 cm L, 15 cm W, 10 cm D
Stage 18	More of fill 4 (C8) was added to the feature, covering the kirby grip (E7)
Stage 19	Earring 1 (E8) was added to fill 4 (C8) at 90 cm L, 35 cm W, 5 cm D
Stage 20	The rest of fill 4 (C8) was added to the feature, covering earring 1 (E8)
Stage 21	Fill 5 (C9) started to be added to the feature overlaying fill 4 (C8)
Stage 22	A packet of cigarette papers (E9) was added to fill 5 (C9) at 10 cm L, 10 cm W, 4 cm D
Stage 23	The rest of fill 5 (C9) was added to the feature, covering the cigarette papers (E9)
Stage 24	The turf (C10) that had been removed during stage 1 was placed back over the feature, overlaying fill 5 (C9)

FIGURE 10: ILLUSTRATES THE STAGES OF THE GRAVE FORMATION PROCESS © EVIS 2016.

Chapter 4 Archaeological Manual/Guideline Analysis

This chapter contains the results from the archaeological manual/guideline analysis. These results are then explored in order to discuss the development, origins and current use of archaeological excavation methods and recording systems in the United Kingdom, Ireland, Australasia, and North America.

Results

Please note that during the analytical assessment of the archaeological manuals/guidelines gathered during the course of this research project, not all of the archaeological manuals/guidelines could be assessed against all of the analytical criteria used. This is because some of the archaeological manuals/guidelines failed to discuss some of the analytical criteria that were being assessed. As a result, some of the analytical figures presented in this results section will not total 100%, as not all of the archaeological manuals/guidelines could be used for that particular assessment.

The development and current use of archaeological excavation methods and recording systems

In order to evaluate the development and current use of archaeological excavation methods and recording systems in the United Kingdom, Ireland, Australasia and North America 153 archaeological manuals and guidelines were analysed against a set of analytical criteria (Appendix B).

General overview of archaeological manuals and guidelines

In terms of the sectors from which the manuals and guidelines that were analysed originated in the United Kingdom, 85% originated from the commercial sector and 15% came from the research sector (Figure 11). In Ireland, 56% came from the commercial sector, 22% came from the research sector and 22% came from the government sector (Figure 11). In Australasia, 67% came from the commercial archaeology sector, 17% came from the research sector and 17% came from the government sector (Figure 11). In North America, 44.3% came from the commercial sector, 34.9% came from the research sector and 20.8%

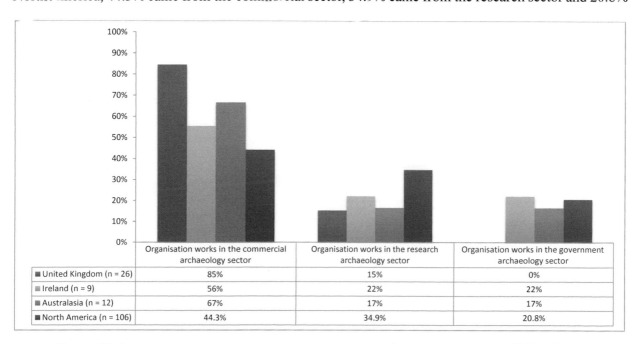

	Organisation works in the commercial archaeology sector	Organisation works in the research archaeology sector	Organisation works in the government archaeology sector
United Kingdom (n = 26)	85%	15%	0%
Ireland (n = 9)	56%	22%	22%
Australasia (n = 12)	67%	17%	17%
North America (n = 106)	44.3%	34.9%	20.8%

FIGURE 11: ARCHAEOLOGICAL SECTOR FROM WHICH THE MANUALS/GUIDELINES ORIGINATE © EVIS 2016.

came from the government sector (Figure 11). In terms of the overall sector distributions from which the manuals and guidelines originated, 54% of the total number of manuals and guidelines analysed originated from the commercial sector, 29% came from the research sector and 17% came from the government sector (Figure 12).

These results indicate that in current archaeological practice, it is the commercial archaeological sector that most often produce and use manuals and guidelines to dictate how archaeological investigations will be conducted. This result can be explained by the fact that commercial units are the archaeological sector that are most often conducting archaeological fieldwork and work on a number of different archaeological sites simultaneously and employ large numbers of staff. Therefore, in order to ensure consistency in the archaeological approaches that are used during such investigations, these companies produce archaeological manuals and guidelines.

The fact that the research sector produce the next highest volume of manuals and guidelines is because such institutions are often responsible for training students in archaeological practice. Consequently, they produce archaeological manuals and guidelines to provide a reference document for students to use in order for the students to become familiar with the principles and practices of archaeological fieldwork.

The fact that the government sector produced the least number of manuals and guidelines is because government organisations are usually office based, and do not conduct archaeological fieldwork themselves. Rather, they are responsible for overseeing the archaeological investigations that are conducted by other archaeological organisations in their jurisdictional area. Those government departments that do produce archaeological manuals or guidelines do so in order to establish a model of archaeological practice for archaeological organisations working within their area, to ensure that such organisations are adhering to local legislation and guidelines.

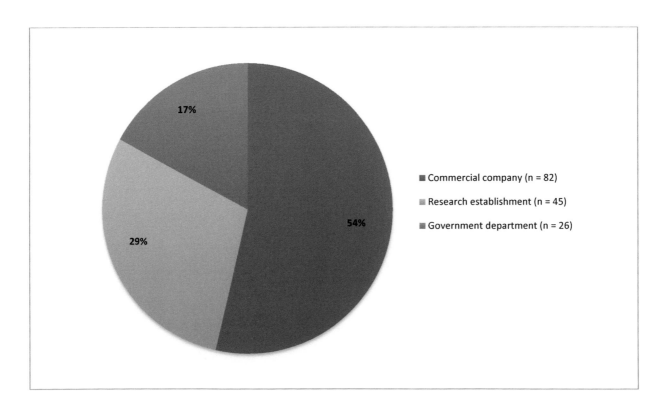

FIGURE 12: OVERALL SECTOR DISTRIBUTION FROM WHICH THE MANUALS/GUIDELINES ORIGINATE © EVIS 2016.

It is evident from Figure 13 that not all organisations have their own archaeological manual or guidelines. Instead, some organisations use another organisation's manual/guideline, or alternatively, develop excavation and recording protocols on a project-by-project basis.

In the United Kingdom, 69% of the organisations have their own archaeological manuals/guidelines, and 31% use another organisation's archaeological manual/guideline (Figure 13). Similarly, in Ireland, 67% of organisations have their own manuals/guidelines and 33% use another organisation's manual/ guideline (Figure 13). Both in the United Kingdom and Ireland, the organisations that don't have their own archaeological manuals/guidelines use the Museum of London Archaeology Service's (MOLAS) excavation manual. The reason why these organisations use the MOLAS manual is because this was one of the first archaeological organisations to formalise their archaeological practice and produce an archaeological manual with the aim of standardising commercial archaeological practice, and it was from this MOLAS manual that many other organisations formed their own archaeological manuals/guidelines. Those who chose not to develop their own archaeological manuals/guidelines used, and continue to use, the MOLAS manual as it provides an accepted approach to archaeological investigations and is available to download for free, thus saving those organisations who use this manual instead of creating their own both money and time in terms of production costs.

In Australasia, 17% of organisations sampled have their own archaeological manual/guidelines and 83% do not, out of the 83% of organisations that don't have their own manual/guidelines 30% use another organisation's manual, which again, is the Museum of London Archaeology Service's excavation manual, and 70% of Australasian organisations develop their excavation and recording protocols on a project-by-project basis (Figure 13). The reason why such a large percentage of organisations develop protocols on a project-by-project basis is due to the fact that these organisations are small and so cannot afford the costs associated with producing a manual. Moreover, due to the small size of the archaeological units

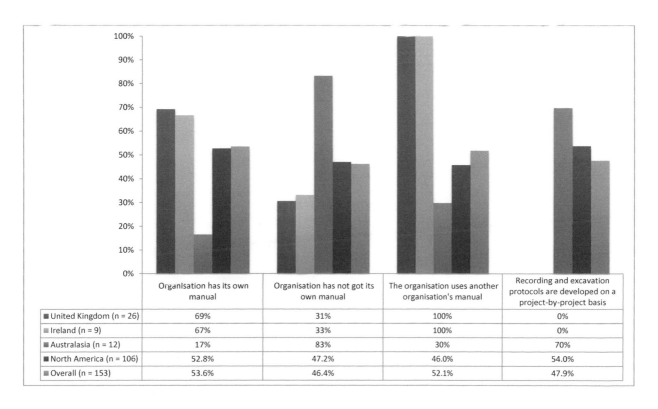

	Organisation has its own manual	Organisation has not got its own manual	The organisation uses another organisation's manual	Recording and excavation protocols are developed on a project-by-project basis
■ United Kingdom (n = 26)	69%	31%	100%	0%
■ Ireland (n = 9)	67%	33%	100%	0%
■ Australasia (n = 12)	17%	83%	30%	70%
■ North America (n = 106)	52.8%	47.2%	46.0%	54.0%
■ Overall (n = 153)	53.6%	46.4%	52.1%	47.9%

Figure 13: Overall manual/guideline usage © Evis 2016.

such organisations tend to work on one project at a time and therefore the archaeological site director can instruct each archaeologist on what methodological approaches will be being used and thereby ensure that a consistent approach is used. In addition, by developing bespoke protocols that are determined by the site that they are excavating, they are able to be more flexible in terms of archaeological approaches than those organisations that follow a set archaeological manual/guideline.

In North America, 52.8% of organisations have their own archaeological manuals/guidelines and 46.4% of organisations do not (Figure 13). Of those organisations that don't have their own archaeological manuals/ guidelines, 46% use another organisation's manual and 54% develop excavation and recording protocols on a project-by-project basis (Figure 13). The organisations that use another organisations archaeological manual/guidelines use the archaeological manuals/guidelines that have been produced by the State department in the State in which they are working. This ensures that archaeological practice is standardised in the State and that they meet the requirements of the local governing body. Those organisations that develop excavation and recording protocols on a project-by-project basis do so again, due to the size of their organisation and the flexibility that this gives them to adapt to the requirements of the archaeological site.

In terms of the time frames in which the archaeological manuals and guidelines analysed in this research project were either last updated or created, in the United Kingdom, 11% of the manuals/guidelines were either created or updated between 1990-1999, 83% between 2000-2009, and 6% between 2010-2013, although, 28% of the organisations are currently in the process of updating their manuals/guidelines (Figure 14). In Ireland, 83% of manuals were either created or updated between 2000-2009, 17% between 2010-2013, although, 17% of the organisations are currently in the process of updating their manuals/guidelines (Figure 14). In Australasia, 50% of the manuals were either created or updated between 2000-2009 and 50% were created or updated between 2010-2013 (Figure 14). In North America, 1.8% of manuals were either created or updated between 1970-1979, 1.8% between 1980-1989, 14.3% between 1990-1999, 71.4% between 2000-2009 and 10.7% between 2010-2013, although, 3.6% of organisations are currently

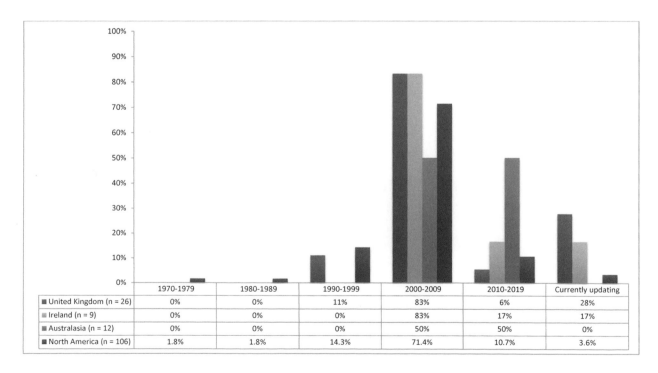

	1970-1979	1980-1989	1990-1999	2000-2009	2010-2019	Currently updating
■ United Kingdom (n = 26)	0%	0%	11%	83%	6%	28%
■ Ireland (n = 9)	0%	0%	0%	83%	17%	17%
■ Australasia (n = 12)	0%	0%	0%	50%	50%	0%
■ North America (n = 106)	1.8%	1.8%	14.3%	71.4%	10.7%	3.6%

FIGURE 14: TIMEFRAME IN WHICH THE MANUALS/GUIDELINES WERE CREATED © EVIS 2016.

in the process of updating their manuals/guidelines (Figure 14). In terms of overall time distribution, 1% of archaeological manuals were created or updated between 1970-1979, 1% between 1980-1989, 12% between 1990-1999, 75% between 2000-2009 and 11% between 2010-2013 (Figure 15).

These results show that the majority of organisations either updated or created their archaeological manuals/ guidelines between 2000-2009 (Figure 15). The reason for this was in part because new technological advances had been made in the field of archaeology and therefore organisations had

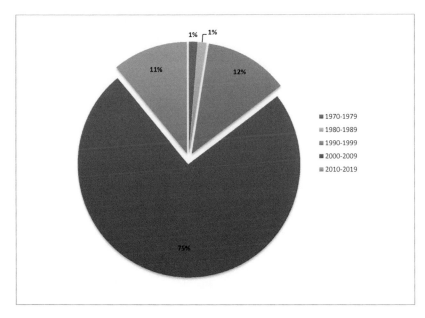

FIGURE 15: OVERALL TIMEFRAME DISTRIBUTION OF THE MANUALS/GUIDELINES © EVIS 2016.

to incorporate these new technological practices into their archaeological manuals/guidelines to instruct the archaeologists using the archaeological manuals/guidelines in how such technology should be utilised during the course of an archaeological investigation. Furthermore, this period saw a significant growth in the commercial archaeology sector, particularly in the United Kingdom and Ireland, due to increased building development projects throughout these countries. As a result, archaeological organisations were employing more archaeologists and working on more archaeological sites simultaneously, therefore, in order to ensure consistency across all of their archaeological investigations, organisations either produced or updated their manuals to ensure that a standardised approach was being utilised by all of their employees.

Similarly, the creation or updating of manuals between 1990-1999 can also be attributed to the increased demand for archaeologists to conduct archaeological investigations prior to building works, and the tightening of legislation in relation to the preservation of heritage sites in all of the countries discussed in this research project. Moreover, the organisations that have updated or created archaeological manuals between 2010-2013 have done so in order to account for new technological advances in the field, primarily in relation to developments in total station and scanning technologies. Those organisations that created or updated their manuals between 1970-1979 and 1980-1989 are small units and believe that their manuals/ guidelines provide enough data to instruct archaeologists on how to conduct archaeological investigations for their organisation. These organisations also stated that if new technology or techniques are to be used they, the organisation directors, can instruct their employees in their use whilst conducting the fieldwork and therefore do not need to update their manuals.

In terms of the archaeological manuals/guidelines general content, in the United Kingdom 19% of archaeological manuals/guidelines stated which excavation methods should be used and 81% suggested which excavation methods should be used, in regards to recording, 96% stated which recording systems should be used and 4% suggested which recording systems should be used (Figure 16). Similarly, in Ireland 11% of archaeological manuals/guidelines stated which excavation methods should be used and 89% suggested which excavation methods should be used, in regards to recording, 78% of the archaeological manuals/guidelines stated which recording systems should be used and 22% suggested which recording systems should be used (Figure 16). In Australasia, 92% of archaeological manuals/guidelines stated which

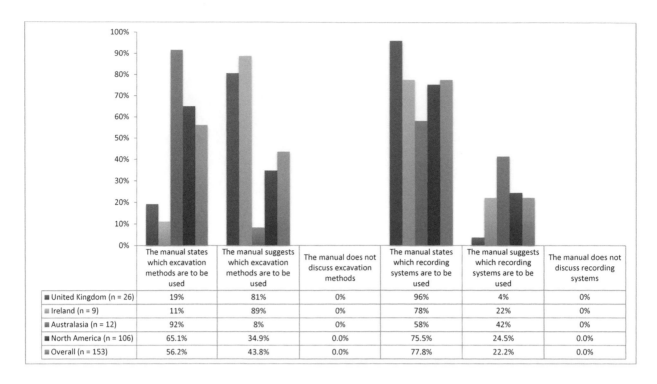

	The manual states which excavation methods are to be used	The manual suggests which excavation methods are to be used	The manual does not discuss excavation methods	The manual states which recording systems are to be used	The manual suggests which recording systems are to be used	The manual does not discuss recording systems
■ United Kingdom (n = 26)	19%	81%	0%	96%	4%	0%
■ Ireland (n = 9)	11%	89%	0%	78%	22%	0%
■ Australasia (n = 12)	92%	8%	0%	58%	42%	0%
■ North America (n = 106)	65.1%	34.9%	0.0%	75.5%	24.5%	0.0%
■ Overall (n = 153)	56.2%	43.8%	0.0%	77.8%	22.2%	0.0%

FIGURE 16: MANUALS/GUIDELINES GENERAL CONTENT © EVIS 2016.

excavation methods should be used and 8% suggested which excavation methods should be used, whereas 58% of the manuals/guidelines stated which recording systems should be used and 42% suggested which should be used (Figure 16). In North America, 65.1% of the archaeological manuals/guidelines stated which excavation methods should be used and 34.9% suggested which excavation methods should be used, in regards to recording, 75.5% of archaeological manuals/guidelines stated which recording systems should be used and 24.5% suggested which recording systems should be used (Figure 16). Overall, 56.2% of the archaeological manuals/guidelines stated which excavation methods should be used and 43.8% suggested which excavation methods should be used, in terms of recording, 77.8% of archaeological manuals/guidelines stated which recording systems should be used and 22.2% suggested which recording systems should be used (Figure 16).

It is clear from these results that the majority of organisations in the United Kingdom and Ireland suggest rather than state outright which excavation methods should be used during an archaeological investigation. This is because these organisations aim to give the site directors in charge of archaeological investigations the opportunity to adapt their excavation approaches to the demands of the individual archaeological site that is being excavated. Thus, if the excavation is taking place in a rural location that does not contain many complex archaeological features, the excavation approach will differ to excavations that are taking place on urban sites with complex intercutting stratigraphy. In comparison, in Australasia and North America the majority of organisations state which archaeological excavation methods should be used, rather than suggesting potential excavation approaches. This means that their excavation methods are less flexible than those of the British and Irish organisations and that they will apply the stated excavation approaches to an archaeological site despite the site type that they are excavating. This can be explained by the fact that a lot of the archaeological investigations that are conducted in Australasia and North America relate to Aboriginal or Native American rural heritage sites. In these sites the archaeological evidence for inhabitation is often widely dispersed and small in volume. Consequently, Australasian and North American archaeologists can apply the same excavation techniques to investigate these sites, as they do not have to be as flexible as British and Irish archaeologists, who deal with both rural, widely dispersed archaeological sites, and urban archaeological sites that have been continually inhabited for hundreds of years.

It is interesting to note that in terms of recording systems the majority of organisations stated specifically what recording systems should be used rather than suggesting which systems should be used. This is due to the fact that recording systems are more standardised than excavation methods within archaeological practice, and need to be, as the records that are produced are all that remains of the archaeological site once it has been excavated. Consequently, in order to ensure that a comprehensive set of records is obtained, all manuals provide blank copies of the recording forms that will be used and set strict guidelines regarding their use.

In terms of the overall objectives of the archaeological manuals and guidelines, apart from one organisation in the United Kingdom, all of the archaeological manuals and guidelines are designed to: instruct archaeologists in what excavation methods and recording systems should be used, explain how recording sheets should be completed, inform archaeologists of what should be recorded and when it should be recorded, and provide archaeologists with a default approach to the excavation and recording of archaeological sites (Figure 17). It is evident from these results that all of the archaeological organisations in the geographical regions under discussion are attempting to produce archaeological manuals/guidelines that standardise archaeological practice in order to ensure that the results that are produced are comprehensive and consistent.

In regards to the applicability of the archaeological manuals/guidelines on different types of archaeological sites, in the United Kingdom, 96% of manuals/guidelines are designed to instruct archaeologists on which excavation and recording techniques can be used for archaeological excavations conducted on both small and large scale sites, but only 92% are designed for rural sites and 96% are designed for urban sites (Figure 18). Whereas, all of the archaeological manuals/guidelines originating from Ireland, Australasia

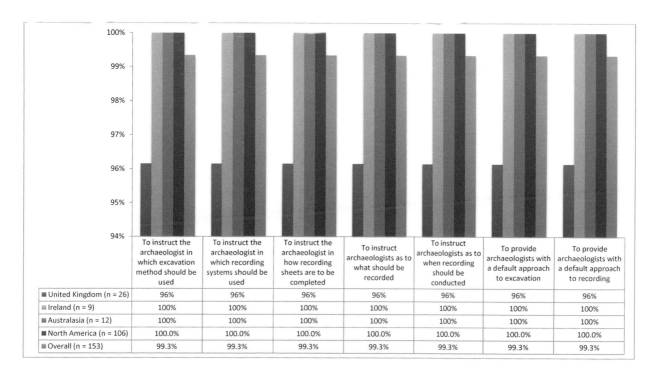

	To instruct the archaeologist in which excavation method should be used	To instruct the archaeologist in which recording systems should be used	To instruct the archaeologist in how recording sheets are to be completed	To instruct archaeologists as to what should be recorded	To instruct archaeologists as to when recording should be conducted	To provide archaeologists with a default approach to excavation	To provide archaeologists with a default approach to recording
United Kingdom (n = 26)	96%	96%	96%	96%	96%	96%	96%
Ireland (n = 9)	100%	100%	100%	100%	100%	100%	100%
Australasia (n = 12)	100%	100%	100%	100%	100%	100%	100%
North America (n = 106)	100.0%	100.0%	100.0%	100.0%	100.0%	100.0%	100.0%
Overall (n = 153)	99.3%	99.3%	99.3%	99.3%	99.3%	99.3%	99.3%

FIGURE 17: OVERALL OBJECTIVES OF THE MANUALS/GUIDELINES © EVIS 2016.

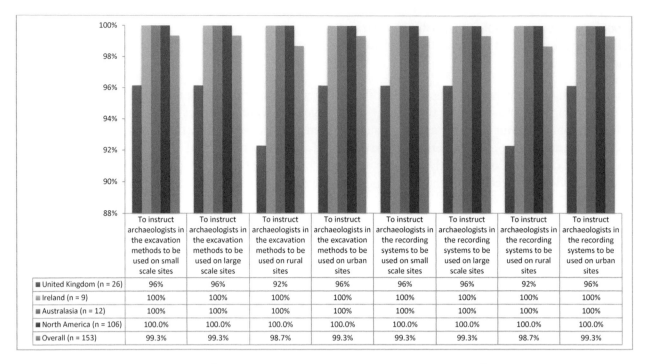

Figure 18: Manuals/guidelines applicability on different site types © Evis 2016.

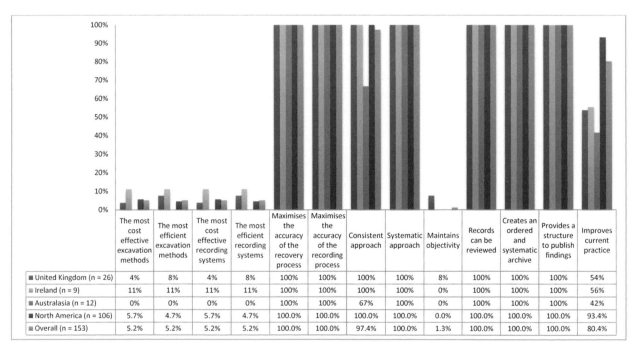

Figure 19: Justifications for the excavation and recording methods advocated in the manuals/guidelines © Evis 2016.

and North America are designed to instruct archaeologists in the excavation and recording techniques to use for small, large, rural and urban archaeological sites (Figure 18). The reason why one British archaeological manual/guideline was deemed to be unsuitable for use on rural sites is due to the fact that this manual/guideline was designed by an organisation that primarily works on urban archaeological sites with very complex stratigraphy, therefore, as they do not conduct archaeological work outside of this context they have not adapted their archaeological manual/guideline for use on rural archaeological sites. The reason why one British archaeological manual/guideline fails to instruct archaeologists on any of the aforementioned variables is due to the fact that this organisation has not stated in their archaeological guideline/manual the circumstances in which their archaeological manual/guideline should be used, and therefore, could not be analysed in the same manner as the other archaeological manuals/guidelines.

In relation to how the organisations have justified the excavation and recording techniques that they have advocated in their archaeological manuals/guidelines. In the United Kingdom, 4% of organisations stated that it was so that the most cost effective excavation and recording methods were used, 8% stated that it was so the most efficient excavation and recording methods are used, 100% stated that it was to ensure accuracy during the recovery and recording process, 100% stated that it was to ensure that a consistent and systematic process was used, 8% stated that it was to make sure that objectivity was maintained, 100% stated that it was so records can be reviewed by interested parties, 100% stated that it was to ensure that an ordered and systematic archive was created so that they are able to publish their findings, and 54% stated that it was to improve current practice (Figure 19).

Similarly, in Ireland, 11% stated that it was to ensure the most cost effective excavation methods and recording systems are used, 11% stated that it was to ensure that the most efficient excavation methods and recording systems were used, 100% stated that it was to maximise the accuracy of the recording and recovery process, 100% stated that it was to ensure that a consistent and systematic approach was used, 100% stated that it was so that records can be reviewed by interested parties, 100% stated that it was to ensure that an ordered and systematic archive was produced that would enable publications to be produced, and 56% thought it would improve current archaeological practice (Figure 19).

Alternatively, in Australasia, 100% of organisations thought the techniques stated in the manual would maximise the accuracy of the recovery and recording process, 67% said it would ensure a consistent approach was used, 100% stated that it would ensure a systematic approach was used, 100% stated that it was so that records could be reviewed by interested parties, 100% stated that it would result in the production of a consistent and structured archive from which they could publish their findings, and 42% thought it would improve current archaeological practice (Figure 19).

In North America, 5.7% of organisations stated that it would ensure that the most cost effective excavation methods and recording systems would be used, 4.7% stated that it would ensure that the most efficient excavation methods and recording systems would be used, 100% stated that it would maximise the accuracy of the recovery and recording process, 100% stated that it would ensure that a consistent and systematic approach would be used, 100% stated that it ensured that records could be reviewed by interested parties, 100% stated that it would result in the creation of a systematic archive from which they can publish their findings, and 93.4% stated that it would improve current archaeological practice (Figure 19).

It is evident from these results that all of the different archaeological organisations analysed in this research project believe that by using the excavation and recording approaches that they advocate in their archaeological manuals/guidelines, archaeologists will use a consistent and systematic archaeological approach that will maximise the accuracy of the recovery and recording process during an archaeological investigation. This in turn, will ensure that the users of this approach produce a systematic and ordered archive that will allow them to publish their findings and interested parties to review the data, and consequently, result in an improvement in the quality of the archaeological investigation. However, the fact that 153 different archaeological organisations have felt the need to produce or use archaeological manuals/guidelines suggests that such justifications cannot be true, as there cannot be 153 correct ways to investigate an archaeological site. Therefore, there is clearly not a universally accepted standardised approach to conducting archaeological investigations.

The identification, definition and recording of archaeological stratigraphy

In regards to who is responsible for identifying and recording archaeological stratigraphy during an archaeological investigation, in the United Kingdom, Ireland and Australasia the archaeologist who has conducted the excavation is responsible for identifying and recording archaeological stratigraphy (Figure 20). In North America, however, 16% of archaeological organisations state that specialist geoarchaeologists are responsible for identifying and recording archaeological stratigraphy (Figure 20).

The fact that some North American organisations employ specialist geoarchaeologists is at odds with archaeological practice in the United Kingdom, Ireland and Australasia, as being able to recognise, define and record archaeological stratigraphy is regarded as one of the fundamental skills of an archaeologist. The reason why some North American organisations feel that they need to employ specialist geoarchaeologists to identify and record archaeological stratigraphy is perhaps due to the fact that in North America archaeology is regarded as a subfield of Anthropology. Therefore, when students graduate from North American universities they have received little practical archaeological training in comparison to archaeology graduates in the United Kingdom, Ireland and Australasia and as a result aren't deemed to be competent enough to effectively recognise and record archaeological stratigraphy. Consequently, some North American organisations believe it is necessary for employees to receive further archaeological training in geoarchaeology (soil science) that in turn, will enable them to recognise and record the archaeological stratigraphy that is present at the archaeological site being investigated.

In terms of what the different archaeological organisations regard as archaeological stratigraphy there are some distinct differences. In the United Kingdom, a positive stratigraphic unit is deemed by 96% of archaeological organisations to be a fill, layer or structure (Figure 21). Likewise, in Ireland a positive stratigraphic unit is deemed by 100% of archaeological organisations to be a fill, layer or structure (Figure 21). Alternatively, in Australasia, a positive stratigraphic unit is deemed by 100% of archaeological organisations to be a fill or a layer, whereas only 8% of Australasian organisations regard structures as positive stratigraphic units with the remaining 92% defining structures as features rather than positive stratigraphic units (Figure 21). Similarly in North America, 74.5% of organisations regard positive stratigraphic units to be a fill or layer, whereas 4.7% of organisations regard structures as positive stratigraphic units with the remaining 70.8% of organisations defining structures as features rather than positive stratigraphic units (Figure 21).

In relation to negative stratigraphic units, in the United Kingdom a negative stratigraphic unit is defined by 100% of archaeological organisations as a cut that has resulted from the removal of material, 96% of British archaeological organisations define each cut that is identified as a separate stratigraphic unit and record them on separate paperwork (Figure 22). Similarly in Ireland, 100% of archaeological organisations regard a negative stratigraphic unit as a cut, and define and record each cut as an individual stratigraphic unit (Figure 22). In Australasia, archaeological organisations also define a negative stratigraphic unit as a cut, however only 8% of organisations define and record each cut as separate stratigraphic units and use separate paperwork for each, instead, the remaining 92% regard cuts as features rather than stratigraphic units (Figure 22). In North America, 74.5% of archaeological organisations define a negative stratigraphic unit as a cut, but only 38.7% of archaeological organisations define and record cuts as distinct stratigraphic units (Figure 22).

It is clear from these results that both British and Irish archaeological organisations define archaeological stratigraphic units using the same criteria. They also record stratigraphic units in the same manner ensuring that each stratigraphic unit is analysed and recorded separately, and that if stratigraphic units share stratigraphic relationships these are collated in order to form feature or unit groups (Figure 21; Figure 22). Unlike archaeological organisations in the United Kingdom and Ireland, the Australasian and

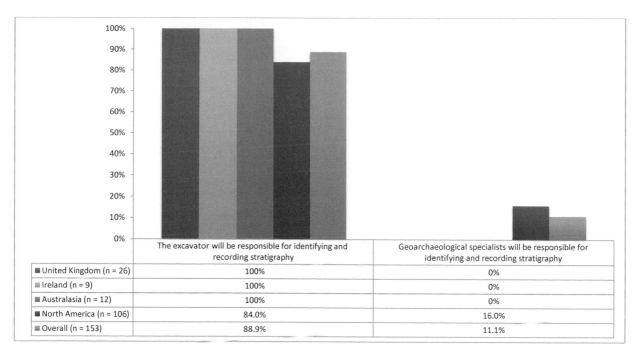

	The excavator will be responsible for identifying and recording stratigraphy	Geoarchaeological specialists will be responsible for identifying and recording stratigraphy
■ United Kingdom (n = 26)	100%	0%
■ Ireland (n = 9)	100%	0%
■ Australasia (n = 12)	100%	0%
■ North America (n = 106)	84.0%	16.0%
■ Overall (n = 153)	88.9%	11.1%

FIGURE 20: MEMBERS OF THE ARCHAEOLOGICAL TEAM RESPONSIBLE FOR IDENTIFYING AND RECORDING STRATIGRAPHY © EVIS 2016.

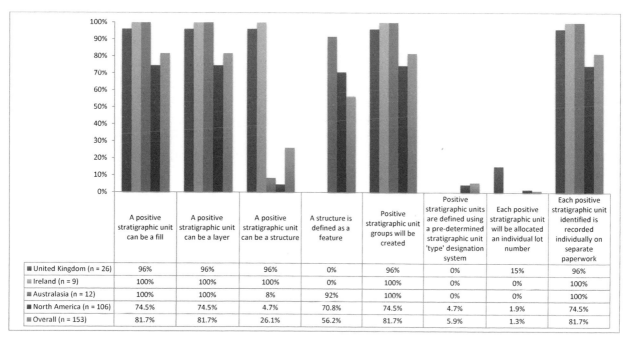

	A positive stratigraphic unit can be a fill	A positive stratigraphic unit can be a layer	A positive stratigraphic unit can be a structure	A structure is defined as a feature	Positive stratigraphic unit groups will be created	Positive stratigraphic units are defined using a pre-determined stratigraphic unit 'type' designation system	Each positive stratigraphic unit will be allocated an individual lot number	Each positive stratigraphic unit identified is recorded individually on separate paperwork
■ United Kingdom (n = 26)	96%	96%	96%	0%	96%	0%	15%	96%
■ Ireland (n = 9)	100%	100%	100%	0%	100%	0%	0%	100%
■ Australasia (n = 12)	100%	100%	8%	92%	100%	0%	0%	100%
■ North America (n = 106)	74.5%	74.5%	4.7%	70.8%	74.5%	4.7%	1.9%	74.5%
■ Overall (n = 153)	81.7%	81.7%	26.1%	56.2%	81.7%	5.9%	1.3%	81.7%

FIGURE 21: MANUALS/GUIDELINES' DEFINITION OF A POSITIVE STRATIGRAPHIC UNIT © EVIS 2016.

North American archaeological organisations define stratigraphic units differently, although similar to each other. The difference is most noticeable in regards to how they define structures and cuts, as they view them as features rather than definable stratigraphic units and will record them on feature forms rather than archaeological stratigraphy forms (Figure 21; Figure 22). One of the most interesting differences in approaches to recording archaeological stratigraphy is that some of the North American organisations (4.7%) define positive stratigraphic units by using a pre-determined stratigraphic unit type designation

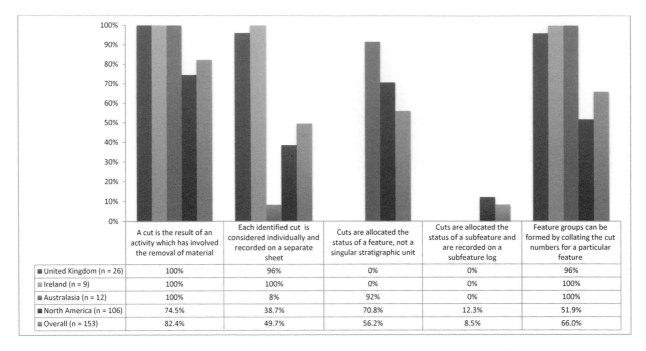

	A cut is the result of an activity which has involved the removal of material	Each identified cut is considered individually and recorded on a separate sheet	Cuts are allocated the status of a feature, not a singular stratigraphic unit	Cuts are allocated the status of a subfeature and are recorded on a subfeature log	Feature groups can be formed by collating the cut numbers for a particular feature
■ United Kingdom (n = 26)	100%	96%	0%	0%	96%
■ Ireland (n = 9)	100%	100%	0%	0%	100%
■ Australasia (n = 12)	100%	8%	92%	0%	100%
■ North America (n = 106)	74.5%	38.7%	70.8%	12.3%	51.9%
■ Overall (n = 153)	82.4%	49.7%	56.2%	8.5%	66.0%

FIGURE 22: MANUALS/GUIDELINES' DEFINITION OF A NEGATIVE STRATIGRAPHIC UNIT © EVIS 2016.

system (Figure 21). When using such a system, if an archaeologist excavates a layer or fill that contains evidence of burning for example, they will refer to the list of strata codes and will allocate that layer the relevant strata code for burnt layer/fill from the archaeological manual/guideline.

In regards to how archaeological organisations record individual stratigraphic units, in the United Kingdom 96% of archaeological organisations allocate stratigraphic units a primary class (cut/fill/deposit etc.), 38% of archaeological organisations will then allocate stratigraphic units a secondary class that will give an indication of the contexts function, and 96% of archaeological organisations encourage their employees to write interpretive comments regarding the stratigraphic unit that has been excavated (Figure 23). Similarly, in Ireland, 100% of archaeological organisations allocate stratigraphic units primary classes, 44% then go on to allocate stratigraphic units secondary classes, and 100% of archaeological organisations encourage their employees to write interpretive comments relating to the stratigraphic unit that has been excavated (Figure 23). In Australasia, 17% of archaeological organisations allocate stratigraphic units primary classes, and 100% encourage their employees to write interpretive comments regarding the stratigraphic unit being dealt with (Figure 23). In North America, 3.8% of archaeological organisations allocate stratigraphic units primary classes and 74.5% encourage their employees to write interpretive comments about the stratigraphic unit that has been excavated (Figure 23). Moreover, in North America, some archaeological organisations will attach suffixes to a stratigraphic unit's identification code, which will indicate the stratigraphic unit's position and function within the stratigraphic sequence (Figure 23).

By reviewing these results it is evident that again, British and Irish organisations tend to adopt the same approaches to recording archaeological stratigraphy, with some archaeological organisations choosing to expand upon their description of a stratigraphic unit by allocating it a secondary class. The reason why the majority of Australasian and North American archaeological units do not record stratigraphy in the same manner as the British and Irish archaeological organisations is due to the fact that they define stratigraphic units in a different manner, as discussed above. Therefore, when recording stratigraphic units, as they do not regard cuts as a distinct unit of archaeological stratigraphy, they do not require recording sheets that will differentiate between cuts, fills and deposits. This also explains why some

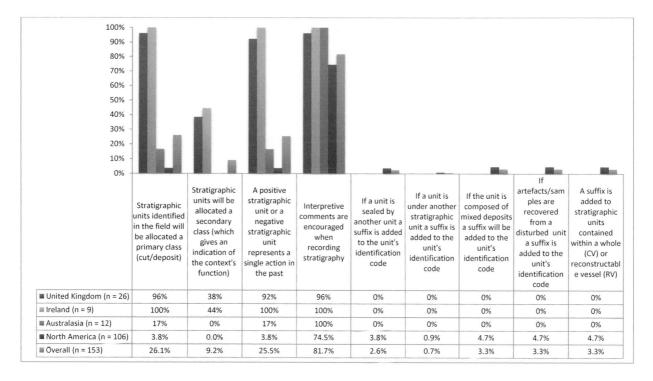

	Stratigraphic units identified in the field will be allocated a primary class (cut/deposit)	Stratigraphic units will be allocated a secondary class (which gives an indication of the context's function)	A positive stratigraphic unit or a negative stratigraphic unit represents a single action in the past	Interpretive comments are encouraged when recording stratigraphy	If a unit is sealed by another unit a suffix is added to the unit's identification code	If a unit is under another stratigraphic unit a suffix is added to the unit's identification code	If the unit is composed of mixed deposits a suffix will be added to the unit's identification code	If artefacts/sam ples are recovered from a disturbed unit a suffix is added to the unit's identification code	A suffix is added to stratigraphic units contained within a whole (CV) or reconstructabl e vessel (RV)
■ United Kingdom (n = 26)	96%	38%	92%	96%	0%	0%	0%	0%	0%
▥ Ireland (n = 9)	100%	44%	100%	100%	0%	0%	0%	0%	0%
■ Australasia (n = 12)	17%	0%	17%	100%	0%	0%	0%	0%	0%
■ North America (n = 106)	3.8%	0.0%	3.8%	74.5%	3.8%	0.9%	4.7%	4.7%	4.7%
▥ Overall (n = 153)	26.1%	9.2%	25.5%	81.7%	2.6%	0.7%	3.3%	3.3%	3.3%

FIGURE 23: MANUALS/GUIDELINES' APPROACHES TO RECORDING STRATIGRAPHIC UNITS © EVIS 2016.

North American organisations use suffixes to outline the function of fills and layers as they do not allocate primary or secondary classes to the stratigraphic units they are excavating. The vast majority of archaeological organisations, do, however, encourage their employees to write interpretive comments regarding the archaeological stratigraphy present. This enables archaeologists to express their ideas about how a stratigraphic unit relates to the archaeological site and to other stratigraphic units present at the site or in the feature that they are excavating, and may help to explain the development of the archaeological site.

In regards to how archaeological organisations record the relationships present within the stratigraphic sequence, in the United Kingdom, 62% of archaeological organisations only record stratigraphic relationships, 4% only record physical relationships, and 31% record both the physical and stratigraphic relationships present (Figure 24). In Ireland, 100% of archaeological organisations record only the stratigraphic relationships present (Figure 24). Likewise, in Australasia, 100% of archaeological organisations record only the stratigraphic relationships present (Figure 24). In North America, 5.7% of archaeological organisations only record the stratigraphic relationships present, and 68.9% record both the physical and stratigraphic relationships present (Figure 24). Moreover, when British or North American archaeological organisations are using the Arbitrary Level Excavation method to excavate an archaeological site, 8% of British and 69.8% of North American archaeological organisations, will determine the stratigraphic accumulation of the archaeological site *post-facto* (Figure 24).

These results indicate that the majority of archaeological organisations in the United Kingdom, Ireland and Australasia only record the stratigraphic relationships present at an archaeological site. This is because it is only the stratigraphic relationships present at an archaeological site that will accurately inform archaeologists of the chronological sequence of events that occurred during the development of the site from its original inception to present day (Figure 26). This in turn, provides such archaeologists with reliable data from which to reconstruct the archaeological site and make informed interpretations regarding the history of the site (Figure 26). The British archaeological organisation that states that only the physical

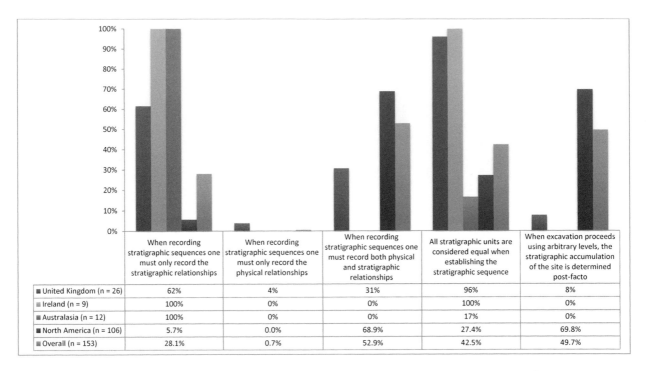

	When recording stratigraphic sequences one must only record the stratigraphic relationships	When recording stratigraphic sequences one must only record the physical relationships	When recording stratigraphic sequences one must record both physical and stratigraphic relationships	All stratigraphic units are considered equal when establishing the stratigraphic sequence	When excavation proceeds using arbitrary levels, the stratigraphic accumulation of the site is determined post-facto
United Kingdom (n = 26)	62%	4%	31%	96%	8%
Ireland (n = 9)	100%	0%	0%	100%	0%
Australasia (n = 12)	100%	0%	0%	17%	0%
North America (n = 106)	5.7%	0.0%	68.9%	27.4%	69.8%
Overall (n = 153)	28.1%	0.7%	52.9%	42.5%	49.7%

FIGURE 24: MANUALS/GUIDELINES' APPROACHES TO RECORDING STRATIGRAPHIC RELATIONSHIPS © EVIS 2016.

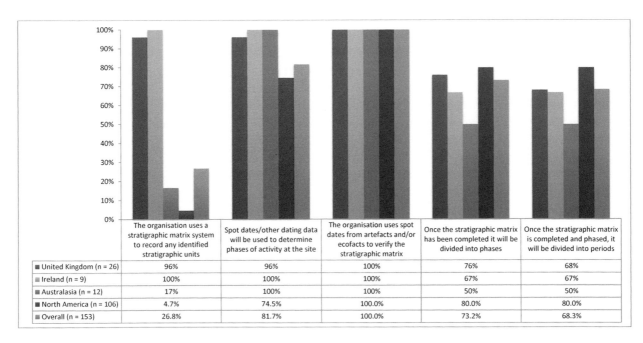

	The organisation uses a stratigraphic matrix system to record any identified stratigraphic units	Spot dates/other dating data will be used to determine phases of activity at the site	The organisation uses spot dates from artefacts and/or ecofacts to verify the stratigraphic matrix	Once the stratigraphic matrix has been completed it will be divided into phases	Once the stratigraphic matrix is completed and phased, it will be divided into periods
United Kingdom (n = 26)	96%	96%	100%	76%	68%
Ireland (n = 9)	100%	100%	100%	67%	67%
Australasia (n = 12)	17%	100%	100%	50%	50%
North America (n = 106)	4.7%	74.5%	100.0%	80.0%	80.0%
Overall (n = 153)	26.8%	81.7%	100.0%	73.2%	68.3%

FIGURE 25: MANUALS/GUIDELINES' APPROACHES TO REPRESENTING AND VERIFYING STRATIGRAPHIC SEQUENCES © EVIS 2016.

relationships present at an archaeological site should be recorded, appears to be an anomaly amongst British archaeological organisations, and is perhaps due to a publication error, as by only recording the physical relationships present at an archaeological site this archaeological company will not be able to accurately interpret or reconstruct the sequence of events that led to the formation of the archaeological site. It is interesting to note that some British and North American archaeological organisations choose to record both stratigraphic and physical relationships. This is due to the fact that by recording physical

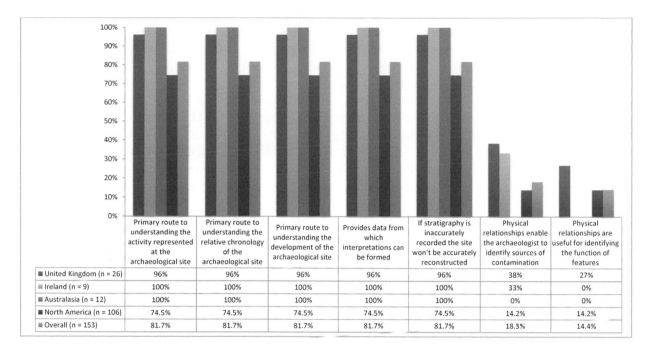

	Primary route to understanding the activity represented at the archaeological site	Primary route to understanding the relative chronology of the archaeological site	Primary route to understanding the development of the archaeological site	Provides data from which interpretations can be formed	If stratigraphy is inaccurately recorded the site won't be accurately reconstructed	Physical relationships enable the archaeologist to identify sources of contamination	Physical relationships are useful for identifying the function of features
United Kingdom (n = 26)	96%	96%	96%	96%	96%	38%	27%
Ireland (n = 9)	100%	100%	100%	100%	100%	33%	0%
Australasia (n = 12)	100%	100%	100%	100%	100%	0%	0%
North America (n = 106)	74.5%	74.5%	74.5%	74.5%	74.5%	14.2%	14.2%
Overall (n = 153)	81.7%	81.7%	81.7%	81.7%	81.7%	18.3%	14.4%

Figure 26: Purpose of recording stratigraphy © Evis 2016.

relationships as well as stratigraphic relationships archaeologists can determine if a stratigraphic unit has become contaminated with artefacts from the stratigraphic units above it due to bioturbation (Figure 26). Additionally, recording physical relationships can also help archaeologists determine the function that a particular archaeological feature served on the archaeological site (Figure 26). The fact that some British and North American organisations record the stratigraphic accumulation of the site *post-facto* after using an Arbitrary Level method of excavation, is due to the fact that if archaeologists are using this technique they have no alternative option, as the process of Arbitrary Level Excavation destroys the dimensions of stratigraphic units and therefore it is not possible to record stratigraphy as the excavation proceeds. Instead, archaeologists must rely on examining the section faces of the excavation units that they have excavated in order to attempt to retrieve stratigraphic data.

In relation to how archaeological organisations represent stratigraphic data, in the United Kingdom, 96% of archaeological organisations use a stratigraphic matrix to represent the stratigraphic sequence, 100% of Irish archaeological organisations use a stratigraphic matrix to represent the stratigraphic sequence, 17% of Australasian archaeological organisations use a stratigraphic matrix to represent the stratigraphic sequence, and 4.7% of North American archaeological organisations use a stratigraphic matrix to represent the stratigraphic sequence (Figure 25). These results indicate that the majority of British and Irish organisations use a stratigraphic matrix to represent stratigraphic data, this is unsurprising as the Harris Matrix, or variations of this matrix system, have been integrated into archaeological investigations in these countries since the mid 1970s, and are now regarded as standard practice. The fact that barely any archaeological organisations in Australasia or North America use a matrix system to represent stratigraphic data is rather concerning, as this suggests that archaeological organisations operating in these areas are behind in archaeological methodological developments by at least 30 years. A possible reason why archaeological organisations operating in these areas have yet to adopt this approach to representing archaeological stratigraphy is because the majority of archaeological investigations that these organisations undertake take place in rural locations in which the archaeological evidence is widely dispersed, small in volume and lack complex stratigraphic sequences. Therefore, as these organisations are not dealing with sites that contain complex stratigraphic sequences that need a logical and structured system in order to represent and interpret them, they are able to understand and reconstruct the stratigraphic sequence by reading the descriptions of the stratigraphic units contained in their recording sheets. Whereas, those archaeological organisations in Australasia and North America that do use a matrix system to represent stratigraphic data are organisations that work on urban sites and need to use the matrix system in order to understand the complex stratigraphic relationships present.

In regards to how the archaeological organisations that use stratigraphic matrices to represent stratigraphic data verify the stratigraphic sequences that they have created and use this data to interpret the archaeological site, in the United Kingdom, 100% of archaeological organisations use spot dates/other dating material to verify the stratigraphic sequence, 96% use spot dates/other dating materials to then determine the phase activity of the site, 76% of archaeological organisations will then divide the stratigraphic matrix into phases, and 68% of archaeological organisations will subsequently divide the stratigraphic matrix into defined periods (Figure 25). In Ireland, 100% of archaeological organisations use spot dates/other dating material to verify the stratigraphic sequence, 100% of archaeological organisations use spot dates/other dating material to determine the phase activity of the site, and 67% of archaeological organisations will then divide the stratigraphic matrix into phases and periods (Figure 25). In Australasia, 100% of archaeological organisations will use spot dates/other dating materials to verify the stratigraphic sequence, 100% of archaeological organisations use spot dates/other dating material to determine the phase activity of the site, and 50% of archaeological organisations will then divide the stratigraphic matrix into phases and periods (Figure 25). In North America, 100% of archaeological organisations use spot dates/other dating materials to verify the stratigraphic sequence, 74.5% will use spot dates/other dating material to divide the site into phases, and 80% of archaeological organisations will divide the stratigraphic matrix into phases and periods (Figure 25).

These results indicate that all of the archaeological organisations that use stratigraphic matrices to represent stratigraphic sequences verify this sequence and phases of activity present at the site by using spot dating and other dating evidence. The majority of organisations then edit their stratigraphic matrices in order to represent phases of activity evident at the site and subsequently split the matrix into periods of inhabitation and abandonment. The fact that not all archaeological organisations choose to phase their matrices or divide them into distinct periods is down to the preferences of that particular organisation, as some organisations feel that as long as they discuss the phases of activity present in the site within their archaeological reports, there is no need to edit the matrix to reflect this.

Recording strategies

In terms of when archaeological organisations use section drawings to document archaeological data, in the United Kingdom 54% of archaeological organisations will only use section drawings when single context planning cannot be used, 4% will create running sections across the archaeological site, 92% will use section drawings to document the long sections of any trenches that have been excavated but only 4% of these organisations will only create long section drawings of trenches if archaeological evidence is present, 96% will use section drawings to record the walls of any test excavation units that have been excavated, and 96% will use section drawings to record any features present at the archaeological site that have been sectioned (Figure 27). In Ireland, 33% of archaeological organisations will only use section drawings if single context planning cannot be used, 100% will record long sections of any trenches that have been excavated regardless of whether archaeological evidence has been identified within them, 100% will record section drawings of the walls of any test excavation units that have been excavated, and 100% will record section drawings of any archaeological features that have been sectioned (Figure 27). In Australasia, 8% of archaeological organisations will only use section drawings if single context planning cannot be used, 100% will record long sections of any trenches that have been excavated regardless of whether archaeological evidence has been identified within them, 100% will record section drawings of the

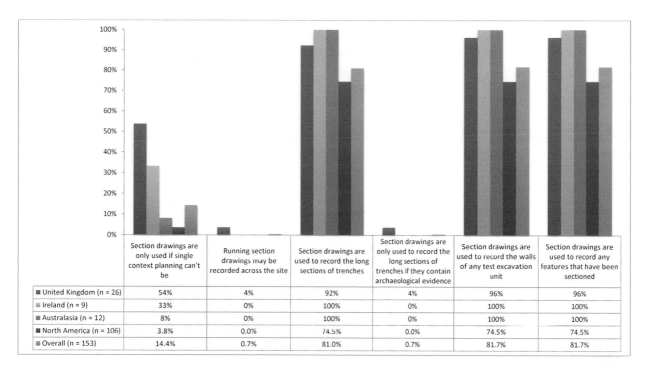

	Section drawings are only used if single context planning can't be	Running section drawings may be recorded across the site	Section drawings are used to record the long sections of trenches	Section drawings are only used to record the long sections of trenches if they contain archaeological evidence	Section drawings are used to record the walls of any test excavation unit	Section drawings are used to record any features that have been sectioned
■ United Kingdom (n = 26)	54%	4%	92%	4%	96%	96%
■ Ireland (n = 9)	33%	0%	100%	0%	100%	100%
■ Australasia (n = 12)	8%	0%	100%	0%	100%	100%
■ North America (n = 106)	3.8%	0.0%	74.5%	0.0%	74.5%	74.5%
■ Overall (n = 153)	14.4%	0.7%	81.0%	0.7%	81.7%	81.7%

FIGURE 27: USE OF SECTION DRAWINGS © EVIS 2016.

walls of any test excavation units that have been excavated, and 100% will record section drawings of any archaeological features that have been sectioned (Figure 27). In North America, 3.8% of archaeological organisations will only use section drawings if single context planning cannot be used, 74.5% will record long sections of any trenches that have been excavated regardless of whether archaeological evidence has been identified within them, 74.5% will record section drawings of any test excavation units that have been excavated, and 74.5% of archaeological organisations will use section drawings to record any features that have been sectioned (Figure 27).

These results indicate that the majority of archaeological organisations use section drawings to document the formation sequence of trenches or excavation units that have been excavated, or archaeological features that have been sectioned in order to document the deposition sequence present. This is because the excavation process used to cut trenches, excavation units, or to section archaeological features results in the deposition sequence present at an archaeological site or in the archaeological feature being cut through, and therefore, the only way in which to record the deposition sequence is to record section drawings, either in the form of a long section or a half section drawing.

The reason why the majority of archaeological organisations in the United Kingdom (54%) and some organisations in Ireland (33%), Australasia (8%) and North America (3.8%) only use section drawings when single context planning cannot be used, is because these organisations prefer to excavate archaeological sites using a Stratigraphic Excavation approach, which is an excavation method that is closely tied with the Single Context Recording system, as a result, these organisations excavate individual stratigraphic units in their entirety and plan each unit as they proceed, that in turn, means that section drawings are unnecessary (Figure 27). This also explains why some organisations choose not to use section drawings at all during the course of an archaeological investigation, as can be seen by the 96% response rate of some British archaeological organisations to section drawing related criteria (Figure 27).

The reason why some British archaeological organisations also only record long section drawings of trenches if they contain archaeological evidence is a money and time saving exercise, as such trenches are not of archaeological interest and so such archaeological organisations deem it to be a waste of resources. The reason why 4% of British archaeological organisations record running section drawings of archaeological sites is because they wish to have additional records available from which they can validate the single context plans that they have created (Figure 27). However, the fact that no other archaeological organisations use this approach indicates that this is not a very common validation practice.

Of those archaeological organisations that do use section drawings to record archaeological data, it is apparent that the data that is included in section drawings is rather generic (Figure 28). All British, Irish, Australasian, and the majority of North American archaeological organisations' section drawings will be photographed and have a unique identification number, they will also contain: a site code and site name, date, scale, cardinal points, datum points, illustrator's name, elevations, keys, section line, artefacts present, disturbances present and grid co-ordinates (Figure 28). The reason why only 82.4% of North American archaeological organisations' section drawings contain the aforementioned data is because not all of the North American organisations that use section drawings stated what data should be included on section drawings (Figure 28). Presumably, however, as the majority of archaeological organisations record the same data, the remaining 17.6% of North American organisations will follow the same approach (Figure 28).

The one area in which the archaeological organisations varied was in whether they included a stratigraphic relationship matrix on the section drawing (Figure 28). In the United Kingdom, 46% of archaeological organisations recorded a stratigraphic relationship matrix on the section drawing, 22% in Ireland, 8% in Australasia, and 4.7% in North America (Figure 28). The reason why some archaeological organisations choose to include a stratigraphic relationship matrix on their section drawings is due to the aesthetic preferences of the organisation, and also because recording this data on the section drawing acts as a back-up procedure, just in case the recording forms associated with the feature being drawn are accidentally lost. In addition, the reason why some Australasian and North American archaeological organisations do not record this data on section drawings is because they do not use a stratigraphic matrix system to record stratigraphic relationships.

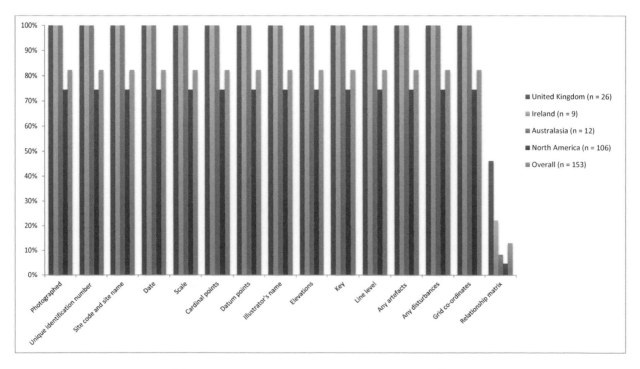

FIGURE 28: DATA THAT IS RECORDED ON SECTION DRAWINGS © EVIS 2016.

There is also a difference between the section drawing conventions that archaeological organisations use (Figure 29). In the United Kingdom, 100% of archaeological organisations' section drawings will illustrate both negative and positive stratigraphic units, that in turn, will illustrate the edge of the cut feature, and 8% will label the section drawing with a specific code that will indicate the feature type/number (Figure 29). In Ireland, 100% of archaeological organisations' section drawings will illustrate both negative and positive stratigraphic units, that in turn, illustrate the edge of the cut feature (Figure 29). In Australasia, 8% of archaeological organisations' section drawings will illustrate positive and negative stratigraphic units, 100% will illustrate each positive stratigraphic unit and the edge of its associated feature, and 92% will label the section drawing with a specific code that will indicate the feature type/number (Figure 29). In North America, 4.7% of archaeological organisations' section drawings will illustrate both negative and positive stratigraphic units, 74.5% will illustrate each positive stratigraphic unit and the edge of its associated feature, 70.8% will label the section drawing with a specific code that will indicate the feature type/number, 69.8% will annotate their section drawing describing the stratum's composition, 44.3% will annotate the stratigraphic boundaries that have been drawn with descriptions of their distinctiveness and topography (Figure 29).

It is evident that in the United Kingdom and Ireland, section drawings will illustrate both positive and negative stratigraphic units. However, in Australasia and North America the majority of archaeological organisations will not illustrate both positive and negative stratigraphic units, but will illustrate the positive units and the edge of the feature that has been excavated. This difference is caused by philosophical differences in what archaeological organisations in the United Kingdom, Ireland, Australasia and North America regard as a stratigraphic unit, as the majority of Australasian and North American archaeological organisations consider 'cuts' to be features rather than negative units of stratigraphy. In the end, however, such differences in terminology make little difference to the production of a section drawing, as all of these archaeological organisations will draw the boundaries of the cut/feature and the fills within.

The inclusion of a separate code relating to the feature number/type by some British, Australasian and North American archaeological organisations is due to the fact that these organisations allocate features specific codes that relate to their function, thus by including this code on the section drawing, individuals who are looking at the section drawing can see what function the feature served without referring to additional data.

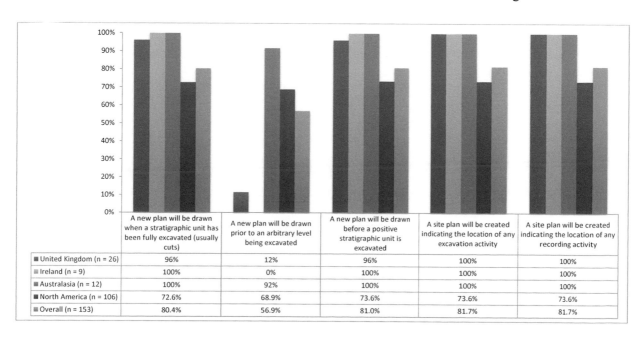

	A new plan will be drawn when a stratigraphic unit has been fully excavated (usually cuts)	A new plan will be drawn prior to an arbitrary level being excavated	A new plan will be drawn before a positive stratigraphic unit is excavated	A site plan will be created indicating the location of any excavation activity	A site plan will be created indicating the location of any recording activity
■ United Kingdom (n = 26)	96%	12%	96%	100%	100%
■ Ireland (n = 9)	100%	0%	100%	100%	100%
■ Australasia (n = 12)	100%	92%	100%	100%	100%
■ North America (n = 106)	72.6%	68.9%	73.6%	73.6%	73.6%
■ Overall (n = 153)	80.4%	56.9%	81.0%	81.7%	81.7%

FIGURE 29: SECTION DRAWING CONVENTIONS © EVIS 2016.

The fact that some North American archaeological organisations annotate their section drawings with descriptions of a stratum's composition and a stratigraphic boundary's distinctiveness and topography, is so that individuals looking at the section drawing can understand the feature without having to look through other recording forms. In addition, it also acts as a back up tool, in case the recording forms relating to the feature are lost accidentally.

In terms of when archaeological organisations will produce plan drawings, in the United Kingdom, 96% of archaeological organisations will produce a plan before a positive stratigraphic unit is excavated, 96% will also produce a plan of a negative stratigraphic unit, 12% will produce a plan before an arbitrary level is excavated, 100% will create a site plan that will indicate the location of any excavation and recording activity that took place at the archaeological site (Figure 30). In Ireland, 100% of archaeological organisations will produce a plan before a positive stratigraphic unit is excavated, 100% will also produce a plan of a negative stratigraphic unit, and 100% will create a site plan that will indicate the location of any excavation and recording activity that took place at the archaeological site (Figure 30). In Australasia, 100% of archaeological organisations will produce a plan before a positive stratigraphic unit is excavated, 100% will also produce a plan of a negative stratigraphic unit/feature, 92% will produce a plan before an arbitrary level is excavated, and 100% will create a site plan that will indicate the location of any excavation and recording activity that took place at the archaeological site (Figure 30). In North America, 73.6% of archaeological organisations will produce a plan before a positive stratigraphic unit is excavated, 72.6% will produce a plan of a negative stratigraphic unit/feature, 68.9% will produce a plan before an arbitrary level is excavated, and 73.6% will create a site plan that will indicate the location of any excavation and recording activity that took place at the archaeological site (Figure 30).

These results indicate that the majority of archaeological organisations will create archaeological site plans that illustrate the locations in which any excavation and/or recording activity has taken place. This is unsurprising, as such plans will demonstrate to individuals reading the archaeological site report where archaeological evidence was found. The fact that not all North American archaeological organisations produce plans is at odds with the majority of archaeological organisations, the reason why they do not produce plans in this manner is probably due to the requirements of the State in which they are working and whether the State recommends that archaeological site plans are produced.

The results also show that the majority of archaeological organisations complete plan drawings of all positive stratigraphic units and negative stratigraphic units/features present, again this is unsurprising as such plans will illustrate how the archaeological site formed. The fact that not all British and North American archaeological organisations choose to plan positive and negative stratigraphic units/features individually is due to the fact that they tend not to excavate using the Stratigraphic Excavation method, and prefer to use alternative excavation approaches that rely on sections rather than plans.

The fact that some British, Australasian and North American archaeological organisations produce plans before excavating an arbitrary level is because when such organisations use the Arbitrary Level Excavation method arbitrary-level plans are expected to be drawn. The small percentage of British archaeological organisations that stated that they create plans before excavating an arbitrary level is due to the fact that very few British archaeological organisations use the Arbitrary Level Excavation method, and therefore, do not discuss arbitrary level planning in their archaeological manuals/guidelines. Conversely, the high percentage of Australasian and North American archaeological organisations that stated that they produce plans before an arbitrary level is excavated is due to the fact that the Arbitrary Level Excavation method is commonly used by Australasian and North American archaeological organisations.

In relation to what data is recorded on plan drawings by archaeological organisations it is evident that it is very consistent between different archaeological organisations (Figure 31). The majority of British, Irish,

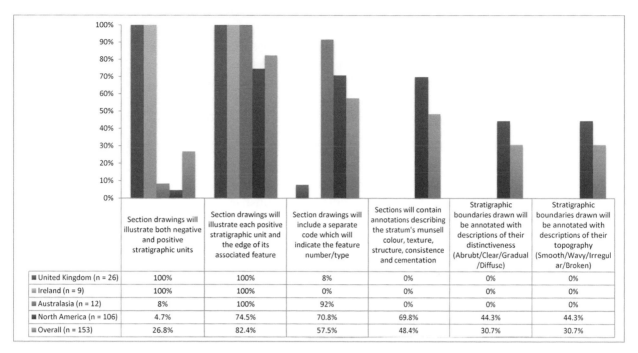

	Section drawings will illustrate both negative and positive stratigraphic units	Section drawings will illustrate each positive stratigraphic unit and the edge of its associated feature	Section drawings will include a separate code which will indicate the feature number/type	Sections will contain annotations describing the stratum's munsell colour, texture, structure, consistence and cementation	Stratigraphic boundaries drawn will be annotated with descriptions of their distinctiveness (Abrubt/Clear/Gradual /Diffuse)	Stratigraphic boundaries drawn will be annotated with descriptions of their topography (Smooth/Wavy/Irregul ar/Broken)
■ United Kingdom (n = 26)	100%	100%	8%	0%	0%	0%
▨ Ireland (n = 9)	100%	100%	0%	0%	0%	0%
■ Australasia (n = 12)	8%	100%	92%	0%	0%	0%
■ North America (n = 106)	4.7%	74.5%	70.8%	69.8%	44.3%	44.3%
▨ Overall (n = 153)	26.8%	82.4%	57.5%	48.4%	30.7%	30.7%

FIGURE 30: USE OF PLAN DRAWINGS © EVIS 2016.

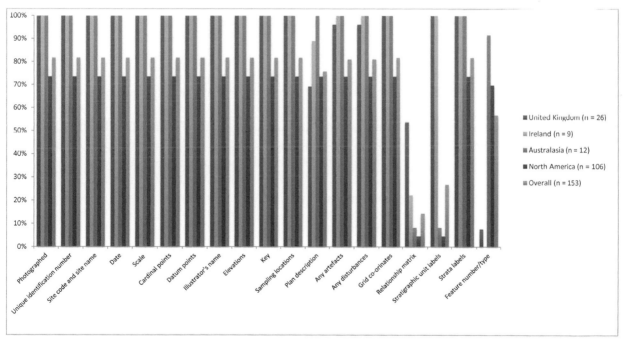

FIGURE 31: DATA THAT IS RECORDED ON PLAN DRAWINGS © EVIS 2016.

Australasian and North American archaeological organisations will photograph their plans and allocate them a unique identification number (Figure 31). The plans will also contain: a site code and site name, date, scale, cardinal points, datum points, illustrator's name, elevations, keys, sampling locations, plan description, artefacts present, disturbances present, grid co-ordinates and strata labels (Figure 31).

The areas in which the archaeological organisations varied in terms of what data was included on plan drawings, included: whether the plan included a stratigraphic relationship matrix, whether stratigraphic units were labelled, and whether the plan included a feature number/type.

The reason why 54% of British archaeological organisations, 22% of Irish archaeological organisations, 8% of Australasian archaeological organisations, and 4.7% of North American archaeological organisations include a stratigraphic relationship matrix on their plan drawings is again due to the aesthetic preferences of the organisation, and also because recording this data on the plan acts as a back-up procedure, just in case any of the recording forms associated with the feature or stratigraphic unit being planned are lost (Figure 31). The reason why some Australasian and North American organisations do not record this data is, as stated before, due to the fact that they do not use a stratigraphic matrix system to record stratigraphic relationships.

The notable differences between whether archaeological organisations label stratigraphic units or feature number/types on their plans is again, down to philosophical differences between how archaeologists operating in these different areas define cuts (Figure 31). As British and Irish archaeological organisations regard cuts as units of archaeological stratigraphy and Australasian and North American archaeological organisations regard them as features. Therefore, when recording plans, British and Irish archaeologists will record the cuts with stratigraphic unit labels and Australasian and North American archaeologists will label them with predetermined feature numbers/types that will give an indication of the function of the feature at the archaeological site. Moreover, as found before, some British archaeological organisations also include feature numbers/types on their plans in order to give individuals reading the plan drawing an idea of what function the archaeological unit being planned served at the archaeological site.

In relation to whether archaeological organisations use pro-formas to document archaeological data, it is evident from Figure 32 that all archaeological organisations do, and therefore it can be stated that pro-formas are a standard recording tool used during archaeological investigations. However, the type of pro-forma that a particular archaeological organisation uses to document archaeological data varies according to their own methodological preferences and the type of archaeological site they are excavating. For example, if an archaeological organisation is excavating an archaeological site using the Arbitrary Level method of excavation they will use a unit level recording form to document each arbitrary level removed, whereas, if they were using the Stratigraphic method of excavation they would use a context recording form to document each stratigraphic unit that was uncovered (Figure 32). However, despite these differences in when a particular type of pro-forma will be used, each archaeological organisation will have a particular pro-forma available to them to deal with the type of archaeological site that they are working on, and the methodological approaches that they will be using. The only variation is in relation to what these forms are called and the terminology used within them.

Excavation strategies

In regards to excavation sampling strategies, in the United Kingdom, 85% of archaeological organisations will excavate a representative sample if a large number of archaeological features are identified, 85% will excavate all archaeological features present if only a small number of archaeological features are identified, and 12% will excavate all units of archaeological stratigraphy present at the site (Figure 33). In Ireland, 89% of archaeological organisations will excavate a representative sample if a large number of archaeological features are identified, 89% will excavate all archaeological features present if only a small number of archaeological features are identified, and 11% will excavate all units of archaeological stratigraphy present at the site (Figure 33). In Australasia, 100% of archaeological organisations will excavate a representative sample if a large number of archaeological features are identified, and 100% of archaeological organisations will excavate all of the archaeological features present if only a small number of archaeological features are identified (Figure 33). In North America, 67% of archaeological organisations will excavate a representative sample if a large number of archaeological features are identified, 67% will excavate all archaeological features present if only a small number of archaeological features are identified, and 8.5% will excavate all units of archaeological stratigraphy present at the site (Figure 33).

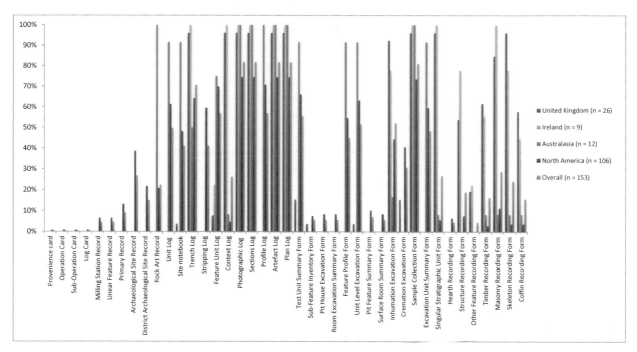

Figure 32: The use of pro-forma recording sheets © Evis 2016.

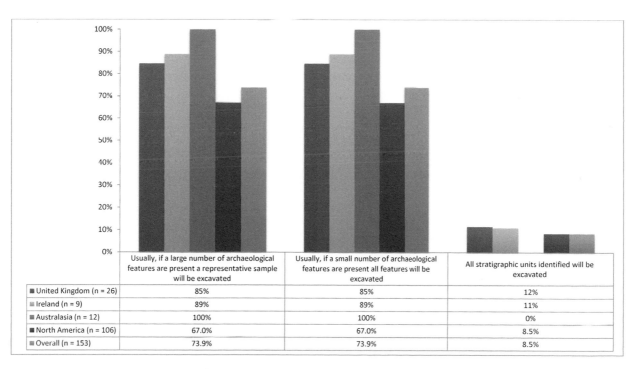

	Usually, if a large number of archaeological features are present a representative sample will be excavated	Usually, if a small number of archaeological features are present all features will be excavated	All stratigraphic units identified will be excavated
■ United Kingdom (n = 26)	85%	85%	12%
▥ Ireland (n = 9)	89%	89%	11%
■ Australasia (n = 12)	100%	100%	0%
■ North America (n = 106)	67.0%	67.0%	8.5%
▥ Overall (n = 153)	73.9%	73.9%	8.5%

Figure 33: Excavation sampling strategies © Evis 2016.

It is evident from this data that the majority of archaeological organisations will adopt a sampling strategy when excavating an archaeological site, which is determined by the time, number of staff, number of archaeological features present, resources and government requirements imposed upon the archaeological investigation. Those that excavate all archaeological stratigraphic units present at the site are either required to by the governing body overseeing the archaeological investigation, or are research organisations that have the time and resources available to do so.

In terms of how archaeological organisations sample archaeological features, in the United Kingdom, 96% of archaeological organisations will excavate a representative slot through the fills of large features, 96% will excavate representative slots through curvilinear features, 8% will use baulks when excavating large or complex structures, and 8% will use test excavation units to sample the site and excavate and record these units individually (Figure 34). In Ireland, 100% of archaeological organisations will excavate a representative slot through the fills of large features and 100% will excavate representative slots through curvilinear features (Figure 34). In Australasia, 100% of archaeological organisations will excavate a representative slot through the fills of large features, 100% will excavate representative slots through curvilinear features, 92% will excavate shovel test pits in areas which lack surface evidence of archaeological inhabitation, 92% will use test excavation units to sample the site and excavate and record these units individually, and 92% will establish feature excavation units when archaeological evidence is found (Figure 34). In North America, 71.7% of archaeological organisations will excavate a representative slot through the fills of large features, 74.5% will excavate representative slots through curvilinear features, 4.7% will use baulks when excavating large or complex structures, 67% will excavate shovel test pits in areas which lack surface evidence of archaeological inhabitation, 69.8% will use test excavation units to sample the site and excavate and record these individually, 59.4% will sample structures using test excavation units prior to excavating the structure in its entirety, and 61.3% of archaeological organisations will establish feature excavation units when archaeological evidence is found (Figure 34).

The results indicate that the majority of archaeological organisations will excavate representative slots through either large or curvilinear features, either by using rectangular or L-shaped slots. This enables archaeologists to determine the sequence of deposition of the features without spending long lengths of time excavating them in their entirety. Moreover, in some cases, this approach is used to determine the chronological relationship between two intercutting features. Usually, when using representative slots in this manner, several slots are excavated at various intervals, in order to confirm that the deposition sequence or chronological relationship that has been identified, is consistent throughout the entire feature, if this is found to be the case then the archaeologists need not excavate the remaining unexcavated sections of the feature.

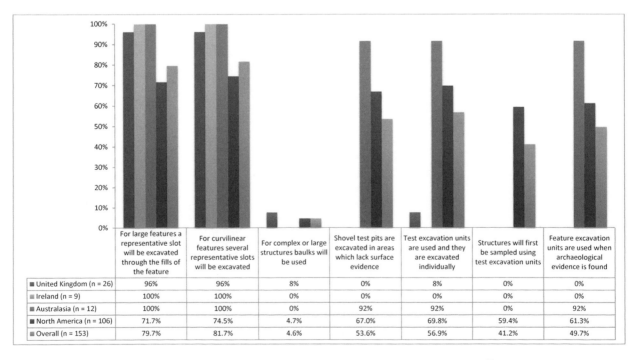

	For large features a representative slot will be excavated through the fills of the feature	For curvilinear features several representative slots will be excavated	For complex or large structures baulks will be used	Shovel test pits are excavated in areas which lack surface evidence	Test excavation units are used and they are excavated individually	Structures will first be sampled using test excavation units	Feature excavation units are used when archaeological evidence is found
United Kingdom (n = 26)	96%	96%	8%	0%	8%	0%	0%
Ireland (n = 9)	100%	100%	0%	0%	0%	0%	0%
Australasia (n = 12)	100%	100%	0%	92%	92%	0%	92%
North America (n = 106)	71.7%	74.5%	4.7%	67.0%	69.8%	59.4%	61.3%
Overall (n = 153)	79.7%	81.7%	4.6%	53.6%	56.9%	41.2%	49.7%

FIGURE 34: EXCAVATION SAMPLING STRATEGIES FOR ARCHAEOLOGICAL FEATURES © EVIS 2016.

The results also show that 8% of British archaeological organisations and 4.7% of North American archaeological organisations use baulks when excavating complex or large structures (Figure 34). Such archaeological organisations use this approach, in order to be able to create a section drawing of the stratigraphic sequence present within the complex or large structures, which can then be compared to the plan drawings created of the area being excavated. By using this approach several archaeologists can excavate in different units or grid squares across the structure at their own pace. They can also confirm, through looking at the section, that the layers that they have identified are the same as their colleagues working on the other sides of the baulks.

The reason why 92% of Australasian and 67% of North American archaeological organisations use shovel test pits in areas that lack evidence of archaeological inhabitation is due to the fact that these archaeological organisations often work on archaeological sites with very sparse archaeological evidence (Figure 34). Therefore, in order to rapidly assess if a site does in fact contain any archaeological evidence, they will excavate shovel test pits, using a shovel, to determine if the site is of archaeological importance.

The fact that 8% of British, 92% of Australasian and 69.8% of North American archaeological organisations use test pits during an archaeological investigation is unsurprising (Figure 34). Such test pits are used to determine whether an area has any archaeological evidence present. The fact that the majority of British and Irish, and some Australasian and North American archaeological organisations do not use test pits is due to the fact that these organisations are commercial archaeological organisations, therefore, they are called in to investigate large swathes of land and use alternative approaches, such as mechanical stripping or large evaluation trenches to determine if there is any archaeological evidence present.

The use of test excavation units to sample structures prior to excavating them in their entirety by 59.4% of North American archaeological organisations is merely a sampling strategy adopted by these organisations (Figure 34). Such test excavation units enables them to gauge the dimensions of the structure that is to be excavated and its complexity. Once such a test excavation unit has been excavated they are then able to adapt their excavation and recording strategies according to what was identified within the test excavation unit.

The results show that 92% of Australasian and 61.3% of North American archaeological organisations use feature excavation units after archaeological evidence has been found (Figure 34). Feature excavation units are shovel test pits or test excavation units that have tested positive for archaeological evidence. Therefore, once archaeological evidence has been found in either a shovel test pit or test excavation unit, the unit or pit is then referred to as a feature excavation unit and the unit is expanded in order to locate any additional archaeological evidence present. Subsequently, any archaeological evidence that is identified from this area is then linked to it by the given feature excavation unit reference number and associated feature excavation unit paperwork, rather than the shovel test pit or test excavation unit reference numbers and paperwork.

In regards to the use of different excavation techniques by archaeological organisations, in the United Kingdom, 12% of archaeological organisations will excavate using fixed arbitrary levels, 96% will excavate using identifiable stratigraphic units, and 58% of archaeological organisations will excavate cut features using the Demirant or Quadrant Excavation methods (Figure 35). In Ireland, 100% of archaeological organisations will excavate using identifiable stratigraphic units and 67% will excavate cut features using the Demirant or Quadrant Excavation methods (Figure 35). In Australasia, 92% of organisations will excavate using fixed arbitrary levels, 100% will excavate using identifiable stratigraphic units, and 92% of organisations will excavate cut features using the Demirant or Quadrant Excavation methods (Figure 35). In North America, 47.2% of archaeological organisations will excavate using fixed arbitrary levels, 70.8% will excavate using identifiable stratigraphic units, and 68.9% of archaeological organisations will use the Demirant or Quadrant Excavation methods to excavate cut features (Figure 35).

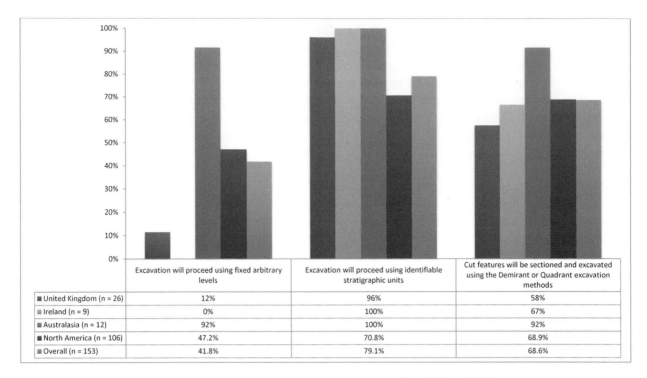

	Excavation will proceed using fixed arbitrary levels	Excavation will proceed using identifiable stratigraphic units	Cut features will be sectioned and excavated using the Demirant or Quadrant excavation methods
■ United Kingdom (n = 26)	12%	96%	58%
■ Ireland (n = 9)	0%	100%	67%
■ Australasia (n = 12)	92%	100%	92%
■ North America (n = 106)	47.2%	70.8%	68.9%
■ Overall (n = 153)	41.8%	79.1%	68.6%

FIGURE 35: USE OF DIFFERENT EXCAVATION METHODS © EVIS 2016.

These results indicate that the majority of archaeological organisations will use a variety of different excavation methods to excavate an archaeological site. The overall decision regarding which excavation method to use will be determined by - the type of archaeological site that is to be excavated and the type of archaeological features present.

It is interesting to note that British and Irish organisations tend not to use the Arbitrary Level Excavation method to excavate archaeological sites. The British archaeological organisations that stated that they do use the Arbitrary Level Excavation technique use it as well as the Stratigraphic, Demirant and Quadrant Excavation methods, and only use the Arbitrary Level Excavation approach if the archaeological site that they are dealing with has particularly thick stratigraphic units or they are unable to define the boundaries of individual stratigraphic units. In comparison, in Australasia and North America it is evident that a larger percentage of archaeological organisations use the Arbitrary Level method of excavation, this is because they are often excavating at archaeological sites that have particularly thick deposits through which they must excavate. Therefore, in order to efficiently deal with such archaeological sites these organisations utilise the Arbitrary Level Excavation method.

Despite the large percentage of Australasian and North American archaeological organisations stating that they do use the Arbitrary Level Excavation approach, the majority of these archaeological organisations as well as the British and Irish archaeological organisations, state that they excavate according to identifiable stratigraphic units, using the Stratigraphic Excavation method. There is a distinction to be made, however, between what British and Irish archaeological organisations and some Australasian and North American archaeological organisations would regard as excavating according to identifiable stratigraphic units or 'stratigraphically'. In the United Kingdom and Ireland, such stratigraphic units include: fills, layers, structures and cuts, whereas, some Australasian and North American archaeological organisations state that stratigraphic units only include: fills and layers. Therefore, when excavating according to stratigraphic units, some of the North American and Australasian organisations will only maintain the boundaries of any fills or layers that they identify, and will excavate through feature boundaries (cuts) as they do not regard them as stratigraphic units, although, they do record the dimensions of such boundaries as they excavate through them, so that a plan/section of the feature/cut can be reconstructed *post-facto*.

The fact that 58% of British, 67% of Irish, 92% of Australasian and 68.9% of North American archaeological organisations choose to section cuts/features using the Demirant or Quadrant Excavation methods is due to the fact that these approaches allow archaeologists to rapidly assess the stratigraphic sequence present. In Australasia and North America, however, there is a difference in how some organisations undertake the sectioning and excavation of a cut/feature. Rather than identifying the limits of the cut/feature, sectioning it, and excavating each section separately, some Australasian and North American archaeological organisations will turn the area in which the cut/feature is present into a feature excavation unit. They will then divide this unit, which includes both the cut/feature and the surrounding sterile soil, into two halves or four quarters, and then excavate both the fills/layers of the cut/feature and the sterile soil within each quarter or half, until the base of the cut/feature has been reached, after which they will record the half section/long section of the cut/feature that they have exposed.

Overall, in terms of the development and current use of archaeological excavation methods and recording systems it is apparent that there are several differences in how the different archaeological organisations evaluated in this research project conduct archaeological investigations.

The most significant differences are between the British and Irish archaeological organisations and the Australasian and North American archaeological organisations. The data analysed in this research project reveals that both British and Irish archaeological organisations adopt similar, if not the same, approaches to conducting archaeological investigations in regards to the recording systems and excavation methods that they use, and in terms of how they define, record and excavate archaeological stratigraphy. Whereas, the majority of Australasian and North American archaeological organisations utilise different excavation methods and recording systems, and define, record and excavate archaeological stratigraphy differently to the British and Irish archaeological organisations, although, use similar approaches to each other.

One reason why the strategies adopted to conduct archaeological investigations differ between archaeological organisations in the United Kingdom and Ireland and in Australasia and North America is because of fundamental differences in how these archaeological organisations define archaeological stratigraphy, as such differences result in these organisations adopting different excavation and recording approaches in order to capture the stratigraphic data that they deem to be significant.

Another reason why is because of the different types of archaeological sites that these different archaeological organisations investigate. In the United Kingdom and Ireland, archaeological investigations are conducted on both urban sites with complex archaeological stratigraphy and rural sites with widely dispersed archaeological evidence with very few complex stratigraphic units present. Consequently, archaeological organisations operating in the United Kingdom and Ireland must ensure that the methodological approaches that they advocate in their archaeological manuals/guidelines can be used and/or adapted for use on both of these archaeological site types.

In comparison, those archaeological organisations operating in Australasia and North America who adopt different archaeological approaches to the archaeological organisations in the United Kingdom and Ireland tend to conduct archaeological investigations on rural Native American or Aboriginal archaeological sites. Such sites contain widely dispersed archaeological evidence and deep stratigraphic deposits, and as a result, such archaeological organisations have developed different archaeological approaches to cater to such sites, and have subsequently published these approaches in their archaeological manuals/guidelines.

Those archaeological organisations in Australasia and North America that do adopt similar archaeological approaches to the archaeological organisations in the United Kingdom and Ireland are archaeological organisations that have begun to excavate colonial historic sites, which are comparative to the urban sites investigated by British and Irish archaeological organisations. Consequently, such organisations have had to adapt their archaeological approaches to cater to the complex stratigraphy present at these sites, and in turn, have looked to British and Irish archaeological publications and investigative strategies to ensure that the methodological approaches that they use will maximise the archaeological evidence that is recovered.

Chapter 5 Archaeological practitioner interviews

This chapter contains the results of the interviews conducted with archaeological practitioners. These interview results are then used to discuss how archaeological practitioners excavate, and explain why and when archaeologists choose to use particular excavation methods and recording systems.

Results

Justifications for the use of archaeological excavation methods and recording systems

In order to obtain data to determine how archaeologists excavate, and why and when archaeologists choose to use particular excavation methods and recording systems, each experimental participant completed an interview in which they answered questions relating to how they conduct archaeological investigations and the factors that influence their choice of excavation method. The participants were also asked to choose an excavation method and the recording techniques that they would use to investigate the grave simulation, and to justify their reasons for doing so.

Archaeological investigations

Through analysing the data obtained through interviewing the participants who excavated the grave simulation, 92% of the participants stated that they followed a set of established archaeological guidelines when conducting archaeological investigations (Figure 36). Such guidelines provide recommendations as to how an archaeological investigation should be undertaken, and outline which archaeological excavation

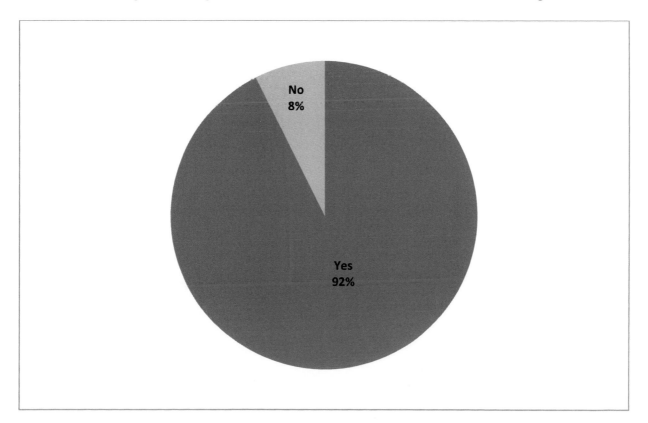

FIGURE 36: INTERVIEW RESULTS FOR QUESTION 7: WHEN CONDUCTING ARCHAEOLOGICAL FIELDWORK, DO YOU, OR THE ORGANISATION WITH WHICH YOU ARE AFFILIATED, FOLLOW A SET OF ESTABLISHED ARCHAEOLOGICAL GUIDELINES? © EVIS 2016.

methods and recording systems should be used. However, 8% of participants stated that they do not follow a set of archaeological guidelines. These archaeologists determine how archaeological investigations will be conducted on a site-by-site basis forming bespoke excavation and recording protocols depending on what the site that is being investigated requires (Figure 36). This provides the archaeologists with the flexibility to adapt their approaches to the needs of the site, rather than being restricted to the methodological approaches advocated in the guidelines. However, such flexibility is only possible in small commercial archaeological units who are working on small-scale archaeological investigations one at a time, as large commercial archaeological units, who are working on multiple sites at the same time, and are dealing with large-scale sites and managing large volumes of staff, many of whom will be rotating between the different archaeological investigations, need to ensure that the methodological approaches utilised are consistent, and that all members of staff are recording and excavating in the same manner, and therefore need archaeological guidelines in place.

When participants were asked whether they are required to report the findings of an archaeological investigation to a governing body 77% of the participants said that they were, and 23% of participants said that they were not (Figure 37). The participants who said that they were required to report their findings to a governing body tended to work in the commercial archaeological sector. When investigations are conducted in this context, local government representatives such as the County Archaeologist, set recommendations as to how the archaeological investigation of a site within their jurisdiction should be conducted, and, as a consequence, the commercial unit dealing with the site are required to report their findings back to the appropriate government representative. The participants who stated that they were not required to report their findings to a governing body tended to conduct most of their archaeological investigations in research or academic contexts. Such archaeological investigations are conducted on private land with the permission of the landowner and therefore the archaeologists are not required to report the findings of the archaeological investigation to a governing body. Instead, archaeologists working

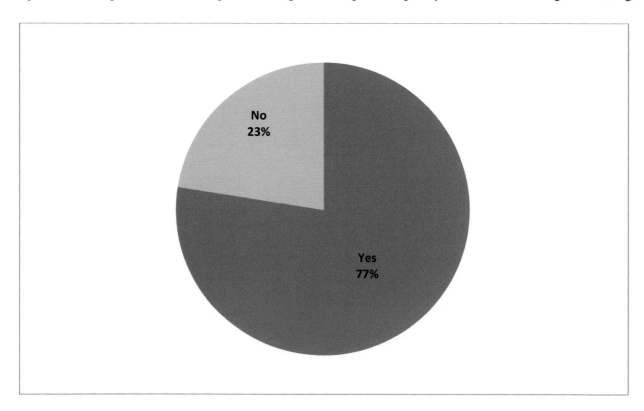

FIGURE 37: INTERVIEW RESULTS FOR QUESTION 8: ARE YOU, OR THE ORGANISATION WITH WHICH YOU ARE AFFILIATED, REQUIRED TO REPORT THE FINDINGS OF AN ARCHAEOLOGICAL INVESTIGATION TO A GOVERNING BODY? © EVIS 2016.

on such archaeological investigations tend to disseminate the findings of the project through publishing them in archaeological journals, speaking at local archaeology society meetings, and writing summary reports for interested parties.

In terms of having a specific excavation manual, which outlines which excavation methods should be used during archaeological investigations 87% of the participants stated that they do, and 13% said that they do not (Figure 39). Similarly, when participants were asked whether they had a specific recording manual, again 87% of participants said that they do, and 13% said that they do not (Figure 43). Interestingly, when one compares the results of these two questions against the results of whether the participants followed a set of archaeological guidelines when conducting archaeological investigations, it is apparent that there was a 5% increase in the number of participants who did not have an excavation or a recording manual available to them. This may be due to the fact that the archaeological guidelines that these participants use do not have detailed excavation methods or recording systems sections within them. Rather, they provide generalised approaches to conducting archaeological investigations, without setting strict rules as to which methodological approaches should be used. Leaving decisions regarding how the site will be excavated and recorded to the discretion of the archaeologists at site and the site director.

Interestingly, however, 92% of the participants stated that the excavation methods that they use vary according to the archaeological site that they are working on, and only 8% said that they do not (Figure 38). This flexibility in the methodological approaches that may be used during archaeological investigations is due to a number of variables which include: the scale of the site that is to be excavated, the time available to conduct the excavation, the type of archaeological site that is being excavated, the number of archaeologists that are available, and the sampling strategy that has been set by the director, each of which will determine which excavation methods will be used. The fact that 92% of participants stated that excavation methods change on a site-by-site basis and 87% of participants stated that they referred to an excavation manual when conducting archaeological fieldwork is not at odds, as excavation manuals provide a variety of excavation approaches to use depending on the impact that the aforementioned variables have been deemed to have on the archaeological investigation (Figure 38; Figure 39).

Despite excavation manuals proffering a variety of excavation methods to use for different types of archaeological investigation only 72% of participants stated that they followed the excavation methods outlined in the excavation manual, whereas 28% said that they do not, and excavate the site according to their own methodological preferences (Figure 40). The reasons why some participants stated that they do not follow the methods outlined in the manual are, firstly, that not all participants have an excavation manual to follow, as the results in Figure 39 highlight. Secondly, some participants consider themselves to have sufficient experience to determine which excavation methods they should use and do not find the excavation manual useful in this respect.

When participants were asked whether the recording techniques that they use varied according to the type of archaeological site they were working on 85% said that they do, and 15% said that they do not (Figure 42). It is interesting to note that the archaeologists adapt their recording approaches to a lesser degree than they do their excavation methods, which were adapted according to the type of site in 92% of cases (Figure 38). This is most likely due to the fact that the majority of archaeological units conducting fieldwork have set pro-formas in place that are used to record archaeological data (Figure 45). Therefore, when an archaeological investigation is undertaken, the same set of pro-formas are used regardless of the type of archaeological site that is being excavated (Figure 45). The participants who stated that they do adapt their recording techniques on a site-by-site basis are again, as with the excavation methods, doing so in accordance with the following factors: the size of the site that is to be recorded, the time available to conduct the recording, the type of archaeological site that is being recorded, the number of archaeologists that are available, and the equipment and resources that are available to complete the

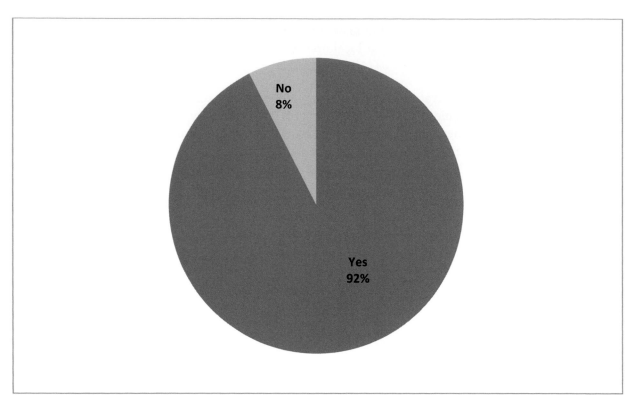

FIGURE 38: INTERVIEW RESULTS FOR QUESTION 10: DO THE EXCAVATION METHODS YOU USE VARY ACCORDING TO THE TYPE OF ARCHAEOLOGICAL SITE YOU ARE WORKING ON? © EVIS 2016.

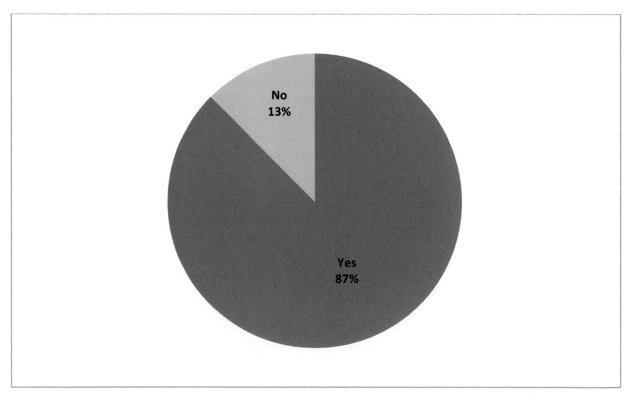

FIGURE 39: INTERVIEW RESULTS FOR QUESTION 11: DO YOU, OR THE ORGANISATION WITH WHICH YOU ARE AFFILIATED, HAVE AN EXCAVATION MANUAL? © EVIS 2016.

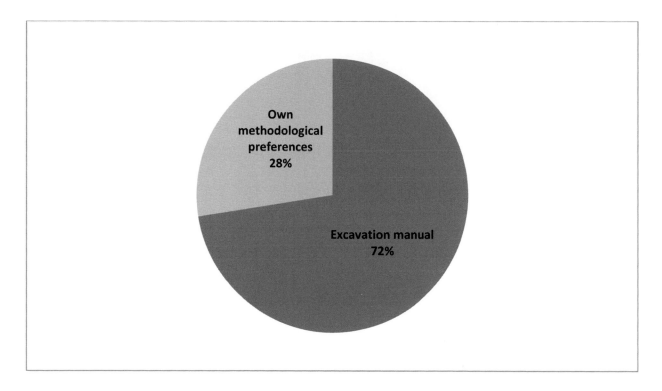

FIGURE 40: INTERVIEW RESULTS FOR QUESTION 12: WHEN EXCAVATING AN ARCHAEOLOGICAL SITE, DO YOU FOLLOW THE EXCAVATION PROCEDURES OUTLINED IN YOUR ORGANISATION'S EXCAVATION MANUAL, OR DO YOU EXCAVATE ACCORDING TO YOUR OWN METHODOLOGICAL PREFERENCES? © EVIS 2016.

recording. This approach, therefore, gives these archaeologists the flexibility to record what is necessary according to the impact that the aforementioned factors have been deemed to have on the archaeological investigation. This reasoning may also explain why when the participants were asked whether they follow the recording procedures outlined in the recording manual 80% said that they do and 20% said that they record according to their own methodological preferences (Figure 44). Furthermore, as stated earlier, not all of the participants had a recording manual available to refer to as the results in Figure 43 show.

Factors that influence excavation method selection

Participants were asked to rank the following factors by the extent to which they influence their choice of excavation method: literary sources, previous archaeological training, the requirements of the local governing body, research aims and objectives, field experience, communication with other archaeologists, site type, and the recording method that will be used. The results from the analysis of this data are displayed in Figure 41.

The results show that the most important factor that influences the participants' choice of excavation method was the aims and objectives of the project. This is perhaps not unsurprising, as the aims and objectives of the project will dictate which archaeological features will be excavated and recorded, and in turn, will state which methodological techniques will be used.

The second most influential factor in determining which excavation methods would be used was the site type that was to be excavated. This is understandable as if an archaeologist was dealing with an urban site with complex, intercutting stratigraphy, they would need to use a different approach than they would if they were dealing with a rural site that had a variety of archaeological features that were vastly spread out from one another.

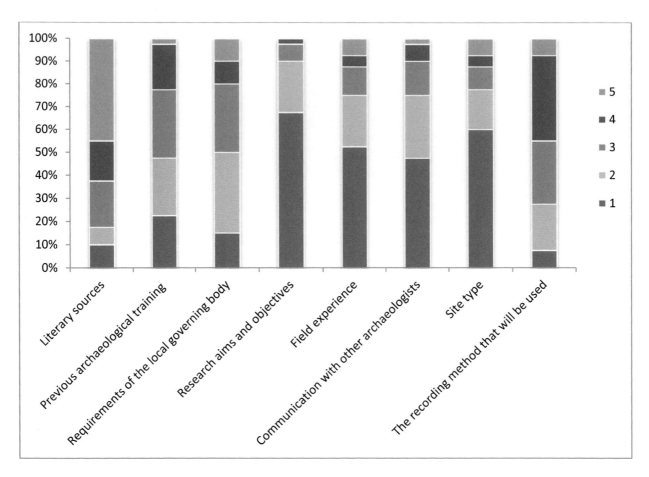

FIGURE 41: INTERVIEW RESULTS FOR QUESTION 13: PLEASE RATE EACH OF THE FOLLOWING FACTORS BY THE EXTENT TO WHICH THEY INFLUENCE YOUR SELECTION OF AN EXCAVATION METHOD 1= MOST INFLUENCE. 5= LEAST INFLUENCE © EVIS 2016.

The third most influential factor in excavation method selection was field experience. This factor was deemed important as through conducting archaeological investigations on a variety of different site types and by excavating a variety of different archaeological features, archaeologists have tried and tested a number of different excavation methods on the same type of feature. This experience then enables the archaeologists to decide which excavation method would be most suitable to use for a particular type of archaeological feature.

The fourth most influential factor in excavation method selection was communication with other archaeologists. Understandably, as with any large-scale project in any industry, discussions amongst team members regarding what approach should be taken to deal with a specific issue, or in the case of archaeology, a specific archaeological feature, contributes greatly to the methodological approach that is eventually taken.

The factor that was ranked fifth was previous archaeological training. Archaeological training provides archaeologists with the knowledge of how each excavation approach is deployed. This knowledge can then be combined with the aforementioned factors to determine which excavation method is best able to deal with the archaeological feature that is to be excavated.

The factor that was ranked sixth was the requirements of the local governing body. Although this factor was ranked as one of the lowest influences on which excavation method will be used, the requirements

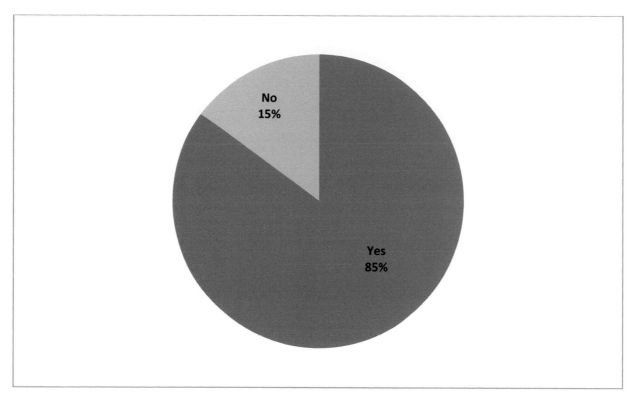

Figure 42: Interview results for question 15: Do the recording techniques you use vary according to the type of archaeological site you are working on? © Evis 2016.

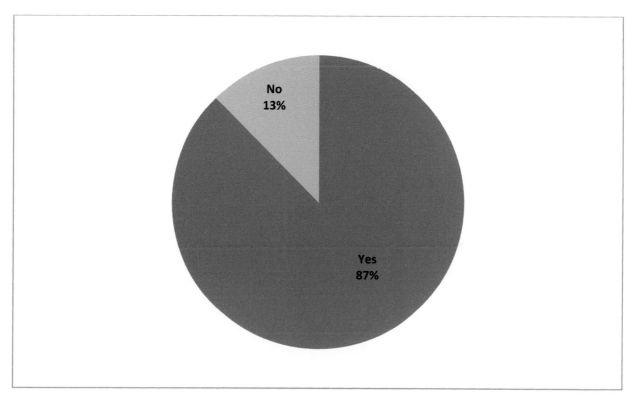

Figure 43: Interview results for question 16: Do you, or the organisation with which you are affiliated, have an archaeological recording manual? © Evis 2016.

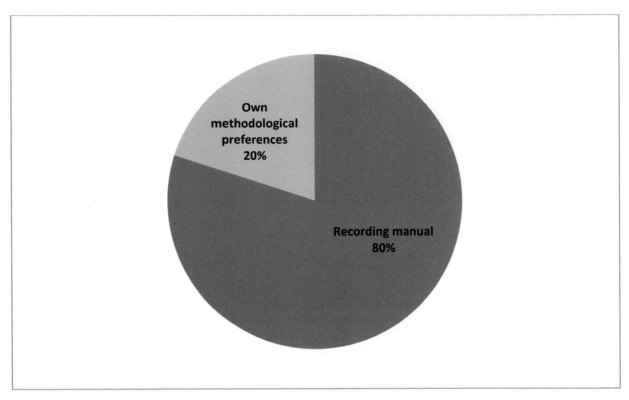

FIGURE 44: INTERVIEW RESULTS FOR QUESTION 17: WHEN RECORDING AN ARCHAEOLOGICAL SITE, DO YOU FOLLOW THE RECORDING PROCEDURES OUTLINED IN YOUR ORGANISATION'S RECORDING MANUAL, OR DO YOU RECORD ACCORDING TO YOUR OWN METHODOLOGICAL PREFERENCES? © EVIS 2016.

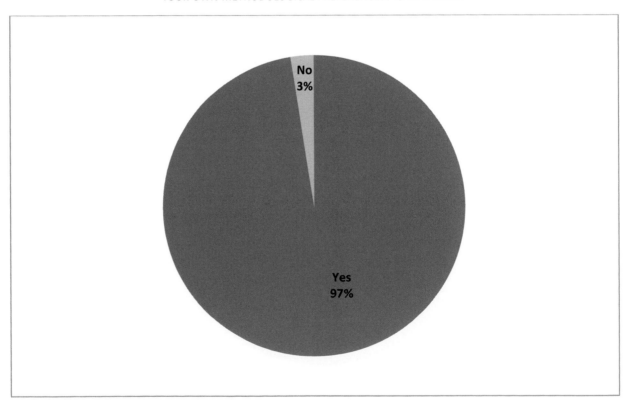

FIGURE 45: INTERVIEW RESULTS FOR QUESTION 18: DO YOU, OR THE ORGANISATION WITH WHICH YOU ARE AFFILIATED, USE PRO-FORMAS WHEN RECORDING ARCHAEOLOGICAL DATA? © EVIS 2016.

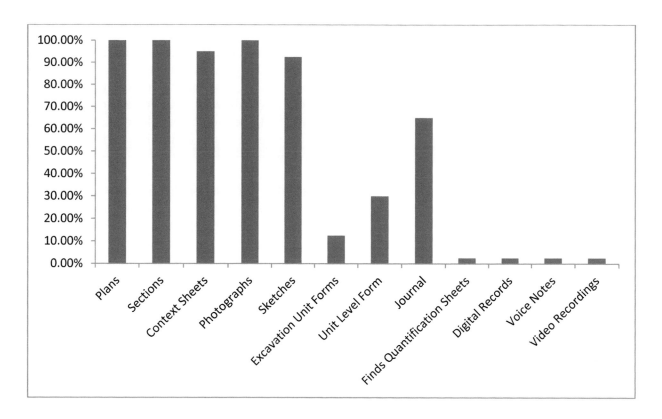

FIGURE 46: INTERVIEW RESULTS FOR QUESTION 19: WHEN RECORDING THE EXCAVATION OF A NEGATIVE FEATURE, WHICH OF THE FOLLOWING RECORDING TECHNIQUES WOULD YOU CHOOSE TO USE? © EVIS 2016.

of the local governing body, especially in commercial archaeological investigations, will be integrated into and help to determine the aims and objectives of the archaeological investigation.

The factor that was ranked seventh was literary sources. It is perhaps unsurprising that this factor was deemed to be one of the least influential factors, as literary sources, apart from archaeological manuals and guidelines, are rarely used in the field to help assess which excavation method should be used. Rather, literary sources tend to be used during the desktop assessment phase and the post-excavation phase of archaeological investigations, to provide some contextual background to the data that could be or was obtained during the course of the excavation.

The factor that was ranked as the least influential factor on excavation method selection was the recording system that was to be used. It is apparent that the participants did not appear to think that recording systems had any impact on how excavations were to be conducted. In fact, it would seem that the participants think that the excavation methods used will determine what recording systems will be used, rather than the other way around.

Excavation method selection

Prior to conducting the excavation experiment participants were asked to choose between four excavation methods to excavate the grave simulation. These methods were: the Stratigraphic Excavation method, the Demirant Excavation method, the Quadrant Excavation method, and the Arbitrary Level Excavation method. Each method was selected by a total of 10 participants. The participants were then asked to justify their choice of excavation method. These justifications are discussed below.

Justifications for the use of the Stratigraphic Excavation method

Participants who chose to use the Stratigraphic Excavation method to excavate the grave simulation justified their selection of this approach by stating the following: the method is widely used in commercial practice, employers expect them to use the Stratigraphic Excavation method, it is what institutions (universities, colleges etc.) and employers have trained them to do, the method allows archaeologists to investigate the archaeological feature of interest in a controlled manner, and, the method enables archaeologists to document and interpret the formation sequence/stratigraphic sequence of the archaeological feature being excavated and the archaeological site as a whole, in a logical manner.

The statement that the Stratigraphic Excavation method is widely used in commercial practice, and as a consequence, is required to be used by employers and also taught in archaeological training institutions, is most likely due to the creation of the Museum of London's Archaeological Site Manual in 1980. This manual introduced a protocol for the excavation and recording of archaeological sites and was the first of its kind to state that the Stratigraphic Excavation method and its associated Single Context Recording system should be used for archaeological investigations, particularly those that are dealing with sites with complex stratigraphy. Since its creation this particular manual has formed the basis for almost all commercial archaeological site manuals in the United Kingdom and Ireland. This has led employers, such as those of the participants, to state that this methodological approach should be used for archaeological investigations. Furthermore, as this is the methodological approach that is widely advocated by the British and Irish archaeological industry, archaeological training institutions train their students to use this method in order to ensure that their pupils are able to find employment after the completion of their courses.

The statement that this method enables archaeologists to investigate the archaeological feature of interest in a controlled manner, and that it allows archaeologists to document and interpret the formation/stratigraphic sequence of archaeological features and archaeological sites in a logical manner, stems from the fact that this method allows archaeologists to excavate and document each stratigraphic context present at a site individually, and as a result, record its relationship both with other contexts present at the site in general, and within any archaeological features being excavated. This provides archaeologists with control over the excavation process, as they focus on a single context at a time, thus ensuring that a full understanding of each context is gathered before moving on to the next, this results in the formation sequence of the site being constructed context-by-context, that in turn, enables archaeologists to identify any discrepancies in the formation sequence immediately, rather than attempting to do it *post-facto*, as other methods do. Thus, leading participants to select this method and highlight the fact that it is both a controlled and logical approach to use in archaeological investigations.

Justifications for the use of the Demirant Excavation method

Participants who chose to use the Demirant Excavation method to excavate the grave simulation justified their selection of this approach by stating the following: this method is the default method used by commercial units on sites where speed of excavation is important, the presence of the half-section allows for the best ratio between information gathered and time spent on the feature, it's the method that the participants have been trained to use, the method enables archaeologists to understand what's happening in the feature rapidly without having to remove all of the fills, and it results in the production of a section drawing that enables one to easily demonstrate the formation process of the feature.

The reason why participants stated that the Demirant Excavation method is the default approach in commercial excavations when time is a critical factor, and that the half section provides the best ratio between information gathered and time spent excavating, stems from the fact that many commercial excavations have a limited time frame in which to investigate an archaeological site. Under such circumstances, archaeologists tend to sample archaeological features rather than excavate them in their entirety. This means that when archaeologists are excavating a pit, for example, they will only excavate half of the pit rather than excavating it in its entirety. This sampling procedure, results in the production of a half section drawing, which archaeologists state, will demonstrate how the pit was constructed – the fills and the cut, and will therefore be sufficient for interpretive purposes, leaving archaeologists with more time to excavate other archaeological features present at the site. This reasoning may also explain why some participants stated that it is the method that they are trained to use for cut features, as this is the methodological approach that they apply on a day-to-day basis during commercial archaeological investigations.

Participants also stated that this methodological approach enables archaeologists to understand what is happening within the archaeological feature rapidly without having to remove all of the fills. This justification stems from the fact that this method requires that half the fill of the feature be removed first. This then enables the archaeologist to view the half section which exhibits the remaining half of the feature's fills in situ, allowing the archaeologist to check the findings of the excavation of the first half by inspecting the half section face to see if the fills that were recorded during the excavation of the first half are present in the half section. This approach, however, relies on the fills of the feature extending into the half section face, which is not always the case. Although, if the fills match the findings from the excavation of the first half, and all extend into the section face, the archaeologist then has a clear understanding of how the feature was formed, and can then, if required, rapidly remove the second half of the feature as they already know the composition of the fills within the feature, and can subsequently update their existing recording forms with any new data gained.

The final justification for the use of this excavation method is that it results in the production of a half section drawing that will demonstrate the formation process of the feature. Again, this justification relies on each fill extending to the point at which the archaeologist set up their section line. However, if they do, the half section drawing displays all of the fills and their relationships with one another at one particular point in the feature, in one drawing. Whereas other approaches rely on the production of multiple plans, or profile drawings, which take time to produce and can be confusing to interpret if they are not correctly labelled, or are being read by layman, such as the general public, or in forensic cases, lawyers and jury members. Thus, having one drawing that can be used to explain how a feature was formed makes it easier to interpret and to explain archaeological findings to interested parties, as long as the section drawing is accurate and each fill within the feature is present in the drawing.

Justifications for the use of the Quadrant Excavation method

Participants who chose to use the Quadrant Excavation method to excavate the grave simulation justified their selection of this approach by stating the following: It's the method that the participants have been trained to use, it provides a highly detailed and clear visual record of the stratigraphic sequence, it allows one to understand the stratigraphic sequence across the length and width of the feature, it allows one to rapidly assess the stratigraphic sequence without disturbing the entire feature, and it captures the most data about the formation process of the feature.

The reason why participants stated that this excavation method was the one which they had been trained to use for cut features, is most likely due to the fact that in commercial archaeology, when dealing with certain types of cut features, such as pits, many archaeologists either excavate them using the Demirant Excavation approach or the Quadrant Excavation approach, depending on the amount of time they have available to dig the feature, the complexity of the feature that is to be excavated, and the preferences of the site director. The Quadrant Excavation method tends to be used on larger cut features, as dividing small features, such as post holes, into four quarters does not allow the archaeologist enough room to remove the fills from within it. This method also allows for multiple archaeologists to excavate opposing quadrants, and thus save time, enabling archaeologists to investigate more features within the time constraints of the commercial archaeological investigation.

The justifications that this method allows one to understand the stratigraphic sequence across the length and width of the feature, and that this method allows one to rapidly assess the stratigraphic sequence without disturbing the entire feature is due to the fact that the method requires that the feature is divided into four quarters, each of which are excavated and recorded separately. If only one archaeologist is working on the archaeological feature, the removal of the first quarter enables the archaeologist to assess the complexity of the stratigraphic sequence, as half of the fill across the length and width of the feature is left in situ. This means that there is minimal disturbance to the archaeological feature as a whole, and that the archaeologist also knows what fills are present within the archaeological feature, from which they can then determine how best to continue the excavation. Having removed the first quarter, the excavation of the opposing quarter then reveals the remaining halves of the feature. This then allows the archaeologist to record the long section of the feature and the half section of the feature. Such data can then be used to reconstruct the stratigraphic sequence across the entire length and width of the archaeological feature, thus ensuring that the dimensions of each fill within the feature are recorded in their entirety, unlike with the Demirant Excavation approach.

This reasoning also accounts for the justifications that this method results in the creation of a highly detailed and clear visual record of the stratigraphic sequence, and that this method captures the most data about the formation process of the feature. As the method leads to the creation of two drawings, one which displays the half section and the other which displays the long section of the archaeological feature, which together, demonstrate the formation sequence of the entire archaeological feature. This visual record again, makes it easier for laymen to understand how the feature was formed, and shows how the archaeological feature was formed across its entirety, rather than at one section, thus resulting in a clear and detailed record of the formation process of the feature.

Justifications for the use of the Arbitrary Level Excavation method

Participants who chose to use the Arbitrary Level Excavation method to excavate the grave simulation justified their selection of this approach by stating the following: it is easy to use, it provides good spatial control of artefacts recovered, it allows for excellent photographs to be taken of artefacts in situ by placing them on pedestals, it's a fast and efficient technique, and it is applicable on archaeological features that lack complex stratigraphy.

The reason why participants stated that this method is an easy, fast and efficient technique to use is because the method itself is simple, all archaeologists need to do is to either define an excavation unit around the archaeological feature of interest, or define the boundaries of the archaeological feature, and then remove spits of 5cm, 10cm, or 15cm, continuously until the base of the feature is reached. This means that no time is spent on defining the boundaries of individual fills contained within the archaeological feature, and individuals with very little archaeological knowledge or training can complete the entire process with ease.

However, as many of the participants stated, this method is only applicable on sites that lack complex stratigraphy, as the method itself destroys the dimensions of individual fills and cuts as the excavation progresses. Despite this, the participants who chose to use this methodological approach stated that as graves usually lack complex stratigraphy, and are normally composed of a singular backfill, the Arbitrary Level Excavation approach would be suitable, as they presumed that the grave simulation would mimic the composition of 'normal' graves.

Participants also justified their use of this technique by stating that the method provides good spatial control of artefacts recovered, and allows for excellent photographs to be taken of artefacts due to the use of soil pedestals. The argument that the method provides good spatial control of artefacts recovered is justifiable if the archaeological feature that is being excavated only has one fill, as the spits allow for a spatial deposition sequence of artefacts found to be formed within the one fill. However, if the archaeological feature has multiple fills, this argument is invalid, as the archaeologist will be able to state that an artefact was found at a certain depth in a certain spit, but will be unable to determine or explain how the artefact came to be in that position, as the contextual information, the fills/deposits of the archaeological feature, have been destroyed during the excavation process. The claim that soil pedestals allow for excellent photographs of artefacts to be taken is due to the fact that the act of pedestalling places each artefact that is found on a soil platform that stands out from the surrounding soil matrix. This ensures that each artefact is visible at a distance and up-close and highlights the artefacts position within the archaeological feature that is being excavated. This platform also enables archaeologists to inspect an artefact in detail before it is lifted, which is useful in the case of fragile artefacts that may become damaged as they are transferred to finds trays or bags.

Recording system selection

Prior to conducting the excavation of the grave simulation each participant was asked to select the various recording techniques that they would choose to use to record a negative archaeological feature, such as a grave. They were then asked to justify their choice of recording techniques.

The results relating to which recording techniques were selected to be used, or at least were made available to be used, by the participants for the excavation of the grave simulation can be found in Figure 46.

The results indicate that all participants thought that plans, section drawings and photographs should be used to record negative archaeological features. However, only 95% of participants thought that context sheets would be of use. Those who chose not to use them preferred to document their findings in journal format. In total, 92.5% of the participants thought that sketches should be used as an aide-memoire as the excavation progressed. In terms of using journals, 65% of the participants advocated their use, stating that using journals enabled them to keep track of their thought processes and aid in the interpretation stage of the investigation. Unit level forms were thought to be important by 30% of the participants and 12.5% of the participants thought excavation unit forms should be used. These two recording techniques tend to be used by archaeologists using the Arbitrary Level Excavation approach and so the reasonably high percentage of participants choosing these two recording techniques can be explained by the fact that 10 participants chose this method to excavate the grave simulation. Finds quantification sheets, digital records, voice notes and video recordings were selected by 2.5% of participants. These results suggest that these types of recording techniques are not deemed to be of critical value for documenting the findings of the excavation of a negative archaeological feature.

It is interesting to note, however, that a number of the participants chose not to use all of the recording techniques that they had said that they would have used when they went on to excavate the grave simulation. Instead, the participants used the recording techniques that were associated with the excavation method that they had chosen to use. Participants who selected to use the Stratigraphic Excavation method used plans, context sheets, photographs, sketches and journals to document their findings. Similarly, the participants who chose to use the Demirant Excavation method and the Quadrant Excavation method used context sheets, section drawings, photographs, sketches and journals to document their findings. Whereas, the participants who chose to use the Arbitrary Level Excavation method used, excavation unit forms, unit level forms, photographs, sketches and journals to document their findings as can be seen in 'Chapter 6 Excavation experiment'.

Justifications for the selection of recording techniques

The reasons given by the participants for the selection of the aforementioned recording techniques were as follows: they are the techniques that the participants have been trained to use and are expected to be used by the companies that they work for, they are standard practice for archaeological investigations, using these techniques enables a variety of detailed data to be collected that allows the site to be reconstructed, interpretations to be formed, and interested parties to reinterpret the archaeological site at a later date, it provides a structured approach to recording that results in a comprehensive and standardised archive being created.

The fact that a number of participants stated that the recording techniques that they selected were those that they had been trained to use and were those that are expected to be used by the companies that they work for can be explained by the fact that the recording techniques that were chosen are the techniques that are most often recommended both in the archaeological literature and archaeological guidelines. Hence, archaeological training institutions train their archaeology students in the use of these methods to ensure that their students are able to find employment within the archaeological industry after graduating. Furthermore, this reasoning may also explain why many of the participants stated that these recording techniques were standard practice.

The statement made by the participants that these recording techniques will enable a variety of detailed data to be collected that allows the site to be reconstructed, interpreted, and interested parties to reinterpret the archaeological site at a later date, makes sense, as this response matches the underlying ethos of archaeological investigations – preservation by record. Thus by using a variety of different techniques, that capture all data, even if it is deemed irrelevant at the time of recording, the site is preserved so that future archaeologists can interrogate the data at a later date and develop new explanations for what was found.

The final justification for the selection of the recording techniques was that it provides a structured approach to recording that results in a comprehensive and standardised archive being formed. This reasoning ties in with the previous justifications, as through using a set of standardised recording techniques to document the findings at the site other archaeologists will be able to interpret the findings, as they will be familiar with the recording techniques that were used. Additionally, the individuals who are working on the post-excavation analysis of the archaeological site under investigation will then have an organised data set from which to form the site archive and create the end of investigation report.

Overall, it is evident from these results that archaeologists use a variety of different archaeological excavation methods and recording systems during the course of an archaeological investigation. The selection of a particular excavation method or recording system is largely determined by the excavation methods and recording systems that are advocated in the archaeological manuals/guidelines that an individual archaeologist has been trained to use, the aims and objectives of the archaeological investigation, and the archaeological site type that is to be excavated. Although, it is interesting to note, that despite the fact that the aims and objectives and site type that was to be excavated during the grave excavation experiment was the same for all participants, different groups of participants chose to utilise different excavation methods and recording systems to complete the experiment. Each group justified their selection by stating that the methodological approach that they chose was one that they had used in the past, that it would recover the stratigraphic units and material evidence present within the grave simulation, and would provide accurate records of the formation sequence of the grave. This finding indicates that there is no standardised archaeological approach to excavating and recording negative archaeological features, such as graves, and that methodological approaches are largely determined by the methodological preferences of individual archaeologists.

Chapter 6 Excavation experiment

This chapter contains the results of the fifty grave excavation experiments. Individual participant's results are presented at the start of the section and are set out in the following manner – participant information, observations of the participant, material evidence recovered by the participant, contexts identified by the participant, and the participant's narrative of the formation sequence of the grave. The results of each of the ten excavation experiments that were completed for each of the methods tested were then integrated to form averages. These averages are presented in Figure 147 - Figure 169. After this, the results relating to how the length of time that individual archaeologists spent excavating and recording the grave simulation impacted the overall recovery of evidence are displayed in Figure 170 - Figure 172. Finally, results relating to how archaeological experience impacted the overall recovery of evidence are displayed in Figure 173 and Figure 174.

Each of the excavation methods and recording systems tested are then comparatively assessed in order to determine how the different methodological approaches impacted the quality and quantity of evidence recovered during the grave excavation experiments. These results are also used to examine how the different methods tested impacted the construction of interpretation-based narratives of the formation sequence of the grave simulation. Additionally, the results are used to discuss the impact that the length of time that archaeologists spent conducting the excavation experiment had on the quality and quantity of evidence recovered. The results are finally used to discuss the impact that archaeological experience had on the quality and quantity of evidence recovered during the grave excavation experiments.

Results

Stratigraphic Excavation method

> **Archaeologist ID:** Archaeologist 01
>
> **Years of experience:** 12 years
>
> **Excavation approach:** Stratigraphic Excavation
>
> **Recording approach:** Single Context Recording
>
> **Tools used to excavate the grave:** Trowel and hand shovel
>
> **Did the participant sieve the fill:** Yes
>
> **Weather conditions:** Warm and slightly overcast
>
> **Time taken:** 4 hours

Observations: The participant chose to use the Stratigraphic Excavation method and Single Context Recording method to excavate and record the grave. The participant started the excavation by removing the replaced turf by hand and then began to excavate the fills contained within the grave.

The participant excavated individual fills by using small spits, approximately 1-2cm in depth, using a trowel, until the participant identified the presence of a new fill. After which the participant carefully discerned the boundary of the underlying context. Once a new fill was identified the participant would plan the fill, fill out a new context form and take photographs.

When recording, the participant also marked on the plans, photographed and noted down on the context forms where material evidence had been found. The participant also sieved all of the spoil removed from the grave as the excavation proceeded, making sure that individual fills were sieved separately. This ensured that any material evidence that was recovered in the sieve could be associated with the fill from which it had originated.

It is interesting to note that this participant classified the dress, placed on the base of the grave, as a separate context as well as a piece of material evidence. The participant justified this approach by stating that the dress covered a large area of the grave and was distinct in nature from the rest of the fills and material evidence contained in the grave, and as a result, acted, in a manner of speaking, as a fill and a unique stratigraphic event.

Evidence	Location	Evidence found	Found in situ	Found out of situ	Evidence not found
E1 Dress	Base of C4	Yes	Yes	No	No
E2 Two pence coin	C5 (Fill 1)	Yes	Yes	No	No
E3 Lighter	C6 (Fill 2)	Yes	Yes	No	No
E4 Fake nail	C7 (Fill 3)	Yes	No	Yes	No
E5 ID card	C7 (Fill 3)	Yes	Yes	No	No
E6 Earring 2	C7 (Fill 3)	Yes	Yes	No	No
E7 Kirby grip	C8 (Fill 4)	No	-	-	Yes
E8 Earring 1	C8 (Fill 4)	Yes	Yes	No	No
E9 Cigarette papers	C9 (Fill 5)	Yes	Yes	No	No
TOTAL:		8	7	1	1
PERCENTAGE:		88.89%	87.50%	12.50%	11.11%

FIGURE 47: MATERIAL EVIDENCE IDENTIFIED BY ARCHAEOLOGIST 01 © Evis 2016.

Context number	Description	Context identified	Context not identified
C1	Natural	Yes	No
C2	Subsoil	Yes	No
C3	Topsoil and turf	Yes	No
C4	Feature cut	Yes	No
C5	Fill 1	Yes	No
C6	Fill 2	Yes	No
C7	Fill 3	Yes	No
C8	Fill 4	Yes	No
C9	Fill 5	Yes	No
C10	Replaced turf	Yes	No
TOTAL:		10	0
PERCENTAGE:		100%	0%

FIGURE 48: CONTEXTS IDENTIFIED BY ARCHAEOLOGIST 01 © Evis 2016.

Archaeologist 01's narrative: A rectangular cut was cut into clayey soil and had straight sides and a flat bottom. A blue and white dress was put on the bottom of the cut and covered with sand. More sand was placed on the western end of the cut. In each of these separate sand deposits modern materials were found. Although neither of these items were in direct connection to the dress or each other. A thick layer of sandy soil, which included modern objects, covered the two separate sand deposits. This deposit was covered by a deposit of charcoal-based compost, with no sign of burned surfaces. The charcoal-based deposit was covered by sand soil and turf.

Most of the objects appear to relate to single or multiple individuals. A matched pair of earrings, a dress, one artificial fingernail, a cigarette lighter, cigarette papers, and a two pence coin may relate to the same individual or another. The ID card may relate to one of the above.

Archaeologist ID: Archaeologist 02

Years of experience: 4 years

Excavation approach: Stratigraphic Excavation

Recording approach: Single Context Recording

Tools used to excavate the grave: Trowel and hand shovel

Did the participant sieve the fill: Yes

Weather conditions: Hot with clear skies

Time taken: 14 hours

Observations: The participant chose to use the Stratigraphic Excavation method and Single Context Recording method to excavate and record the grave. The participant began by removing the replaced turf by hand and then proceeded to excavate the fills of the grave.

The participant was particularly methodical when excavating the fills, removing millimetres of fill at a time, ensuring that the boundaries of each fill were maintained as each new fill was uncovered, this might explain why it took this participant so long to excavate the grave.

As a new fill was identified the participant would record it in plan and then fill out a new context sheet and take photographs, which would include both the location of any evidence that had been recovered and the fills contained in the grave. The participant also created a long section drawing of the fills contained in the grave at the end of the excavation process. The participant did this by measuring the remnants of each of the fills that were present in the grave, as they had adhered to the walls of the grave cut.

During the excavation process the participant also sieved all of the spoil that had been removed from the grave. The participant ensured that each fill was sieved separately so that any material evidence recovered could be reassociated with the fill from which it had originated.

Archaeologist 02's narrative: The cut was dug first.

The dress and two pence coin from the first sand fill and the lighter from the second sand fill were deposited first. These contexts may have been formed contemporaneously, based on their near identical depths and soil types.

Evidence	Location	Evidence found	Found in situ	Found out of situ	Evidence not found
E1 Dress	Base of C4	Yes	Yes	No	No
E2 Two pence coin	C5 (Fill 1)	Yes	No	Yes	No
E3 Lighter	C6 (Fill 2)	Yes	Yes	No	No
E4 Fake nail	C7 (Fill 3)	Yes	Yes	No	No
E5 ID card	C7 (Fill 3)	Yes	Yes	No	No
E6 Earring 2	C7 (Fill 3)	Yes	No	Yes	No
E7 Kirby grip	C8 (Fill 4)	No	-	-	Yes
E8 Earring 1	C8 (Fill 4)	No	-	-	Yes
E9 Cigarette papers	C9 (Fill 5)	Yes	Yes	No	No
TOTAL:		7	5	2	2
PERCENTAGE:		77.78%	71.43%	28.57%	22.22%

FIGURE 49: MATERIAL EVIDENCE IDENTIFIED BY ARCHAEOLOGIST 02 © Evis 2016.

Context number	Description	Context identified	Context not identified
C1	Natural	Yes	No
C2	Subsoil	Yes	No
C3	Topsoil and turf	Yes	No
C4	Feature cut	Yes	No
C5	Fill 1	Yes	No
C6	Fill 2	Yes	No
C7	Fill 3	Yes	No
C8	Fill 4	Yes	No
C9	Fill 5	Yes	No
C10	Replaced turf	Yes	No
TOTAL:		10	0
PERCENTAGE:		100%	0%

FIGURE 50: CONTEXTS IDENTIFIED BY ARCHAEOLOGIST 02 © Evis 2016.

The two sand deposits may have originally been one deposit as they are both composed of what happens to be identical sand. It is possible that the grave was dug into at a later date, during which one deposit was split into two. Since this later cut was not identified during the excavation, the relationship between these two sand deposits must be classified separately and artefacts from each of these sand deposits were recorded as coming from separate deposits (stratigraphic events); although they may have been deposited at the same time.

A sandy soil deposit was placed on top of the two sand deposits and slumped down in the middle, following the sloping profile of the two sand deposits. This deposit contained a fake nail, an ID card and an earring. The colour, composition and components of this deposit were very similar to the natural earth from this area and so probably originated from this site.

The next deposit resembled potting soil in its colour, composition and components and was placed on top of the sandy soil deposit. It differed from the natural earth from the area and so was probably brought in from elsewhere.

Finally, another deposit was placed on top of the potting soil along with a packet of cigarette papers. The composition of this deposit again matched the natural earth from the local area. This deposit may have been removed during the cutting of the grave and then re-deposited as a fill. Alternatively, this deposit may have been formed through natural taphonomic processes.

Further analysis of context soils and the material evidence is needed to determine the time frame of this depositional sequence and to examine the possible interrelatedness of the material evidence found in the same depositional deposits. At present it is unclear how much time elapsed between the depositions of artefacts in the same deposit.

No human remains were found.

Archaeologist ID: Archaeologist 03

Years of experience: 1 year

Excavation approach: Stratigraphic Excavation

Recording approach: Single Context Recording

Tools used to excavate the grave: Trowel and hand shovel

Did the participant sieve the fill: Yes

Weather conditions: Hot with clear skies

Time taken: 14 hours

Observations: The participant chose to use the Stratigraphic Excavation method and the Single Context Recording method to excavate and record the grave. The participant lacked any real structure in their approach to the excavation.

The participant started the excavation by removing the replaced turf by hand and then began to excavate the grave. The participant stated that on previous excavations when the participant had dealt with graves, the graves lacked complexity and were usually filled with one fill. Therefore, the participant removed the fifth (last) fill rapidly, using both the hand shovel and trowel. This resulted in the participant failing to recognise the presence of the fourth fill in the grave. This might explain why the participant failed to recover items from this fourth fill. When the participant began to excavate the third fill the participant slowed down, having discovered that the fill was overlaying two sand fills, and thus, needed closer examination. After this, the participant carefully uncovered and discerned the boundaries of these fills.

When the participant successfully identified a new fill a new context form and plan would be drawn and photographs taken, both of which illustrated the dimensions of each fill and the presence of artefacts found within it. However, the accuracy of the early plans was questionable, as the participant had failed to remove all of the fills from the grave walls. This meant that the dimensions of the fills recorded in plan were smaller than in reality. It was not until the end of the excavation that the participant realised that there was still some fill adhering to the walls of the feature. In order to account for this earlier error, the participant noted down on the plan drawings the mistake and drew new plans to rectify the identified error. As the excavation proceeded the participant sieved each of the identified fills separately. However, as the participant failed to identify the fourth fill, if the participant had successfully recovered material evidence from this fill it would have been mistakenly attributed to the fifth fill. The participant did, however, fail to recover any of the material evidence from this fill, and therefore this did not pose a problem.

In terms of sieving, the participant did not spend much time inspecting the sieve for the presence of evidence; rather the participant would shake the sieve a few times and then throw the spoil remaining in it on the spoil heap. This might explain why the participant failed to recover the smaller sized items during the experiment.

Evidence	Location	Evidence found	Found in situ	Found out of situ	Evidence not found
E1 Dress	Base of C4	Yes	Yes	No	No
E2 Two pence coin	C5 (Fill 1)	Yes	Yes	No	No
E3 Lighter	C6 (Fill 2)	Yes	Yes	No	No
E4 Fake nail	C7 (Fill 3)	Yes	Yes	No	No
E5 ID card	C7 (Fill 3)	Yes	Yes	No	No
E6 Earring 2	C7 (Fill 3)	No	-	-	Yes
E7 Kirby grip	C8 (Fill 4)	No	-	-	Yes
E8 Earring 1	C8 (Fill 4)	No	-	-	Yes
E9 Cigarette papers	C9 (Fill 5)	Yes	Yes	No	No
TOTAL:		6	6	0	3
PERCENTAGE:		66.67%	100%	0%	33.33%

FIGURE 51: MATERIAL EVIDENCE IDENTIFIED BY ARCHAEOLOGIST 03 © Evis 2016.

Context number	Description	Context identified	Context not identified
C1	Natural	Yes	No
C2	Subsoil	Yes	No
C3	Topsoil and turf	Yes	No
C4	Feature cut	Yes	No
C5	Fill 1	Yes	No
C6	Fill 2	Yes	No
C7	Fill 3	Yes	No
C8	Fill 4	No	Yes
C9	Fill 5	Yes	No
C10	Replaced turf	Yes	No
TOTAL:		9	1
PERCENTAGE:		90%	10%

FIGURE 52: CONTEXTS IDENTIFIED BY ARCHAEOLOGIST 03 © Evis 2016.

Archaeologist 03's narrative: The feature was excavated in stratigraphic order with several plans showing the relationships between the different contexts. Artefacts were recorded with the context in which they were found.

Some care was taken in digging of the grave cut. Judging from the horizontal clarity of the grave fill, it would appear that the fills of the grave came from elsewhere, since normally fills would have been mixed in burial.

No human bones were found within the grave, however, the clothing found at the bottom of the grave suggests that the occupant was female in gender, the position of the clothing within the grave cut implies that the body was oriented roughly east to west with the head pointing west. There is no evidence of a coffin or any other items of clothing.

The first layer deposited in the burial was light greyish yellow sand, which was deposited in two separate parts at the head and foot end of the grave. They were therefore treated as two different contexts, but classified as potentially contemporary. Context 5, was deposited at the head end of the grave, a 2p coin was found in the deposit, and gives the contexts the date of 1994, as the earliest possible date of burial. The dress found at the bottom of context 5 was made by a designer company called "Atmosphere" and was size 14 (UK measurements), its position at the bottom of the context suggests that it belonged to the occupant of the grave. Context 6 was deposited at the foot end of the grave. It was composed of the same material as context 5, which implies that the two contexts were deposited at the same time. A pink cigarette lighter made by "Romon" was found in context 6. It may not have belonged to the occupant due to its position within the context and the deposit would have been covering the feet of the occupant. There is a gap between context 6 and context 5 which is filled by context 7. This context (context 7) lies on top of context 5 and context 6.

Context 7 is re-deposited topsoil which contains slate, which is local geology and also contains roots. It could be suggested that the topsoil is re-deposited from the cutting of the grave cut. An acrylic nail found near the bottom of the context in the area which covers the bottom of the grave cut (between context 5 and context 6) suggests it may have belonged to the occupant of the grave, giving further evidence that the occupant was female in gender. An ID card was found closer to the top of context 7, suggests that it was dropped during burial and did not belong to the occupant.

The re-deposited topsoil was then covered over by a greyish brown material (context 9) , which, like context 5 and context 6, was brought in from elsewhere. Close to the surface was a packet of empty cigarette papers, which is probably not related to the lighter found in context 6 due to the cigarette lighter being full of lighter fuel.

The overall good condition of the artefacts suggests that the burial was recent enough for decomposition not to have had too much effect on the artefacts, most likely more recently than 1994 (as given by the 2p coin).

In regards to the acrylic nail found at the bottom of context 7, there is a possibility that it belonged to the occupant.

The occupant of the grave is most likely female in gender. Judging by the dress found in context 5, the individual would be young but without human remains a rough age is difficult to determine.

The individual responsible for the burial may be female in gender, based on the acrylic fingernail found in context 7. The individual may also be a smoker based on the cigarette lighter found in context 6, whether the packet of cigarette papers belonged to the individual is unclear, due to it being found in the top layer (context 9). However, that does not rule out the possibility that the individual had more than one cigarette lighter and had simply dropped one.

In conclusion, a possible young female in a blue and white striped dress was buried by an over 18 female, who smoked. The burial was dug with some care, but covered the head, toes and feet with sand brought from elsewhere, dropping a cigarette lighter and a coin from 1994. They then covered the sand and the rest of the body with topsoil, possibly the material removed to make the grave, and finally this layer was covered with another deposit brought from elsewhere. The grave was then covered over with the turf that had been removed when the grave was cut.

Archaeologist ID: Archaeologist 04

Years of experience: 18 years

Excavation approach: Stratigraphic Excavation

Recording approach: Single Context Recording

Tools used to excavate the grave: Trowel and hand shovel

Did the participant sieve the fill: No

Weather conditions: Overcast and warm

Time taken: 2 hours

Observations: The participant chose to use the Stratigraphic Excavation method and the Single Context Recording method to excavate and record the grave. Before beginning the excavation the participant noted that due to the size of the burial the grave might contain the remains of an infant. The participant then trowelled around the edges of the replaced turf to ensure that the edge of the grave cut matched the edge of the replaced turf.

Having defined the grave edge, the participant began to remove the fills using a trowel, excavating 1-2cm of soil at a time until a different fill was identified. This ensured that when the participant identified a new fill the boundaries of this fill were defined carefully.

Rather than sieving the excavated spoil the participant ran their fingers through the spoil. This was made possible because the participant only removed a small amount of soil at a time. The participant ensured that each fill was inspected separately in order to ensure that any artefacts that were recovered were able to be reassociated with the context from which they came.

When each new fill was identified the participant would fill out a new context sheet and then photograph and draw a plan of the fill, ensuring to note down and photograph the location of any evidence that was found. Interestingly, when the participant came down onto the second fill, the dark colour and charcoal inclusions led the participant to believe that this layer may represent evidence of burning. Therefore, the participant took a sample and stated that it should be looked at by a specialist to confirm whether this colouration was the result of burning.

Evidence	Location	Evidence found	Found in situ	Found out of situ	Evidence not found
E1 Dress	Base of C4	Yes	Yes	No	No
E2 Two pence coin	C5 (Fill 1)	Yes	No	Yes	No
E3 Lighter	C6 (Fill 2)	Yes	Yes	No	No
E4 Fake nail	C7 (Fill 3)	Yes	Yes	No	No
E5 ID card	C7 (Fill 3)	Yes	Yes	No	No
E6 Earring 2	C7 (Fill 3)	Yes	Yes	No	No
E7 Kirby grip	C8 (Fill 4)	No	-	-	Yes
E8 Earring 1	C8 (Fill 4)	Yes	Yes	No	No
E9 Cigarette papers	C9 (Fill 5)	Yes	Yes	No	No
TOTAL:		8	7	1	1
PERCENTAGE:		88.89%	87.50%	12.50%	11.11%

FIGURE 53: MATERIAL EVIDENCE IDENTIFIED BY ARCHAEOLOGIST 04 © Evis 2016.

Context number	Description	Context identified	Context not identified
C1	Natural	Yes	No
C2	Subsoil	Yes	No
C3	Topsoil and turf	Yes	No
C4	Feature cut	Yes	No
C5	Fill 1	Yes	No
C6	Fill 2	Yes	No
C7	Fill 3	Yes	No
C8	Fill 4	Yes	No
C9	Fill 5	Yes	No
C10	Replaced turf	Yes	No
TOTAL:		10	0
PERCENTAGE:		100%	0%

FIGURE 54: CONTEXTS IDENTIFIED BY ARCHAEOLOGIST 04 © Evis 2016.

Once the participant was confident that the base of the grave had been reached and had completed recording, they excavated the base and edges of the grave to see if there was evidence of any bullets that had penetrated the grave edges, and to confirm that the base and edges of the grave had been correctly defined.

Archaeologist 04's narrative: The feature presents as a small 1.10m x 0.4m SW/NE oriented cut feature containing very loose backfill. Probably material derived from the initial excavation cut. This loose material was the final event in the filling of this grave, this deposit contained a packet of cigarette papers. The second deposit encountered was topsoil like soil, again could be derived from the initial excavation and filled separately. It was quite dark suggesting burning, and contained an earring. A third layer of backfill again derived from the excavation was found, and contained an ID card, an earring and a false nail. Next, two lenses of sand, clearly derived from the south and the north. These covered a dress, which was at the base of the cut. One contained a two pence coin dated to 1994 and the other contained a lighter. This feature is clearly a murder victim's grave, but there was no evidence of a skeleton. The lenses of sand present the primary deposits in this sequence.

Archaeologist ID: Archaeologist 05

Years of experience: 16 years

Excavation approach: Stratigraphic Excavation

Recording approach: Single Context Recording

Tools used to excavate the grave: Trowel and hand shovel

Did the participant sieve the fill: Yes

Weather conditions: Overcast and warm

Time taken: 4 hours

Observations: The participant chose to use the Stratigraphic Excavation method and the Single Context Recording method to excavate and record the grave. Before starting the excavation the participant searched the surrounding area for any potential forensic evidence and conducted a fingertip search over the replaced turf in case it contained any evidence.

The participant then drew the replaced turf's dimensions on a pre-excavation plan and gave each of the separate turf clumps an evidence number in case the turfs could be used by other experts to identify the perpetrator.

The participant then gridded up a tarpaulin so that different contexts could be stored in the different grid squares of the tarpaulin. This made sure that fills were not mixed and could be inspected separately.

Having finished this task, the participant then removed the replaced turf and began excavating the grave, using 1-2cm spits at a time, until the boundary of the next fill was identified. Before beginning to excavate a new fill the participant would gently trowel the surface of the fill in order to see if there was any evidence of additional cuts.

As each new fill was uncovered the participant would fill out a new context form and plan and photograph the fill, noting down and photographing the location of any evidence that was found.

Interestingly, when the participant found an item of evidence they would block lift it, putting soil in the evidence bag along with the evidence item. This was due to the fact that the participant thought that the soil might contain additional evidence.

When the participant came across the imported soil fills, the participant took samples and stated that the police should use these samples to check against any soil or sand bags contained at a suspect's house as it may link them to the scene. In order to ensure that these soils were in fact imported, the participant excavated some exploratory sondages 10m out from each of the corners of the feature. Having inspected these sondages the participant was then able to confirm that this soil did not originate from the site.

When the participant had finished excavating each fill, the fill would then be sieved and any artefacts associated with that fill were recorded on the context sheets photographed, and noted on the plans.

Archaeologist 05's narrative: Rectangular North-South oriented cut. Contained six modern finds.

Formation: Sod layers deliberately cut into a rectangular shape.

Fill sequence from the base: Sand (context 5 and context 6) divided in two at the north and south ends of the cut. They did not join at any point. Then context 7, backfill from the natural local subsoil. Then context 8, imported non-local charcoal based topsoil. Then, context 9, backfill of natural local subsoil. Finally, sod layers placed back into their original positions over the feature. No human remains were found.

Evidence	Location	Evidence found	Found in situ	Found out of situ	Evidence not found
E1 Dress	Base of C4	Yes	Yes	No	No
E2 Two pence coin	C5 (Fill 1)	Yes	No	Yes	No
E3 Lighter	C6 (Fill 2)	Yes	Yes	No	No
E4 Fake nail	C7 (Fill 3)	Yes	No	Yes	No
E5 ID card	C7 (Fill 3)	Yes	Yes	No	No
E6 Earring 2	C7 (Fill 3)	No	-	-	Yes
E7 Kirby grip	C8 (Fill 4)	No	-	-	Yes
E8 Earring 1	C8 (Fill 4)	No	-	-	Yes
E9 Cigarette papers	C9 (Fill 5)	Yes	Yes	No	No
TOTAL:		6	4	2	3
PERCENTAGE:		66.67%	66.67%	33.33%	33.33%

FIGURE 55: MATERIAL EVIDENCE IDENTIFIED BY ARCHAEOLOGIST 05 © Evis 2016.

Context number	Description	Context identified	Context not identified
C1	Natural	Yes	No
C2	Subsoil	Yes	No
C3	Topsoil and turf	Yes	No
C4	Feature cut	Yes	No
C5	Fill 1	Yes	No
C6	Fill 2	Yes	No
C7	Fill 3	Yes	No
C8	Fill 4	Yes	No
C9	Fill 5	Yes	No
C10	Replaced turf	Yes	No
TOTAL:		10	0
PERCENTAGE:		100%	0%

FIGURE 56: CONTEXTS IDENTIFIED BY ARCHAEOLOGIST 05 © Evis 2016.

Archaeologist ID: Archaeologist 06

Years of experience: 12 years

Excavation approach: Stratigraphic Excavation

Recording approach: Single Context Recording

Tools used to excavate the grave: Trowel, hand shovel and spade

Did the participant sieve the fill: No

Weather conditions: Overcast, with occasional rainy spells

Time taken: 2 hours

Observations: The participant chose to use the Stratigraphic Excavation method and the Single Context Recording method to excavate and record the grave. Before starting the excavation the participant recorded the dimensions of the grave. The participant then removed the replaced turf by hand.

The participant then began to excavate the grave using a trowel, working from one end of the grave to the other in approximately 1-2cm spits. As the participant excavated, spoil would be placed on a tarpaulin using a hand shovel and would be briefly inspected with a trowel to see if it contained any evidence. However, after removing approximately two spits of the first fill, the participant changed approach and used a spade to remove the fills and place them onto the tarpaulin. This resulted in large volumes of soil being removed at a time. As a consequence, the participant failed to identify the fourth fill. Moreover, as the participant came across the sand fills at the bottom of the grave, they continued to use the spade. This resulted in the participant intermixing these two separate fills, as the gap between them became contaminated with sand, and the participant mistakenly classified them as being one fill.

Having adopted this approach the participant then ceased inspecting the extracted spoil, and relied on spotting material evidence in situ. This approach also led to the participant inaccurately defining the boundaries of the grave, leaving at least 5cm of soil in situ at either end of the grave. The participant only realised this error, when the excavation was towards the end and the fills that had been left in place collapsed in, contaminating the rest of the fills that were in the grave, and hindering the participant's attempts to determine whether the two sand fills were in fact separated.

In terms of recording, as the participant identified different fills and material evidence a new context form would be filled out and a plan would be drawn which would illustrate where evidence was found.

Archaeologist 06's narrative:

Events:

The grave was dug.

A blue dress was carefully laid out at the bottom of the grave.

Very clean sand, presumably from a builders' yard, was brought in, covering the dress.

The grave was backfilled with topsoil mixed with sand. This suggests that the topsoil was mixed with the lower layer, but this was possibly not intentional.

Evidence	Location	Evidence found	Found in situ	Found out of situ	Evidence not found
E1 Dress	Base of C4	Yes	Yes	No	No
E2 Two pence coin	C5 (Fill 1)	No	-	-	Yes
E3 Lighter	C6 (Fill 2)	Yes	Yes	No	No
E4 Fake nail	C7 (Fill 3)	No	-	-	Yes
E5 ID card	C7 (Fill 3)	Yes	Yes	No	No
E6 Earring 2	C7 (Fill 3)	No	-	-	Yes
E7 Kirby grip	C8 (Fill 4)	No	-	-	Yes
E8 Earring 1	C8 (Fill 4)	No	-	-	Yes
E9 Cigarette papers	C9 (Fill 5)	Yes	Yes	No	No
TOTAL:		4	4	0	5
PERCENTAGE:		44.44%	100%	0%	55.56%

FIGURE 57: MATERIAL EVIDENCE IDENTIFIED BY ARCHAEOLOGIST 06 © Evis 2016.

Context number	Description	Context identified	Context not identified
C1	Natural	Yes	No
C2	Subsoil	Yes	No
C3	Topsoil and turf	Yes	No
C4	Feature cut	Yes	No
C5	Fill 1	No	Yes
C6	Fill 2	Yes	No
C7	Fill 3	Yes	No
C8	Fill 4	No	Yes
C9	Fill 5	Yes	No
C10	Replaced turf	Yes	No
TOTAL:		8	2
PERCENTAGE:		80%	20%

FIGURE 58: CONTEXTS IDENTIFIED BY ARCHAEOLOGIST 06 © Evis 2016.

Backfill soil, presumably from when the grave was dug, was placed into the grave.

The turf was carefully put back.

It is important to note that the backfilling happened very quickly, probably within a couple of hours. Though differences between layers/contexts are obvious, there is good reason to believe that there was no time between the infilling of the different layers (absence of silting, sand clinging to the sides of the grave). The infilling of the grave was conducted in one action, and the layers are essentially of the same date (though stratigraphically distinct).

The artefacts recovered from event 3 and 4 were probably thrown in, rather than deliberately placed. I would assume that the items found in the upper layers were dropped, not rubbish. The lighter presumably works, and has lighter fluid in it, and would indicate that the person digging/infilling the grave is a smoker. Cigarette papers are signs of low status (or rather, poor economy), where as the ID card (from House of Fraser), is possibly a sign of the opposite. The juxtaposition of these two items is intriguing, but can't be resolved. It's worth noting though that the card is not signed, probably indicating that it was never used.

Archaeologist ID: Archaeologist 07

Years of experience: 4 years

Excavation approach: Stratigraphic Excavation

Recording approach: Single Context Recording

Tools used to excavate the grave: Trowel, shovel and mattock

Did the participant sieve the fill: No

Weather conditions: Overcast and cold

Time taken: 2 1/2 hours

Observations: The participant chose to use the Stratigraphic Excavation method and the Single Context Recording method to excavate and record the grave. Prior to starting the excavation the participant measured the dimensions of the grave. The participant then removed the replaced turf by hand.

The participant then proceeded to excavate the fills of the feature using a mattock and a shovel. The participant justified this approach by stating that this is the approach that would be used in the industry to excavate cut features. Once the participant removed a shovelful of spoil the participant would place it on the tarpaulin.

The participant made no attempt to sieve the soil, either with the sieve provided or with hands. The participant justified this approach by stating that it was not normal practice to sieve spoil in commercial fieldwork.

Through using a mattock to excavate the fills of the grave the participant failed to identify the presence of the fourth fill, and also dug through both of the two sand fills (context 5 and context 6), resulting in the participant misclassifying these two separate fills as one fill. Additionally, through using this approach, the participant failed to delineate the grave walls resulting in the grave structure being recorded as being 10cm larger on all edges, and unrelated material evidence becoming integrated into the fills contained in the grave. This approach also resulted in the vast majority of material evidence contained within the grave being knocked out of its original position and its origin being unable to be ascertained, as the participant

Evidence	Location	Evidence found	Found in situ	Found out of situ	Evidence not found
E1 Dress	Base of C4	Yes	Yes	No	No
E2 Two pence coin	C5 (Fill 1)	Yes	No	Yes	No
E3 Lighter	C6 (Fill 2)	Yes	No	Yes	No
E4 Fake nail	C7 (Fill 3)	No	-	-	Yes
E5 ID card	C7 (Fill 3)	Yes	Yes	No	No
E6 Earring 2	C7 (Fill 3)	No	-	-	Yes
E7 Kirby grip	C8 (Fill 4)	No	-	-	Yes
E8 Earring 1	C8 (Fill 4)	No	-	-	Yes
E9 Cigarette papers	C9 (Fill 5)	Yes	Yes	No	No
TOTAL:		5	3	2	4
PERCENTAGE:		55.56%	60%	40%	44.44%

FIGURE 59: MATERIAL EVIDENCE IDENTIFIED BY ARCHAEOLOGIST 07 © Evis 2016.

Context number	Description	Context identified	Context not identified
C1	Natural	Yes	No
C2	Subsoil	Yes	No
C3	Topsoil and turf	Yes	No
C4	Feature cut	Yes	No
C5	Fill 1	Yes	No
C6	Fill 2	No	Yes
C7	Fill 3	Yes	No
C8	Fill 4	No	Yes
C9	Fill 5	Yes	No
C10	Replaced turf	Yes	No
TOTAL:		8	2
PERCENTAGE:		80%	20%

FIGURE 60: CONTEXTS IDENTIFIED BY ARCHAEOLOGIST 07 © Evis 2016.

had not carefully defined the boundaries of each context. Although, when the participant identified a change in soil composition the participant would use a trowel to confirm the difference in fill type, but would then revert to using a shovel to remove the fill.

In terms of recording, the participant chose not to complete context sheets or take notes in the field, rather the participant decided to fill out context forms post-facto, in the site hut, and relied on memory to document the composition of the different contexts that were identified and the location in which material evidence was recovered. The participant also did not take soil samples into the site hut to aid in the context form recording process. However, the participant did record plans in the field of the different fills contained in the grave, but these plans weren't annotated and so provided no assistance to the participant when completing context forms in the site hut, and were inaccurate as the participant had overcut the feature.

Archaeologist 07's narrative: The uppermost fill (context 9) appeared recent judging by how loosely compacted the soil was. Close to the surface I uncovered a blue cigarette lighter still containing some lighter fluid (more than half full). At the same level was found some cigarette papers.

Slightly further down an ID card was found in context 7.

There was a shallow layer of sand (context 5) that was unlike anything else within the feature, suggesting it was a deliberate deposit. Directly below this layer was an article of clothing. It looked to be a nearly new blue and white striped dress. Slightly below this was a two pence piece, which appears to be out of context.

The dress was deposited first, along with the two pence coin. Then the lighter, ID card and cigarette papers were deposited later. This was perhaps a last minute attempt to hide further evidence.

Archaeologist ID: Archaeologist 08

Years of experience: 18 years

Excavation approach: Stratigraphic Excavation

Recording approach: Single Context Recording

Tools used to excavate the grave: Trowel and hand shovel

Did the participant sieve the fill: No

Weather conditions: Overcast and cold

Time taken: 3 1/2 hours

Observations: The participant chose to use the Stratigraphic Excavation method and the Single Context Recording method to excavate and record the grave. The participant started the excavation by removing the replaced turf by hand. After which the participant scraped the surrounding area of the grave with a trowel to confirm the dimensions of the grave and to ensure that any photographs would clearly display the grave fills against the undisturbed natural surrounding it.

Having completed this the participant stated that due to the dimensions of the feature, if an individual was buried in the grave, they would most likely be a child. The participant then set up a baseline to use as a reference point for finds and plans. The participant then measured the dimensions of the grave and began to excavate it.

As each new fill was identified the participant would complete a new context form, photograph it and plan it, documenting the location of any material evidence found. The participant also kept a personal log to note down additional notes to aid in interpreting the feature. The participant was very careful when excavating taking time to delineate the boundaries of each of the fills contained within the grave using a trowel and a hand shovel.

The participant did not, however, sieve the extracted spoil, as the participant stated that sieving was rarely conducted in a commercial context. However, the participant did store each of the different fills on different tarpaulins, so that context specific sampling could be conducted and, if required, sieving to recover missed material evidence. Despite the careful approach to excavating the participant failed to identify many of the material evidence items contained within the grave. This might be due to the fact that the participant's area of interest was stratigraphy, and therefore the participant concentrated on recovering the stratigraphic units in the feature rather than the material evidence contained within it.

Interestingly, as Archaeologist 01 did, the participant recorded the dress at the bottom of the grave as a separate context, justifying this by stating that the placement and size of the dress represented a significant stratigraphic event and so should be recorded as such.

Evidence	Location	Evidence found	Found in situ	Found out of situ	Evidence not found
E1 Dress	Base of C4	Yes	Yes	No	No
E2 Two pence coin	C5 (Fill 1)	No	-	-	Yes
E3 Lighter	C6 (Fill 2)	Yes	Yes	No	No
E4 Fake nail	C7 (Fill 3)	No	-	-	Yes
E5 ID card	C7 (Fill 3)	Yes	Yes	No	No
E6 Earring 2	C7 (Fill 3)	No	-	-	Yes
E7 Kirby grip	C8 (Fill 4)	No	-	-	Yes
E8 Earring 1	C8 (Fill 4)	No	-	-	Yes
E9 Cigarette papers	C9 (Fill 5)	Yes	Yes	No	No
TOTAL:		4	4	0	5
PERCENTAGE:		44.44%	100%	0%	55.56%

FIGURE 61: MATERIAL EVIDENCE IDENTIFIED BY ARCHAEOLOGIST 08 © Evis 2016.

Context number	Description	Context identified	Context not identified
C1	Natural	Yes	No
C2	Subsoil	Yes	No
C3	Topsoil and turf	Yes	No
C4	Feature cut	Yes	No
C5	Fill 1	Yes	No
C6	Fill 2	Yes	No
C7	Fill 3	Yes	No
C8	Fill 4	Yes	No
C9	Fill 5	Yes	No
C10	Replaced turf	Yes	No
TOTAL:		10	0
PERCENTAGE:		100%	0%

FIGURE 62: CONTEXTS IDENTIFIED BY ARCHAEOLOGIST 08 © Evis 2016.

Archaeologist 08's narrative: The primary cut (context 4) was regular, measuring 110cm in length and 40cm wide, with a straight sided profile, the upper edges were well defined and met the base at right angles, and the base of the cut was flat.

On the base of the cut, a 'fashion item' (dress) identified as a Primark product, had been laid out, with the edges of the dress extended to the edges of the cut. The dress was laid out at the end of the feature closest to pin 'B'.

A deposit of clean sand (context 5) had been laid over the dress, the deposit extended to the edge of the cut, suggesting it was a deliberate backfill. The sandy deposit had been divided into two sections by an interface (context 5 and context 6). This interface might have been a recut, but was wholly confined in the primary cut, so may have been an attempt to expose the dress (or meaty occupant) by the person who laid the dress in, or an animal attempting to burrow to the base of the cut. A cigarette lighter was found in the other sand deposit (context 6) closest to pin 'A'.

Context 7 overlaid the interface, the deposit extends over the whole area of the cut, suggesting it was carefully placed in the cut over the interface, suggesting an attempt to refill the grave that had been revisited. This deposit contained an ID card, near the centre of the deposit in the middle of the grave cut. The deposit was thicker where it lay in the interface.

Context 7 was overlaid by a fine silt (context 8) with high humic content (much like potting compost), which also extended across the whole area of the cut. The deposit was consistent and sloped up towards pin 'A'.

Above context 7, a deposit of sandy, gritty clay with frequent pebble inclusions and regularly distributed stones and flint (context 9) was laid. This covered the whole area of the grave and was thicker nearer pin 'B' and sloped up towards pin 'A'. A packet of cigarette rolling papers was found at the end of the trench closest to pin 'B'.

All of the deposits were wholly contained within the primary cut, the sequence appears to have been formed fairly quickly as the cut is easily visible and well defined. If the interface was created by an attempt to recover material from the base, the deposits were deposited in a secondary filling process, but contained within the original cut.

Archaeologist ID: Archaeologist 09

Years of experience: 12 years

Excavation approach: Stratigraphic Excavation

Recording approach: Single Context Recording

Tools used to excavate the grave: Trowel and hand shovel

Did the participant sieve the fill: Yes

Weather conditions: Overcast and warm

Time taken: 3 1/2 hours

Observations: The participant chose to use the Stratigraphic Excavation method and the Single Context Recording method to excavate and record the grave. The participant began the excavation by measuring the dimensions of the feature and then removed the replaced turf by hand.

The participant then proceeded to excavate the fills of the grave with a trowel and a hand shovel. The participant would dig down a few centimetres at one end and then work backwards removing the fills in a mechanical fashion, until the boundary of the next fill was identified. This resulted in the participant maintaining the boundaries of each of the fills with a high level of accuracy.

As the participant removed each fill a new context form was filled out, a plan was drawn, and the participant also drew a long section drawing of the fill in order to capture the dimensions of the fills along the length of the feature, making later interpretation easier.

The participant also took photographs of each of the fills and any material evidence recovered from the grave.

The participant also sieved all extracted spoil, making sure to sieve each fill separately so that material evidence could be reassociated with the context from which it had originated.

Evidence	Location	Evidence found	Found in situ	Found out of situ	Evidence not found
E1 Dress	Base of C4	Yes	Yes	No	No
E2 Two pence coin	C5 (Fill 1)	Yes	Yes	No	No
E3 Lighter	C6 (Fill 2)	Yes	Yes	No	No
E4 Fake nail	C7 (Fill 3)	Yes	Yes	No	No
E5 ID card	C7 (Fill 3)	Yes	Yes	No	No
E6 Earring 2	C7 (Fill 3)	No	-	-	Yes
E7 Kirby grip	C8 (Fill 4)	No	-	-	Yes
E8 Earring 1	C8 (Fill 4)	Yes	Yes	No	No
E9 Cigarette papers	C9 (Fill 5)	Yes	Yes	No	No
TOTAL:		7	7	0	2
PERCENTAGE:		77.78%	100%	0%	22.22%

FIGURE 63: MATERIAL EVIDENCE IDENTIFIED BY ARCHAEOLOGIST 09 © Evis 2016.

Context number	Description	Context identified	Context not identified
C1	Natural	Yes	No
C2	Subsoil	Yes	No
C3	Topsoil and turf	Yes	No
C4	Feature cut	Yes	No
C5	Fill 1	Yes	No
C6	Fill 2	Yes	No
C7	Fill 3	Yes	No
C8	Fill 4	Yes	No
C9	Fill 5	Yes	No
C10	Replaced turf	Yes	No
TOTAL:		10	0
PERCENTAGE:		100%	0%

FIGURE 64: CONTEXTS IDENTIFIED BY ARCHAEOLOGIST 09 © Evis 2016.

Archaeologist 09's narrative: Rectangular cut with sharp sides and edges, probably dug by a spade (with roots sliced through).

Woman's dress deposited and then backfilled with sand (context 5) in which a coin was incorporated. At a similar point context 6 was laid down, similar sand incorporating a cigarette lighter. The sand does not appear to have been from the grave cut, probably imported.

Those two contexts were covered with clay soil (context 7) that appears to have originated from the dug feature. This contained a number of finds – ID card and a fake nail. Context 7 was covered by a dark soil (context 8), which contained an earring. This was then covered by more soil (context 9) that contained cigarette papers.

Finally, the perpetrator replaced the turf that had been removed when the grave was initially dug.

Archaeologist ID: Archaeologist 10

Years of experience: 19 years

Excavation approach: Stratigraphic Excavation

Recording approach: Single Context Recording

Tools used to excavate the grave: Trowel and hand shovel

Did the participant sieve the fill: Yes

Weather conditions: Overcast and warm

Time taken: 4 hours

Observations: The participant chose to use the Stratigraphic Excavation method and the Single Context Recording method to excavate and record the grave. Before beginning the excavation the participant searched the surrounding area for any potential material evidence or other disturbances that might relate to the grave, and to inspect open trenches in the surrounding area to record the natural stratigraphy. After which, the participant used a trowel to dig around the area of the grave to confirm the boundaries of the feature and confirm that the grave had not been disturbed by any later cut events.

Having confirmed that this was not the case, the participant then measured the feature and removed the replaced turf by hand. Interestingly, the participant numbered each turf clump separately in case they could provide significant evidence at a later time. The participant then began to excavate the grave using a trowel and a hand shovel.

When excavating the grave the participant would remove approximately 1-2 cm of soil at a time from only half of the feature, until the boundary of the next fill had been reached and defined. This ensured that the participant was able to record a running section of how the grave had been filled in. The participant was careful to maintain the boundaries of each of the fills. The participant did however note that between the fills there was approximately 1cm of soil where the overlying fill had intermixed with the fill below. Therefore, any material evidence identified in this interface zone would be classified as being 'most likely' from a particular context, and notes were made highlighting the possibility that this evidence could have originated from the layer above or beneath the one within which the material evidence had been associated.

As fills were removed from the grave, each fill was sieved and kept separately from another, to ensure that any material evidence could be reassociated with the correct context. In addition, this provided the opportunity for the participant/investigators to sample the soil for further analysis.

As each fill was uncovered a new context form would be filled out which would also contain information regarding the location of any material evidence that was found. In addition, the participant would record a new plan for each fill and a section drawing mid-way through the excavation of a fill. Photographs were also taken to document the section, plans and location of material evidence. The participant also kept a notebook to record their thoughts and findings throughout the excavation and recording process, and to aid with creating the interpretation of the grave.

As with previous participants, this participant also recorded the dress as a separate context as it represented a significant stratigraphic event.

Evidence	Location	Evidence found	Found in situ	Found out of situ	Evidence not found
E1 Dress	Base of C4	Yes	Yes	No	No
E2 Two pence coin	C5 (Fill 1)	Yes	No	Yes	No
E3 Lighter	C6 (Fill 2)	Yes	Yes	No	No
E4 Fake nail	C7 (Fill 3)	Yes	Yes	No	No
E5 ID card	C7 (Fill 3)	Yes	Yes	No	No
E6 Earring 2	C7 (Fill 3)	Yes	No	Yes	No
E7 Kirby grip	C8 (Fill 4)	Yes	No	Yes	No
E8 Earring 1	C8 (Fill 4)	Yes	No	Yes	No
E9 Cigarette papers	C9 (Fill 5)	Yes	Yes	No	No
TOTAL:		9	5	4	0
PERCENTAGE:		100%	55.56%	44.44%	0%

FIGURE 65: MATERIAL EVIDENCE IDENTIFIED BY ARCHAEOLOGIST 10 © Evis 2016.

Context number	Description	Context identified	Context not identified
C1	Natural	Yes	No
C2	Subsoil	Yes	No
C3	Topsoil and turf	Yes	No
C4	Feature cut	Yes	No
C5	Fill 1	Yes	No
C6	Fill 2	Yes	No
C7	Fill 3	Yes	No
C8	Fill 4	Yes	No
C9	Fill 5	Yes	No
C10	Replaced turf	Yes	No
TOTAL:		10	0
PERCENTAGE:		100%	0%

FIGURE 66: CONTEXTS IDENTIFIED BY ARCHAEOLOGIST 10 © Evis 2016.

Archaeologist 10's narrative: All archaeological contexts were identified and recorded stratigraphically. Initial examination of the area indicated an area of turf with a clear area of disturbance; turf blocks appearing to stick out of the feature. Open trenches were examined to record the natural stratigraphy of the area. The areas of turf appeared to be trampled. Loose soil covered the turf in places. The same loose soil appeared to lip over the edge of the feature. Footprints were visible pressed into the loose soil. Compression of the turf could be seen.

The loose soil was spread around the south and north sides of the feature. Some areas on the south and north side of the feature were compacted, with turf absent.

The loose soil appears to be consistent to backfill. It was spread outside the grave, and continued unbroken to the fill within the grave. In the grave it ran under the numerous blocks of turf. Each block of turf was numbered separately, but the blocks were given one context number to simplify stratigraphic phasing (context 10). The turf blocks represent the highest (uppermost) fill deposit filling the feature. Some of the blocks of turf overlaid other turf blocks. However, no pattern could be discerned as to laying. The blocks were quite dried in places.

Tool marks were apparent on the edges of the blocks. Some were curved and indicative of spade or shovel cut marks. All blocks were individually numbered and laid out in lifting sequence. The underlying sides of the blocks were checked for trace or other evidence. Some of these blocks had sedimentation patters indicting that they had been exposed to rain. The blocks appeared to have been placed to fill the uppermost volume of the feature. Once removed, the outline of a rectangular feature could be discerned.

The loose soil that composed context 9 had caused the turf over which it lay to die off on the outside of the feature. It was excavated outside and inside to show its dimensions and depth. It filled the whole feature and surrounding area. This soil is similar to the upper sandy soil seen in section in the nearby trenches. It is re-deposited, and had flint and stone inclusions.

Under this loose soil was a layer similar in appearance and texture to garden compost (context 8). This had inclusions of plastic, rubber and plastic bag.

Both contexts 9 and 8 had fresh grass strands within, which appear to have been pressed into the feature by the addition of sods/deposits. This is indicative of the recent filling of the feature. This also contrasts with the grass sods that were found to have died off earlier in the investigation.

Contexts 9 and 8 were both recorded and removed. Both filled the whole length of the feature and sloped west to east. This may be indicative of filling the east end first, then the west, where the deposits lipped up.

Removal of these deposits exposed a cut edge to the feature. Tool marks could be seen. The curved profile of these were indicative of spade or shovel marks.

Under context 8 was a layer of sandy soil (context 7) that was very similar to context 9.

Evidence was found within each of the aforementioned contexts; some found in situ, some found in sieving. All were pinpointed to the deposits in question, and were not on interfaces.

Soil excavated was separated into 1cm spits (within each context) so that they could be examined and recorded from known spatial locations.

Under context 7 was a deposit of yellow/white sand. This had no inclusions. The sand formed two separate contexts in the east and west ends of the feature (context 6 and context 5). They did not physically connect, and so were numbered as separate contexts. They are similar in appearance, friability and content. Both sand deposits sloped towards the centre of the linear feature. This may indicate that these deposits were tipped in from each end. Both of these deposits contained evidence.

One of the sand deposits (context 5) covered a dress. Therefore it was removed to expose the dress, so that it would be able to be planned and photographed.

The dress was lifted. It formed a separate layer that separated the cut of the feature from the sand (context 5).

The cut of the grave was examined for evidence, and then cleaned to demonstrate that it truncated part of the natural soil sequence. The shape of the feature was recorded, as were tool marks. No body or human remains were found.

The feature was filled with a sequence of deposits, which were recorded stratigraphically. The feature appeared to have been recently dug, as grass found in the fills was fresh. Evidence recovered may provide dating evidence. The feature was deliberately filled. There was no evidence of water action or weathering. The die off of the turf used to cover the feature does not match what appears to be a recent backfilling. The feature may have lain open for some time, then rapidly filled, with dead turf placed on the top. There is no evidence of the feature being re-cut.

There are no indications that the fills have been re-cut or removed. Due to the presence of grass in the feature, it is recommended that a botanist examine the vegetation to provide a time frame for the creation of the feature. There was no evidence that there had been a time-lag between deposits being placed in the feature; no surfaces appeared to have been formed between deposits, by rain action or drying. The friable nature of the deposits meant that they could not be easily separated during excavation. The area of 'interface' between deposits was collected and examined separately. No evidence was found within these 'interface' areas. All evidence that was found was found within firm contexts within the stratigraphic sequence.

Demirant Excavation method

Archaeologist ID: Archaeologist 11

Years of experience: 1 year

Excavation approach: Demirant Excavation

Recording approach: Standard Context Recording

Tools used to excavate the grave: Trowel and hand shovel

Did the participant sieve the fill: Yes

Weather conditions: Hot with clear skies

Time taken: 14 hours

Observations: The participant chose to use the Demirant Excavation method and the Standard Context Recording method to excavate and record the grave. The participant started the investigation by measuring the dimensions of the grave, after which the participant divided the feature in two across its width, marking this boundary with test pegs, string, and a line level. The participant then removed the turf from one half of the feature and began excavating.

The participant carefully defined each fill in the half that was being excavated, and would verify whether the fill being excavated was a new context by inspecting the section face. As the participant came upon context 5, one of the angled sand fills at the bottom of the grave, the participant found that this context did not extend into the section face. This meant that the section drawing would not include this fill. Therefore, the participant planned context 5 separately, in order to ensure that there was a record of it. Similarly, when the participant excavated context 6, which was the other angled sand fill, the same procedure was followed.

As each new fill was discovered in the first half of the grave, the participant would complete a new context form and document where material evidence was found on these forms. Additionally, the participant took photos of each new fill and material evidence items that were discovered. In addition, the participant took samples of each of the fills that were used to verify that the fills uncovered on the other side of the grave were the same as the fills on the first side. Having completed the excavation of the first side the participant then recorded the section. Once this had been recorded, the participant went on to excavate the second half of the grave.

The second half of the grave was excavated more quickly than the first half as the participant appeared to be comfortable with how the grave was constructed and had soil samples to assess changes to the fills in the grave. When excavating the second half the participant would go back to the context forms that had been recorded earlier, and provide more information about the context that was being excavated and include information about any further material evidence items that had been discovered.

The participant also kept all of the fills separately from one another, and even split fills that were classified as one context but excavated from the two halves apart from one another. This was to ensure that when the participant sieved the spoil at the end of the excavation, any material evidence items could not only be reassociated with the context from which they came but spatially from one of the two excavation areas of the grave.

Evidence	Location	Evidence found	Found in situ	Found out of situ	Evidence not found
E1 Dress	Base of C4	Yes	Yes	No	No
E2 Two pence coin	C5 (Fill 1)	Yes	Yes	No	No
E3 Lighter	C6 (Fill 2)	Yes	Yes	No	No
E4 Fake nail	C7 (Fill 3)	Yes	Yes	No	No
E5 ID card	C7 (Fill 3)	Yes	Yes	No	No
E6 Earring 2	C7 (Fill 3)	No	-	-	Yes
E7 Kirby grip	C8 (Fill 4)	No	-	-	Yes
E8 Earring 1	C8 (Fill 4)	No	-	-	Yes
E9 Cigarette papers	C9 (Fill 5)	Yes	Yes	No	No
TOTAL:		6	6	0	3
PERCENTAGE:		66.67%	100%	0%	33.33%

FIGURE 67: MATERIAL EVIDENCE IDENTIFIED BY ARCHAEOLOGIST 11 © Evis 2016.

Context number	Description	Context identified	Context not identified
C1	Natural	Yes	No
C2	Subsoil	Yes	No
C3	Topsoil and turf	Yes	No
C4	Feature cut	Yes	No
C5	Fill 1	Yes	No
C6	Fill 2	Yes	No
C7	Fill 3	Yes	No
C8	Fill 4	Yes	No
C9	Fill 5	Yes	No
C10	Replaced turf	Yes	No
TOTAL:		10	0
PERCENTAGE:		100%	0%

FIGURE 68: CONTEXTS IDENTIFIED BY ARCHAEOLOGIST 11 © Evis 2016.

Archaeologist 11's narrative: Cut (context 4), rectangular, dug with straight and very regular sides. The feature appears to have been cut using a shovel. After, the dress was placed on the base of the cut. Then sand was placed on both sides of the feature, but does not meet in the middle and are not recorded on the section as the section line fell in the middle of these deposits. A lighter (in context 6) and a two pence coin (in context 5) were disposed of. Context 7 was then placed in the grave, on top of the sand deposits and the natural. This deposit seems to be have originated from the cutting of the grave as it contains local slate. A fingernail and an ID card were found in this fill. The ID card may have been placed carefully just under the surface of this fill as it was level and central in orientation within the feature. Context 8 appears to be shop bought potting soil, due to the presence of twigs and roots and it covered context 7 in its entirety. Context 9 was composed of topsoil, and had a packet of cigarette papers in, which were found in the corner. Finally, turf was placed back on top of the feature after it had been filled in.

Archaeologist ID: Archaeologist 12

Years of experience: 9 years

Excavation approach: Demirant Excavation

Recording approach: Standard Context Recording

Tools used to excavate the grave: Trowel and hand shovel

Did the participant sieve the fill: Yes

Weather conditions: Overcast and warm

Time taken: 4 hours

Observations: The participant chose to use the Demirant Excavation method and the Standard Context Recording method to excavate and record the grave. Prior to starting the excavation the participant measured the dimensions of the grave and divided the grave into two halves across its width using test pegs, string and a line level. The participant then removed the turf off of the first half of the feature and proceeded to excavate the fills of the feature with a trowel and a hand shovel.

The participant would remove approximately 1-2cm of soil at a time until the participant noticed the presence of a new fill. The participant would then carefully define the boundaries of the underlying fill and confirm the change in context through inspecting the section face. When material evidence was identified, it was photographed and put in a finds bag. The participant did not fill out context sheets during the excavation of the first half of the grave; rather the participant relied on using the section to describe the fills and the origin of material evidence and added information to the context sheets as the second half of the grave was excavated.

As with the previous participant, the section point was placed in the middle of the grave, thus meaning that context 5 did not extend into the section face. The participant therefore recorded context 5 and context 6 in plan and made notes on the section drawing to explain that the section point missed these two contexts. The participant actually commented that the quadrant excavation method would have captured this data more reliably, and therefore wished that this alternative approach had been used. Having finished the excavation of the first half, the participant recorded the section and completed context forms. The participant then proceeded to excavate the second half of the grave using the same approach as above.

In terms of sieving, the participant sieved each fill separately as it was excavated, making sure to store each fill in a separate area.

Archaeologist 12's narrative: Cut of feature suggested deliberate/careful excavation, with deposits that were very carefully deposited within. All contexts suggest that the feature was very much a deliberate act.

Uncertain why not all of the material within the cut was material derived from the original excavation of the feature. There is a disturbing absence of human remains. The feature was possibly used to dispose of material evidence rather than a grave cut. Although, extensive sampling might recover some remnants of human bone not observed during excavation.

All deposits appeared to be the result of a deliberate backfill.

Evidence	Location	Evidence found	Found in situ	Found out of situ	Evidence not found
E1 Dress	Base of C4	Yes	Yes	No	No
E2 Two pence coin	C5 (Fill 1)	Yes	Yes	No	No
E3 Lighter	C6 (Fill 2)	Yes	Yes	No	No
E4 Fake nail	C7 (Fill 3)	Yes	Yes	No	No
E5 ID card	C7 (Fill 3)	Yes	Yes	No	No
E6 Earring 2	C7 (Fill 3)	Yes	No	Yes	No
E7 Kirby grip	C8 (Fill 4)	No	-	-	Yes
E8 Earring 1	C8 (Fill 4)	Yes	No	Yes	No
E9 Cigarette papers	C9 (Fill 5)	Yes	Yes	No	No
TOTAL:		8	6	2	1
PERCENTAGE:		88.89%	75%	25%	11.11%

FIGURE 69: MATERIAL EVIDENCE IDENTIFIED BY ARCHAEOLOGIST 12 © Evis 2016.

Context number	Description	Context identified	Context not identified
C1	Natural	Yes	No
C2	Subsoil	Yes	No
C3	Topsoil and turf	Yes	No
C4	Feature cut	Yes	No
C5	Fill 1	Yes	No
C6	Fill 2	Yes	No
C7	Fill 3	Yes	No
C8	Fill 4	Yes	No
C9	Fill 5	Yes	No
C10	Replaced turf	Yes	No
TOTAL:		10	0
PERCENTAGE:		100%	0%

FIGURE 70: CONTEXTS IDENTIFIED BY ARCHAEOLOGIST 12 © Evis 2016.

Archaeologist ID: Archaeologist 13

Years of experience: 4 years

Excavation approach: Demirant Excavation

Recording approach: Standard Context Recording

Tools used to excavate the grave: Trowel and hand shovel

Did the participant sieve the fill: Yes

Weather conditions: Overcast and warm

Time taken: 4 hours

Observations: The participant chose to use the Demirant Excavation method and the Standard Context Recording method to excavate and record the grave. The participant started the excavation by measuring the dimensions of the grave and then divided the grave across its width using test pegs, string and a line level.

The participant then removed the turf from the first half of the grave and began excavating the fills within. The participant would excavate down in one area until a change in fill was noticed and then would dig across to the section point, being careful to maintain the boundaries of the underlying fill. As with the previous participant the material evidence that was recovered was photographed and placed in separate bags according to the fill from which it had originated.

As with the previous participants, the section point fell at the mid-point of the grave and therefore context 5 and context 6 (the sand fills) did not reach the section point, resulting in the participant recording plans of these fills, and noting on this drawing that these fills were recorded in this fashion as these fills would not appear in the section drawing.

As with the previous participant, the participant did not record context forms until the first half of the feature had been excavated, and relied on the section face to describe and record the fills on the context forms. The participant also stated that the quadrant excavation method would have captured the two sand fills in the long section, but stated that the method was too time consuming to use. Having completed the excavation of the first half, the participant used the same approach as before to excavate the second half of the grave.

In terms of sieving, the participant sieved each fill as they were removed, ensuring that any material evidence recovered could be reassociated with the fill from which it had originated.

Evidence	Location	Evidence found	Found in situ	Found out of situ	Evidence not found
E1 Dress	Base of C4	Yes	Yes	No	No
E2 Two pence coin	C5 (Fill 1)	Yes	No	Yes	No
E3 Lighter	C6 (Fill 2)	Yes	Yes	No	No
E4 Fake nail	C7 (Fill 3)	Yes	Yes	No	No
E5 ID card	C7 (Fill 3)	Yes	Yes	No	No
E6 Earring 2	C7 (Fill 3)	Yes	Yes	No	No
E7 Kirby grip	C8 (Fill 4)	No	-	-	Yes
E8 Earring 1	C8 (Fill 4)	Yes	No	Yes	No
E9 Cigarette papers	C9 (Fill 5)	Yes	Yes	No	No
TOTAL:		8	6	2	1
PERCENTAGE:		88.89%	75%	25%	11.11%

FIGURE 71: MATERIAL EVIDENCE IDENTIFIED BY ARCHAEOLOGIST 13 © Evis 2016.

Context number	Description	Context identified	Context not identified
C1	Natural	Yes	No
C2	Subsoil	Yes	No
C3	Topsoil and turf	Yes	No
C4	Feature cut	Yes	No
C5	Fill 1	Yes	No
C6	Fill 2	Yes	No
C7	Fill 3	Yes	No
C8	Fill 4	Yes	No
C9	Fill 5	Yes	No
C10	Replaced turf	Yes	No
TOTAL:		10	0
PERCENTAGE:		100%	0%

FIGURE 72: CONTEXTS IDENTIFIED BY ARCHAEOLOGIST 13 © Evis 2016.

Archaeologist 13's narrative: The grave cut (context 4) appears to have been dug with precision, which may be an indication of intent as straight edges, particularly within a slate filled matrix, would take time to prepare. The grave is cut through a layer of probable hill wash and bottoms at the natural. The base of the grave follows the slope of the hill it is positioned on. The two lower fills (context 5 and 6), are most likely due to slumping of material into the grave cut shortly after the initial excavation of the grave. Context 7 appears to be a purposeful backfilling event most likely associated with the primary function of the grave – i.e. the deposition of a body or the partial remains of one. It is important to note that no human or animal bone was observed during excavation. Context 8 appears to be imported garden soil. Context 9 is probably re-deposited natural. Finally turf was placed back over the grave. It's important to note that the turf is standing proud of the grave cut meaning that the cut contains more material than was excavated during the 'cutting process'.

The finds within suggest that either the excavator of the grave was a female and a smoker, or the body within the grave was a female, and the excavator was a smoker.

Archaeologist ID: Archaeologist 14

Years of experience: 5 years

Excavation approach: Demirant Excavation

Recording approach: Standard Context Recording

Tools used to excavate the grave: Trowel, shovel and hand shovel

Did the participant sieve the fill: No

Weather conditions: Overcast and warm

Time taken: 4 hours

Observations: The participant chose to use the Demirant Excavation method and the Standard Context Recording method to excavate and record the grave. The participant started the excavation by recording the dimensions of the grave, after which the participant divided the feature into two halves across its width using test pegs, string and a line level.

The participant then removed the overlying turf on the first half of the feature by hand and began to excavate the feature. The participant started excavating the fills using a shovel, however, having uncovered the cigarette papers the participant altered their approach and used a trowel and a hand shovel. In order to keep track of the finds and fills the participant kept a notebook, to assist in creating an interpretation and filling out the context forms after the first half of the grave had been excavated.

The participant uncovered each fill carefully, and maintained the boundaries of each of the fills in the first half of the grave by excavating down in each fill until the underlying fill was revealed, then the participant would work back to the section point. The participant would store the different fills on a tarpaulin, but instead of sieving would scrape the spoil heaps with a trowel to search for evidence. If items of material evidence were found, the participant would store them in different bags according to the fill from which they had come and take photographs of each recovered item.

Evidence	Location	Evidence found	Found in situ	Found out of situ	Evidence not found
E1 Dress	Base of C4	Yes	Yes	No	No
E2 Two pence coin	C5 (Fill 1)	No	-	-	Yes
E3 Lighter	C6 (Fill 2)	Yes	Yes	No	No
E4 Fake nail	C7 (Fill 3)	Yes	Yes	No	No
E5 ID card	C7 (Fill 3)	Yes	Yes	No	No
E6 Earring 2	C7 (Fill 3)	No	-	-	Yes
E7 Kirby grip	C8 (Fill 4)	No	-	-	Yes
E8 Earring 1	C8 (Fill 4)	No	-	-	Yes
E9 Cigarette papers	C9 (Fill 5)	Yes	Yes	No	No
TOTAL:		5	5	0	4
PERCENTAGE:		55.56%	100%	0%	44.44%

FIGURE 73: MATERIAL EVIDENCE IDENTIFIED BY ARCHAEOLOGIST 14 © Evis 2016.

Context number	Description	Context identified	Context not identified
C1	Natural	Yes	No
C2	Subsoil	Yes	No
C3	Topsoil and turf	Yes	No
C4	Feature cut	Yes	No
C5	Fill 1	No	Yes
C6	Fill 2	Yes	No
C7	Fill 3	Yes	No
C8	Fill 4	Yes	No
C9	Fill 5	Yes	No
C10	Replaced turf	Yes	No
TOTAL:		9	1
PERCENTAGE:		90%	10%

FIGURE 74: CONTEXTS IDENTIFIED BY ARCHAEOLOGIST 14 © Evis 2016.

As with the other participants, the participant found that the sand fill didn't reach the section point. However, having discovered the other sand fill on the other side of the grave, the participant assumed that they must have dug through the sand that had connected the two and mistakenly classified the two separate sand fills as one. Having completed excavating the first half of the grave the participant completed the relevant context forms and drew a section drawing. The participant then proceeded to excavate the second half of the feature using the same approach as described earlier.

Archaeologist 14's narrative: The feature was dug. The dress was placed on the bottom of the feature. Sand was spread over the bottom of the feature but done unevenly, so that more sand was present at each end of the feature (context 6). A lighter was added to this fill. This fill was covered by local topsoil/subsoil (context 7). This was covered by dark topsoil (context 8). Then the dark topsoil was covered by local topsoil/subsoil (context 9). Then the cut turf was placed back over the feature (context 10).

Archaeologist ID: Archaeologist 15

Years of experience: 15 years

Excavation approach: Demirant Excavation

Recording approach: Standard Context Recording

Tools used to excavate the grave: Trowel and hand shovel

Did the participant sieve the fill: Yes

Weather conditions: Overcast and warm

Time taken: 4 hours

Observations: The participant chose to use the Demirant Excavation method and the Standard Context Recording method to excavate and record the grave. The participant started the investigation by setting up a 1m x 1.5m grid around the feature. The participant then dug the area within the grid around the grave in order to inspect if there were any other cuts intercutting the feature. Having confirmed that this was not the case the participant set up a line across the width of the grave to mark its mid-point. The participant kept a notebook throughout the process of the excavation to document findings and assist with the interpretation of the grave.

The participant then began to excavate the first half of the feature. The participant would dig down in one area until a new fill was identified. Then the participant would excavate backwards, to ensure that the boundaries of the underlying fill were maintained. Whilst excavating the participant would not only fill out context forms and maintain a notebook, but also drew plans and took photographs of the fills identified whilst excavating. This meant that when the participant discovered that the two separate sand fills did not meet the section point the planning system would ensure that they were recorded. Having reached the base of the first half of the feature, the participant drew a section drawing and then followed the same excavation and recording procedures to remove the second half of the grave.

Any material evidence that was identified was kept in separate finds bags, with labels stating where each item had come from. The participant sieved as the excavation progressed, and ensured that each fill was sieved and kept separately so that material evidence could be reassociated with where it had originated.

Archaeologist 15's narrative: The feature was excavated and was rectangular in form with straight sides and a flat bottom. A dress was then placed on the base of the feature. Following this, two separate sand deposits were added to the feature at either end of the feature (context 5 and context 6). These deposits sloped down to the base of the feature and did not meet at any discernable point. Context 5 contained a two pence coin and context 6 contained a lighter. A light brown deposit (context 7) was then placed above these two sand deposits and sloped down to fill the space left by the sand deposits. This deposit contained a fake nail, an ID card and an earring. A dark brown deposit (context 8) was then placed on top of the light brown deposit. This deposit contained another earring. Finally, another light brown deposit (context 9) was placed on top of context 8. This deposit contained a packet of cigarette papers. The turf that had been removed when the feature had initially been dug was then placed back over the feature.

Evidence	Location	Evidence found	Found in situ	Found out of situ	Evidence not found
E1 Dress	Base of C4	Yes	Yes	No	No
E2 Two pence coin	C5 (Fill 1)	Yes	No	Yes	No
E3 Lighter	C6 (Fill 2)	Yes	Yes	No	No
E4 Fake nail	C7 (Fill 3)	Yes	Yes	No	No
E5 ID card	C7 (Fill 3)	Yes	Yes	No	No
E6 Earring 2	C7 (Fill 3)	Yes	No	Yes	No
E7 Kirby grip	C8 (Fill 4)	No	-	-	Yes
E8 Earring 1	C8 (Fill 4)	Yes	Yes	No	No
E9 Cigarette papers	C9 (Fill 5)	Yes	Yes	No	No
	TOTAL:	8	6	2	1
	PERCENTAGE:	88.89%	75%	25%	11.11%

FIGURE 75: MATERIAL EVIDENCE IDENTIFIED BY ARCHAEOLOGIST 15 © Evis 2016.

Context number	Description	Context identified	Context not identified
C1	Natural	Yes	No
C2	Subsoil	Yes	No
C3	Topsoil and turf	Yes	No
C4	Feature cut	Yes	No
C5	Fill 1	Yes	No
C6	Fill 2	Yes	No
C7	Fill 3	Yes	No
C8	Fill 4	Yes	No
C9	Fill 5	Yes	No
C10	Replaced turf	Yes	No
TOTAL:		10	0
PERCENTAGE:		100%	0%

FIGURE 76: CONTEXTS IDENTIFIED BY ARCHAEOLOGIST 15 © Evis 2016.

Archaeologist ID: Archaeologist 16

Years of experience: 7 years

Excavation approach: Demirant Excavation

Recording approach: Standard Context Recording

Tools used to excavate the grave: Trowel and hand shovel

Did the participant sieve the fill: No

Weather conditions: Overcast and warm

Time taken: 3 1/2 hours

Observations: The participant chose to use the Demirant Excavation method and the Standard Context Recording method to excavate and record the grave. The participant started by dividing the grave across its width using test pegs, string and a line level.

The participant then removed the turf on the first half of the feature by hand and began excavating. The participant wrote a notebook and filled in context forms as the excavation proceed in order to make interpretation of the grave easier at the end. Interestingly, the participant filled out two different sets of context forms for each half of the grave. When excavating the second half, the participant would record contexts as being the 'same as' the others from the other side of the grave. The participant excavated the first half of the grave by removing small spits approximately 1-2cm in depth until a new fill was identified. Then the participant would carefully uncover the boundaries of the underlying fill and complete context forms and write notes in the notebook.

The participant did not sieve the fills but would inspect each fill separately by running fingers through it whilst transferring it to the tarpaulin. As with the previous excavations using this method, the sand fill did not extend into the section, therefore the participant wrote this in the notebook, completed a separate context form and photographed it.

All material evidence identified was kept in separate finds bags and documented in the context forms, notebooks and photographed. Once the participant had finished excavating the first half of the grave a section drawing was completed. The participant then excavated the remaining half of the grave following the same approach as stated earlier.

Evidence	Location	Evidence found	Found in situ	Found out of situ	Evidence not found
E1 Dress	Base of C4	Yes	Yes	No	No
E2 Two pence coin	C5 (Fill 1)	Yes	Yes	No	No
E3 Lighter	C6 (Fill 2)	Yes	Yes	No	No
E4 Fake nail	C7 (Fill 3)	Yes	Yes	No	No
E5 ID card	C7 (Fill 3)	Yes	Yes	No	No
E6 Earring 2	C7 (Fill 3)	Yes	Yes	No	No
E7 Kirby grip	C8 (Fill 4)	No	-	-	Yes
E8 Earring 1	C8 (Fill 4)	Yes	Yes	No	No
E9 Cigarette papers	C9 (Fill 5)	Yes	Yes	No	No
TOTAL:		8	8	0	1
PERCENTAGE:		88.89%	100%	0%	11.11%

FIGURE 77: MATERIAL EVIDENCE IDENTIFIED BY ARCHAEOLOGIST 16 © Evis 2016.

Context number	Description	Context identified	Context not identified
C1	Natural	Yes	No
C2	Subsoil	Yes	No
C3	Topsoil and turf	Yes	No
C4	Feature cut	Yes	No
C5	Fill 1	Yes	No
C6	Fill 2	Yes	No
C7	Fill 3	Yes	No
C8	Fill 4	Yes	No
C9	Fill 5	Yes	No
C10	Replaced turf	Yes	No
TOTAL:		10	0
PERCENTAGE:		100%	0%

FIGURE 78: CONTEXTS IDENTIFIED BY ARCHAEOLOGIST 16 © Evis 2016.

Archaeologist 16's narrative: Feature, known also as cut 4 was identified with loose topsoil. This was measured and investigated for disturbance.

Context 9 – top layer of fill of cut 4 is a mid-brown crumbly context. Within this layer a cigarette paper packet was recovered.

Context 9 merged into context 8 – the merge was easily delineated as mid-brown blended into black/ brown over 1cm into a clear horizon. Context 8 had an earring in.

Context 8 merged again into context 7 – over a very clear horizon with little blending.

Context 7 was composed of red/brown soil, it was very crumbly and contained an ID card, fingernail and earring.

Context 5, a sand layer was under context 7 and there was a very clear context change between the two. Context 5 contained a two pence coin. Context 5 also overlaid a piece of blue material that was found to be a dress.

Context 6, underlying context 7 was found to be constructed of the same sand as context 5 and sloped to the base of the feature, but did not join with context 8. Within this context a lighter was found.

Archaeologist ID: Archaeologist 17

Years of experience: 7 years

Excavation approach: Demirant Excavation

Recording approach: Standard Context Recording

Tools used to excavate the grave: Trowel and hand shovel

Did the participant sieve the fill: No

Weather conditions: Overcast and warm

Time taken: 3 1/2 hours

Observations: The participant chose to use the Demirant Excavation method and the Standard Context Recording method to excavate and record the grave. The participant started the investigation by measuring the dimensions of the grave. The participant then divided the grave across its width using test pegs, string and a line level. Unfortunately, the section line was angled and so did not reflect a true horizontal cut through the feature. Despite this, the participant began by removing the turf that was overlaying the first half of the feature by hand.

The participant then began to excavate the fills in the first half of the grave. The participant did this by removing small spits of soil approximately 1-2cm in depth from across the fill up unto the section point, the participant would continue using this approach until a different fill was identified. The participant verified this change by inspecting the section face. When a new fill was identified the participant would complete a new context form and would write down thoughts and additional information in a notebook.

Any material evidence that was found was recorded on the context form, in notes, and had photographs taken of it. Each item found was stored in a separate labelled finds bag. The participant did not sieve the fills removed from the grave, instead, the participant would inspect the hand shovel by running a trowel through it to locate any out of situ material evidence.

Evidence	Location	Evidence found	Found in situ	Found out of situ	Evidence not found
E1 Dress	Base of C4	Yes	Yes	No	No
E2 Two pence coin	C5 (Fill 1)	Yes	Yes	No	No
E3 Lighter	C6 (Fill 2)	Yes	Yes	No	No
E4 Fake nail	C7 (Fill 3)	Yes	No	Yes	No
E5 ID card	C7 (Fill 3)	Yes	Yes	No	No
E6 Earring 2	C7 (Fill 3)	No	-	-	Yes
E7 Kirby grip	C8 (Fill 4)	No	-	-	Yes
E8 Earring 1	C8 (Fill 4)	Yes	Yes	No	No
E9 Cigarette papers	C9 (Fill 5)	Yes	Yes	No	No
TOTAL:		7	6	1	2
PERCENTAGE:		77.78%	85.71%	14.29%	22.22%

FIGURE 79: MATERIAL EVIDENCE IDENTIFIED BY ARCHAEOLOGIST 17 © Evis 2016.

Context number	Description	Context identified	Context not identified
C1	Natural	Yes	No
C2	Subsoil	Yes	No
C3	Topsoil and turf	Yes	No
C4	Feature cut	Yes	No
C5	Fill 1	Yes	No
C6	Fill 2	Yes	No
C7	Fill 3	Yes	No
C8	Fill 4	Yes	No
C9	Fill 5	Yes	No
C10	Replaced turf	Yes	No
TOTAL:		10	0
PERCENTAGE:		100%	0%

FIGURE 80: CONTEXTS IDENTIFIED BY ARCHAEOLOGIST 17 © Evis 2016.

As with previous participants, when the participant got to the sand fill (context 5) it did not reach into the section. The participant therefore recorded a separate context sheet for this fill and documented its presence in notes and photographs. Having reached the base of the first half of the grave, the participant recorded a section drawing and then proceeded to excavate the other half of the feature using the aforementioned approach. When additional material evidence was identified, the participant would update the notes, context forms and take additional photographs.

Archaeologist 17's narrative: Following the initial deposition of a blue dress across half of the pit, sand was deposited from both edges of the pit (context 5 and context 6). At the beginning of the sand deposition a coin was left above the dress (context 5) and a lighter on the other side (in the other sand deposit) (context 6).

The next layer (context 7), a light brown greyish sediment contained a fake nail, and later an ID card.

The next layer was a dark brown topsoil (context 8). It extended across the area of the pit and contained an earring.

The last deposition was a brown sediment with sand spots (context 9) which contained many stones, the only find belonging to this layer, at the edge of the pit, was a packet of cigarette rolling papers.

Turf was then placed on top of the pit covering the fills.

The layers slope towards the middle of the pit most strongly at the beginning of the sequence. The pit was therefore probably filled from the short sides towards the middle.

Archaeologist ID: Archaeologist 18

Years of experience: 3 years

Excavation approach: Demirant Excavation

Recording approach: Standard Context Recording

Tools used to excavate the grave: Trowel and hand shovel

Did the participant sieve the fill: No

Weather conditions: Cold, overcast with rainy intervals

Time taken: 5 1/2 hours

Observations: The participant chose to use the Demirant Excavation method and the Standard Context Recording method to excavate and record the grave. The excavation took place over two days due to weather conditions. The participant started the investigation by dividing the grave in two across its width using test pegs, string and a line level.

The participant then began to excavate the fills in the first half of the grave using a trowel and a hand shovel. The participant would remove 1-2cm spits of soil at a time until a new fill was identified, then the participant would follow the boundaries of the underlying fill until the section point was reached. This resulted in the participant maintaining the boundaries of each of the fills with a high level of accuracy. When a new fill was identified the participant would make notes in a notebook, fill out a context form and take photographs.

Any material evidence identified was stored in separate labelled finds bags and was noted down in the notebook and on the context forms. The participant did not sieve the extracted fills. However, the participant did inspect the fills as they were transferred to the tarpaulin, and the participant stored each fill separately on the tarpaulin so that they could be inspected/sampled at a later time if required.

As with the previous archaeologists using this method, the sand fill (context 5) did not extend into the section, and so the participant planned it and took photographs to explain why this fill did not appear in the section. The participant also followed this approach for the second sand fill (context 6). The participant finished excavating the first half of the grave on the first day and ended the day by recording the section. On the second day, the participant excavated the second half of the grave following the same approach as on the first day of the experiment.

Archaeologist 18's narrative: From this excavation, the following information can be determined about the creation of the feature: the cut of the feature was dug to a depth of approximately 30cm, and 40cm in width and 110cm in length.

At some point after it was dug a long dress (blue/white stripes with a tie at the top) was placed flat on the bottom of one half of the feature.

After the feature was cut, a small layer of light sand (context 6) was placed. This could have been before or at the same time as the dress was left. Again, after the cut, and after the dress was left in the feature,

Evidence	Location	Evidence found	Found in situ	Found out of situ	Evidence not found
E1 Dress	Base of C4	Yes	Yes	No	No
E2 Two pence coin	C5 (Fill 1)	No	-	-	Yes
E3 Lighter	C6 (Fill 2)	Yes	Yes	No	No
E4 Fake nail	C7 (Fill 3)	No	-	-	Yes
E5 ID card	C7 (Fill 3)	Yes	Yes	No	No
E6 Earring 2	C7 (Fill 3)	No	-	-	Yes
E7 Kirby grip	C8 (Fill 4)	No	-	-	Yes
E8 Earring 1	C8 (Fill 4)	No	-	-	Yes
E9 Cigarette papers	C9 (Fill 5)	Yes	Yes	No	No
TOTAL:		4	4	0	5
PERCENTAGE:		44.44%	100%	0%	55.56%

FIGURE 81: MATERIAL EVIDENCE IDENTIFIED BY ARCHAEOLOGIST 18 © Evis 2016.

Context number	Description	Context identified	Context not identified
C1	Natural	Yes	No
C2	Subsoil	Yes	No
C3	Topsoil and turf	Yes	No
C4	Feature cut	Yes	No
C5	Fill 1	Yes	No
C6	Fill 2	Yes	No
C7	Fill 3	Yes	No
C8	Fill 4	Yes	No
C9	Fill 5	Yes	No
C10	Replaced turf	Yes	No
TOTAL:		10	0
PERCENTAGE:		100%	0%

FIGURE 82: CONTEXTS IDENTIFIED BY ARCHAEOLOGIST 18 © Evis 2016.

another layer of light sand was put into the feature (context 5). This again, could have been placed at the same time or after context 6. Context 6 contained a plastic lighter which could be dated using manufacturer information, but context 5 contained a two pence coin dated to 1994.

At some point after both contexts 5 and 6 a sandy loam layer was placed into the feature (context 7). Within this context a fake fingernail, an ID card, and a gold earring were found. These artefacts could have already been in the soil or were dropped in as the feature was being filled. Since the coin from context 5 was dated to 1994 the sandy loam (context 7) must have been placed in or after the year 1994.

After the sandy loam layer (context 7), a sandy dark brown layer (context 8) was placed. In this layer another gold earring was found. The gold earring was exactly the same as the one found in context 7, suggesting the two layers either came into contact with one another or the same person. Again, this context must date to 1994 or later, based on the two pence coin found in context 8.

After context 8, a rocky soil layer was placed (context 9). This layer contained a packet of cigarette papers. Again, this layer would have to date to 1994 or later.

Finally, a layer of turf (context 10) was placed on top of context 9, noting again that this was at or after the year 1994.

Archaeologist ID: Archaeologist 19

Years of experience: 3 years

Excavation approach: Demirant Excavation

Recording approach: Standard Context Recording

Tools used to excavate the grave: Trowel and hand shovel

Did the participant sieve the fill: No

Weather conditions: Cold, overcast with rainy intervals

Time taken: 4 hours

Observations: The participant chose to use the Demirant Excavation method and the Standard Context Recording method to excavate and record the grave. The participant started the excavation process by dividing the grave in two across its width using test pegs, string and a line level. The participant then removed the turf overlying the first side of the grave.

The participant then began excavating the fills using a trowel and a hand shovel. The participant would remove fills by excavating down at one end of the half until a change in fill was noticed, and then excavated backwards to uncover the dimensions of the underlying fill. The participant would place all excavated spoil onto a tarpaulin and did not try to inspect it for the presence of evidence, or to keep fills separate for later analysis. Each time a new fill was identified the participant would complete a context form and photograph it.

Any material evidence that was identified was stored in separate bags according to the fill from which it had originated. As with the previous excavations using this method, the section point being in the middle of the grave, meant that the sand contexts (contexts 5 and 6) did not appear in the section. Therefore, the participant filled in the context forms for these fills and noted on them that the sand fills didn't extend to the section point. Having completed the excavation of the first half of the grave, the participant recorded the section. The participant then went on to excavate the second half of the feature. It's interesting to note however, that the participant sped up when excavating the second half of the feature and used the hand shovel to remove the fills, as the participant seemed to be confident that no additional fills would be identified. When the participant reached the second sand fill (context 6) they altered this approach and reverted back to using a trowel as this fill had yet to be recorded.

Evidence	Location	Evidence found	Found in situ	Found out of situ	Evidence not found
E1 Dress	Base of C4	Yes	Yes	No	No
E2 Two pence coin	C5 (Fill 1)	No	-	-	Yes
E3 Lighter	C6 (Fill 2)	Yes	Yes	No	No
E4 Fake nail	C7 (Fill 3)	No	-	-	Yes
E5 ID card	C7 (Fill 3)	Yes	Yes	No	No
E6 Earring 2	C7 (Fill 3)	No	-	-	Yes
E7 Kirby grip	C8 (Fill 4)	No	-	-	Yes
E8 Earring 1	C8 (Fill 4)	No	-	-	Yes
E9 Cigarette papers	C9 (Fill 5)	Yes	Yes	No	No
TOTAL:		4	4	0	5
PERCENTAGE:		44.44%	100%	0%	55.56%

FIGURE 83: MATERIAL EVIDENCE IDENTIFIED BY ARCHAEOLOGIST 19 © Evis 2016.

Context number	Description	Context identified	Context not identified
C1	Natural	Yes	No
C2	Subsoil	Yes	No
C3	Topsoil and turf	Yes	No
C4	Feature cut	Yes	No
C5	Fill 1	Yes	No
C6	Fill 2	Yes	No
C7	Fill 3	Yes	No
C8	Fill 4	Yes	No
C9	Fill 5	Yes	No
C10	Replaced turf	Yes	No
	TOTAL:	10	0
	PERCENTAGE:	100%	0%

FIGURE 84: CONTEXTS IDENTIFIED BY ARCHAEOLOGIST 19 © Evis 2016.

Archaeologist 19's narrative: The feature that has been dug to the natural was commenced by the deposition of the dress on the top of the natural layer. Then two sand layers context 5 and context 6 were deposited in this feature. In the sand layer that was furthest away from the dress (context 6) a lighter was found. The next layer, context 7, was composed of clay, with many stone inclusions, these stones were perhaps used to seal the evidence from the previous layers, it contained an ID card. Context 7 was sealed by context 8, which was a sandy dark brown layer, no evidence was found in this layer. Context 8 was completely covered by another layer, context 9, which had cigarette papers within it. At the very top of the feature, completing the sequence was context 10, the replaced turf that was placed back over the grave sealing the feature and the evidence within it.

Archaeologist ID: Archaeologist 20

Years of experience: 11 years

Excavation approach: Demirant Excavation

Recording approach: Standard Context Recording

Tools used to excavate the grave: Trowel and hand shovel

Did the participant sieve the fill: No

Weather conditions: Sunny and warm

Time taken: 4 hours

Observations: The participant chose to use the Demirant Excavation method and the Standard Context Recording method to excavate and record the grave. The participant started the excavation process by dividing the grave in half. Interestingly, this participant divided the grave in half across its length using test pegs, string and a line level. The participant then began to excavate the first half of the grave.

The participant excavated by removing small spits of fill approximately 1-2cm in depth across the length of the area being excavated until a new fill was identified. Then the participant would carefully follow the boundaries of the underlying fill. As each new fill was identified the participant would complete a new context form and take photographs.

Any material evidence items recovered were placed in labelled finds bags. As the participant was excavating along the length of the feature, you would have expected the participant not to have the same problems as the other participants, with the sand fills not appearing in the section face. However, as the participant failed to clean the section face properly, and had moved sand across the surface of the section when removing halves of the sand fills (contexts 5 and 6), the separation between the two fills was not evident, and therefore the participant misclassified context 5 and 6 as being one fill.

The participant did not sieve as the excavation progressed. The participant would occasionally inspect the hand shovel for the presence of any material evidence, but as the excavation progressed the participant ceased doing this. When the participant reached the base of the first half of the grave, they recorded the section. The participant then used the same approach to excavate the second half of the feature. Like one of the other participants, the participant recorded different context forms for the fills in the second half of the grave and would describe fills as being the 'same as' the fills in the first half.

Evidence	Location	Evidence found	Found in situ	Found out of situ	Evidence not found
E1 Dress	Base of C4	Yes	Yes	No	No
E2 Two pence coin	C5 (Fill 1)	No	-	-	Yes
E3 Lighter	C6 (Fill 2)	Yes	No	Yes	No
E4 Fake nail	C7 (Fill 3)	Yes	Yes	No	No
E5 ID card	C7 (Fill 3)	Yes	Yes	No	No
E6 Earring 2	C7 (Fill 3)	No	-	-	Yes
E7 Kirby grip	C8 (Fill 4)	No	-	-	Yes
E8 Earring 1	C8 (Fill 4)	No	-	-	Yes
E9 Cigarette papers	C9 (Fill 5)	No	-	-	Yes
TOTAL:		4	3	1	5
PERCENTAGE:		44.44%	75%	25%	55.56%

FIGURE 85: MATERIAL EVIDENCE IDENTIFIED BY ARCHAEOLOGIST 20 © Evis 2016.

Context number	Description	Context identified	Context not identified
C1	Natural	Yes	No
C2	Subsoil	Yes	No
C3	Topsoil and turf	Yes	No
C4	Feature cut	Yes	No
C5	Fill 1	Yes	No
C6	Fill 2	No	Yes
C7	Fill 3	Yes	No
C8	Fill 4	Yes	No
C9	Fill 5	Yes	No
C10	Replaced turf	Yes	No
TOTAL:		9	1
PERCENTAGE:		90%	10%

FIGURE 86: CONTEXTS IDENTIFIED BY ARCHAEOLOGIST 20 © Evis 2016.

Archaeologist 20's narrative: A rectangular feature was excavated, under the assumption of it being a grave of recent date.

Following the removal of the covering turfs, and what appeared to be a layer of sandy silt associated with the topsoil (context 9), another fill was revealed. This was highly organic and appeared to be garden compost (context 8). Underneath another fill was identified (context 7) and contained an ID card. It appeared to be carefully placed not casually thrown or deposited during backfill. An artificial fingernail was found further down in this fill. The nail appeared to have been tossed in during backfilling. It was not associated with any other finds.

Further excavation revealed a deposit of pure sand (context 5) – clearly extraneous to the location. It contained a plastic lighter in the corner. Unfortunately, this item was recovered from the spoil heap, but its location in the grave was sufficiently constrained so that an approximate location point could be recorded. This item was also most likely tossed into the sandy fill.

Underneath the sand at the other end of the grave was a blue piece of material that was found on the base of the cut. There was no sand underneath and the item seemed to have been carefully placed.

The removal of this final deposit and material concluded the excavation.

The cut was dug into local subsoil.

Interpretation – On the basis of the extraneous soil deposits and the arrangements of the finds – apparently deliberate- this seems to be an intentional deposition of materials. No bones were found. The feature is therefore a pit with a number of deliberately placed artefacts.

Weather conditions were fine and there were no time constraints.

Quadrant Excavation method

Archaeologist ID: Archaeologist 21

Years of experience: 4 years

Excavation approach: Quadrant Excavation

Recording approach: Standard Context Recording

Tools used to excavate the grave: Trowel and hand shovel

Did the participant sieve the fill: Yes

Weather conditions: Overcast and warm

Time taken: 3 1/2 hours

Observations: The participant chose to use the Quadrant Excavation method and the Standard Context Recording method to excavate and record the grave. The participant began the investigation by measuring the dimensions of the grave and then divided it into four quarters, which were labelled 'A', 'B', 'C' and 'D'. The participant then removed the turf from quarter 'A' and began to excavate. The participant would excavate by removing 1-2cm of soil at a time until the boundary of the next fill was uncovered. This meant that the participant maintained the boundaries of each context. Each time the participant believed they had uncovered a new fill they would check against both the long section and section of the grave to confirm that this assumption was correct.

As each new fill was uncovered the participant would complete a new context form, on which the participant would record the placement of any material evidence identified and provide context descriptions and sketch drawings. The participant would also take photographs of each of the new fills, evidence, and both section faces. This process was continued until the participant reached the base of the first quadrant. The participant then recorded the long section and section of the remaining grave fills, and annotated the diagrams with relevant descriptive information.

As with the Demirant approach, the sand fills (context 5 and context 6) did not reach the section point, however, as the long section was still intact, a record of their existence was recorded on the long section drawing. Having completed this recording, the participant then moved on to the second quadrant. Unlike some archaeologists, this participant decided to excavate the adjacent quadrant rather than the one in the opposite corner, this meant that the long section face was never fully visible, and could only be seen by looking at the section drawings once the excavation was completed. This, however, did not hinder the archaeologist from correctly identifying all fills present in the grave, as the participant fully recorded each section face as it became unveiled through the excavation process. In terms of excavation, the participant used the aforementioned approach as the excavation continued, and as the participant uncovered more of previous contexts and additional material evidence the context records were updated and more photographs were taken.

The participant also sieved all extracted spoil, and made sure to sieve each fill separately so that any evidence found out of situ could be reassociated with the fill from which it had come. At some points, however, the participant would inspect the spoil with hands, rather than the sieve, this is due to the fact that the participant's arms had begun to ache from sieving and so this alternative, less labour intensive, approach was used.

Evidence	Location	Evidence found	Found in situ	Found out of situ	Evidence not found
E1 Dress	Base of C4	Yes	Yes	No	No
E2 Two pence coin	C5 (Fill 1)	Yes	Yes	No	No
E3 Lighter	C6 (Fill 2)	Yes	Yes	No	No
E4 Fake nail	C7 (Fill 3)	Yes	Yes	No	No
E5 ID card	C7 (Fill 3)	Yes	Yes	No	No
E6 Earring 2	C7 (Fill 3)	No	-	-	Yes
E7 Kirby grip	C8 (Fill 4)	No	-	-	Yes
E8 Earring 1	C8 (Fill 4)	Yes	Yes	No	No
E9 Cigarette papers	C9 (Fill 5)	Yes	Yes	No	No
TOTAL:		7	7	0	2
PERCENTAGE:		77.78%	100%	0%	22.22%

FIGURE 87: MATERIAL EVIDENCE IDENTIFIED BY ARCHAEOLOGIST 21 © Evis 2016.

Archaeologist 21's narrative: The feature has been dug and soil re-deposited demonstrated by the presence of multiple contexts and the deposition of modern material.

Inclusions are found throughout the feature and have been recovered through hand collection and sieving. The evidence recorded in all contexts suggests that the feature has been backfilled with five different contexts.

Each quadrant contained evidence, although this was mainly focused towards the centre of the feature, also suggesting that it was part of a single depositional event.

Context number	Description	Context identified	Context not identified
C1	Natural	Yes	No
C2	Subsoil	Yes	No
C3	Topsoil and turf	Yes	No
C4	Feature cut	Yes	No
C5	Fill 1	Yes	No
C6	Fill 2	Yes	No
C7	Fill 3	Yes	No
C8	Fill 4	Yes	No
C9	Fill 5	Yes	No
C10	Replaced turf	Yes	No
TOTAL:		10	0
PERCENTAGE:		100%	0%

FIGURE 88: CONTEXTS IDENTIFIED BY ARCHAEOLOGIST 21 © Evis 2016.

Archaeologist ID: Archaeologist 22

Years of experience: 5 years

Excavation approach: Quadrant Excavation

Recording approach: Standard Context Recording

Tools used to excavate the grave: Trowel and hand shovel

Did the participant sieve the fill: Yes

Weather conditions: Overcast and warm

Time taken: 4 hours

Observations: The participant chose to use the Quadrant Excavation method and the Standard Context Recording method to excavate and record the grave. The participant began the process by marking out the grave into four different quarters. The participant allocated each quarter a different name 'Q1', 'Q2', 'Q3' and 'Q4'. Having completed this the participant then began to excavate the first quadrant 'Q1'. The participant started by removing the turf by hand and then excavated the fill using a trowel, excavating down in the zone closest to the cut until a new context was identified, and then moving across the fill to the section points being careful to maintain the boundaries of the underlying fill.

Each time a new fill was identified the participant would complete a notebook, context form, take photographs and draw sketches. This helped the participant keep good records of what was being found. Each context form, sketch, and the notebook was updated after the participant had identified material evidence, photographs were also taken. As with previous participants, the sand contexts (5 and 6) did not reach the section point, however, as with the previous participant, as the method left in situ one quarter of the grave on each side, these fills were present and recorded in the long section drawings. After the participant had completed excavating the first quadrant 'Q1', the participant moved onto the second quadrant 'Q2' which was in the opposing corner. As this quadrant was being excavated, the same excavation and recording procedure was followed, with all records being updated when more of the same context was uncovered. Having excavated this particular quadrant the whole structure of the grave was

121

apparent, as the long section and the section faces were still intact. This made understanding the grave structure and the recording process a lot easier, as the participant could easily draw both the long section and the section and take photographs to demonstrate the findings. However, due to the fact that the soils were recently deposited in the grave, and had yet to compact, the participant found that the standing quadrants left in situ started to crumble. Although, this situation was rectified by the participant angling the section faces. This provided enough support for the standing quadrants for recording to be completed. The participant then excavated the remaining quadrants following the aforementioned excavation and recording procedures.

In terms of sieving, the participant would sieve all extracted spoil as the grave was being excavated, to make sure that any finds could be reassociated with their contexts. Although, due to the careful approach that the participant adopted, all but one find was found in situ.

Evidence	Location	Evidence found	Found in situ	Found out of situ	Evidence not found
E1 Dress	Base of C4	Yes	Yes	No	No
E2 Two pence coin	C5 (Fill 1)	Yes	Yes	No	No
E3 Lighter	C6 (Fill 2)	Yes	Yes	No	No
E4 Fake nail	C7 (Fill 3)	Yes	Yes	No	No
E5 ID card	C7 (Fill 3)	Yes	Yes	No	No
E6 Earring 2	C7 (Fill 3)	Yes	Yes	No	No
E7 Kirby grip	C8 (Fill 4)	No	-	-	Yes
E8 Earring 1	C8 (Fill 4)	Yes	Yes	No	No
E9 Cigarette papers	C9 (Fill 5)	Yes	Yes	No	No
	TOTAL:	8	8	0	1
	PERCENTAGE:	88.89%	100%	0%	11.11%

FIGURE 89: MATERIAL EVIDENCE IDENTIFIED BY ARCHAEOLOGIST 22 © Evis 2016.

Context number	Description	Context identified	Context not identified
C1	Natural	Yes	No
C2	Subsoil	Yes	No
C3	Topsoil and turf	Yes	No
C4	Feature cut	Yes	No
C5	Fill 1	Yes	No
C6	Fill 2	Yes	No
C7	Fill 3	Yes	No
C8	Fill 4	Yes	No
C9	Fill 5	Yes	No
C10	Replaced turf	Yes	No
	TOTAL:	10	0
	PERCENTAGE:	100%	0%

FIGURE 90: CONTEXTS IDENTIFIED BY ARCHAEOLOGIST 22 © Evis 2016.

Archaeologist 22's narrative: The feature contained no human remains. It did, however, contain modern artefacts that may be of interest. These include: a blue and white striped dress, a two pence coin (1994), a lighter, an acrylic nail, an ID card, a pair of earrings, and a packet of cigarette papers.

The creation of the feature is as follows:

The feature was dug, probably with a spade or shovel. A blue and white striped dress was laid on the base of the feature. Two sand deposits were placed in at either end of the feature, one containing a lighter, the other a two pence coin (1994). The sand from which these deposits were made was clearly imported from elsewhere, and may lead investigators to the suspect. These two sand deposits did not have any physical connection with one another, but given their placement and similarities in composition, are likely to have been placed in the feature contemporaneously. These deposits also sloped down to the base of the feature, suggesting that the feature was filled from either end.

Above these two deposits was a third, constructed from the topsoil and subsoil that had been removed when the feature was dug. This deposit contained the acrylic nail, ID card, and one of the pair of earrings. This deposit slumped in the middle filling the space left by the two sand deposits.

Above this deposit was another fourth deposit, which appeared to be composed from garden potting soil. This again, is foreign to the area and could lead investigators to the suspect. An earring was found in this fill. As with the previous fills, this deposit sloped upwards nearly filling to the top of the grave at the northern end. It was also thicker in the middle than at either end.

There was a final deposit placed in the feature, again, composed of the topsoil and subsoil removed during the excavation of the feature. This contained a packet of cigarette papers. This deposit also sloped upwards at the northern end. After this deposit was placed in the feature the turf was placed back over the feature.

Archaeologist ID: Archaeologist 23

Years of experience: 6 years

Excavation approach: Quadrant Excavation

Recording approach: Standard Context Recording

Tools used to excavate the grave: Trowel and hand shovel

Did the participant sieve the fill: Yes

Weather conditions: Clear and hot

Time taken: 6 hours

Observations: The participant chose to use the Quadrant Excavation method and the Standard Context Recording method to excavate and record the grave. The participant began the process by measuring the dimensions of the grave and then divided it into four different sections. The participant then started excavating the first quadrant. The participant started to excavate the quadrant with a trowel removing approximately 1-2cm spits of soil at a time, and would transfer it to the sieve and inspect it.

However, the participant then altered their approach and started using the hand shovel as the main digging tool to remove the fills contained within the quadrant. This resulted in the participant failing to identify the fourth fill (context 8) and misclassifying the third fill (context 7) as the same as fill five (context 9) as the participant had not identified the middle context. However, the participant did notice the sand context (context 5) at the base of the cut and changed back to using the trowel in order to correctly define its structure. Having failed to identify context 8 and 7 during the excavation, one would have thought the participant would notice these contexts presence in the long section and section faces, however, the participant failed to do so. Therefore, when the participant moved on to excavate the other quadrants, first the opposing corner and then the others, they again used a hand shovel to remove them, as the participant thought they had identified all of the contexts correctly. It was not until the participant reached the second sand fill (context 6), that they went back to using a trowel to uncover its dimensions correctly.

In terms of recording, when the participant successfully identified a new fill, a context form was completed and photographs were taken, and updated, as more of the same fill or material evidence was located. In terms of sieving, the participant sieved all of the spoil as the excavation progressed.

Evidence	Location	Evidence found	Found in situ	Found out of situ	Evidence not found
E1 Dress	Base of C4	Yes	Yes	No	No
E2 Two pence coin	C5 (Fill 1)	Yes	No	Yes	No
E3 Lighter	C6 (Fill 2)	Yes	Yes	No	No
E4 Fake nail	C7 (Fill 3)	Yes	Yes	No	No
E5 ID card	C7 (Fill 3)	Yes	Yes	No	No
E6 Earring 2	C7 (Fill 3)	Yes	Yes	No	No
E7 Kirby grip	C8 (Fill 4)	No	-	-	Yes
E8 Earring 1	C8 (Fill 4)	Yes	No	Yes	No
E9 Cigarette papers	C9 (Fill 5)	Yes	Yes	No	No
TOTAL:		8	6	2	1
PERCENTAGE:		88.89%	75%	25%	11.11%

FIGURE 91: MATERIAL EVIDENCE IDENTIFIED BY ARCHAEOLOGIST 23 © Evis 2016.

Context number	Description	Context identified	Context not identified
C1	Natural	Yes	No
C2	Subsoil	Yes	No
C3	Topsoil and turf	Yes	No
C4	Feature cut	Yes	No
C5	Fill 1	Yes	No
C6	Fill 2	Yes	No
C7	Fill 3	No	Yes
C8	Fill 4	No	Yes
C9	Fill 5	Yes	No
C10	Replaced turf	Yes	No
TOTAL:		8	2
PERCENTAGE:		80%	20%

FIGURE 92: CONTEXTS IDENTIFIED BY ARCHAEOLOGIST 23 © Evis 2016.

124

Archaeologist 23's narrative: Ground soil/bedrock was cut by feature 1 (grave) which is approximately 40cm wide, 110cm long, and 30cm deep.

A clothing item was deliberately placed at the south end of the grave.

Fine sand context 5 and context 6 covers both ends of the grave. One of these deposits contained a lighter and the other contained a two pence coin.

The rest of the grave was backfilled with context 9 (sandy silt). Cigarette papers, a false nail, an ID card, and a pair of stud earrings appear to have been accidentally deposited during the filling of the grave.

The turf that was removed during the cutting of the feature was then placed over the feature.

Archaeologist ID: Archaeologist 24

Years of experience: 28 years

Excavation approach: Quadrant Excavation

Recording approach: Standard Context Recording

Tools used to excavate the grave: Trowel and hand shovel

Did the participant sieve the fill: No

Weather conditions: Overcast and warm

Time taken: 7 1/2 hours

Observations: The participant chose to use the Quadrant Excavation method and the Standard Context Recording method to excavate and record the grave. Prior to starting the excavation the participant took photographs of the site and the surrounding area so that the grave would have some contextual information to aid in interpretation. The participant also inspected open trenches at the site in order to check what the natural stratigraphy was like, so that it could be compared against any fills identified during the excavation. The participant then searched the surrounding area for any other evidence that might assist with interpreting the feature. The participant documented the findings of this in a notebook.

The participant then recorded the individual turf clumps overlaying the grave, allocating each a finds number as they may have evidence on them. Rather than establishing a grid with string and pegs, the participant divided the feature to just under a third and split that third into two. The participant then excavated one half of that third. The participant used miniature spits no larger than 1cm to remove the fill within the grave until a new fill was identified, after which, the participant would carefully follow the boundary of the underlying fill.

As the participant identified a new fill they would fill out a new context form, take photographs, add notes to the notebook, and draw the section of each fill before excavating the next fill in the sequence. When material evidence was identified the participant would update the context forms and notebook, take photographs, and store the evidence in labelled finds bags. The participant would confirm that a piece of evidence originated from a particular context by excavating approximately 1cm underneath the item to check that it belonged in that particular fill. The participant would also store the spoil from each fill in different containers so that any evidence found at a later date could be associated with the fill from which it had come. The participant repeated the outlined procedures for the second half of the third that had already been excavated. The participant then extended the excavation by another third and split that in two, and then the final third was again split in two, and the participant followed the same excavation and recording procedure for each.

The participant decided not to sieve in the field, as this is not normal practice, and stated that, as the fills were stored in different containers they could be sieved in sterile conditions, if needed, at a later date.

Evidence	Location	Evidence found	Found in situ	Found out of situ	Evidence not found
E1 Dress	Base of C4	Yes	Yes	No	No
E2 Two pence coin	C5 (Fill 1)	Yes	Yes	No	No
E3 Lighter	C6 (Fill 2)	Yes	Yes	No	No
E4 Fake nail	C7 (Fill 3)	Yes	Yes	No	No
E5 ID card	C7 (Fill 3)	Yes	Yes	No	No
E6 Earring 2	C7 (Fill 3)	Yes	Yes	No	No
E7 Kirby grip	C8 (Fill 4)	No	-	-	Yes
E8 Earring 1	C8 (Fill 4)	Yes	Yes	No	No
E9 Cigarette papers	C9 (Fill 5)	Yes	Yes	No	No
TOTAL:		8	8	0	1
PERCENTAGE:		88.89%	100%	0%	11.11%

FIGURE 93: MATERIAL EVIDENCE IDENTIFIED BY ARCHAEOLOGIST 24 © Evis 2016.

Context number	Description	Context identified	Context not identified
C1	Natural	Yes	No
C2	Subsoil	Yes	No
C3	Topsoil and turf	Yes	No
C4	Feature cut	Yes	No
C5	Fill 1	Yes	No
C6	Fill 2	Yes	No
C7	Fill 3	Yes	No
C8	Fill 4	Yes	No
C9	Fill 5	Yes	No
C10	Replaced turf	Yes	No
TOTAL:		10	0
PERCENTAGE:		100%	0%

FIGURE 94: CONTEXTS IDENTIFIED BY ARCHAEOLOGIST 24 © Evis 2016.

Archaeologist 24's narrative: Feature cut 4 into 3, 2 and 1 (current soil profile). Primary fills (alien to the site) 5 and 6 deposited. Fill five has a terminus post quem of 1994. Fill 7 deposited (mix of original soils), followed by fill 8 (imported alien soil). Then fill 9, which was redeposited topsoil, then turf (context 10). The ID card gives a terminus post quem for fill 7, and the fills above. The feature was filled from both sides, or rather there was no side preference apparent for fills 5 and 6. Fill 7 may represent settling or compression. Fill 8 may have been filled from the right, potentially.

Archaeologist ID: Archaeologist 25

Years of experience: 6 years

Excavation approach: Quadrant Excavation

Recording approach: Standard Context Recording

Tools used to excavate the grave: Trowel and hand shovel

Did the participant sieve the fill: No

Weather conditions: Overcast and warm

Time taken: 3 hours

Observations: The participant chose to use the Quadrant Excavation method and the Standard Context Recording method to excavate and record the grave. The participant started by measuring the dimensions of the grave and then divided the grave into four quarters. The participant then removed the turf on the first quadrant and began excavating.

The participant excavated the fill using a trowel by digging down in the corner until a new fill was identified, then the participant removed the rest of the fill within the quadrant, being careful to maintain the boundaries of the underlying fill. The participant would regularly check the section faces to ensure that the fills were being identified correctly. Once a new fill was identified the participant would complete a context form, make notes in a notebook, and draw sketches and take photographs.

As material evidence was identified the participant would update these records and take additional photographs. As the excavation proceeded the participant would store each fill in separate piles on the tarpaulin. The participant did not sieve as the excavation progressed but would inspect the spoil with hands to determine if it contained any additional evidence. As the excavation progressed however, the participant ceased checking the soil and focused on uncovering and recording the different fills in the grave. Once the participant completed excavating the first quadrant, the participant recorded both the section and the long section. As with the other participants, the sand fill did not reach the section point, but was recorded in the long section drawing. After this was recorded, the opposing quarter was excavated and the section and long section drawing was completed.

The participant continued to update all recording forms as the excavation progressed, and continued to excavate in the manner described above. One issue the participant had was again with the crumbling section faces, however, the participant sloped them and this prevented this problem causing any further issues.

Archaeologist 25's narrative: The grave was constructed as follows:

Latest event

Context 10- Replaced turf.

Context 9 – Rocky, orange/brown clumped soil with a grainy texture. Contained: Cigarette papers.

Context 8 – Dark brown soil with high organic context.

Context 7 – Loose, light brown sandy soil. Contained: ID card.

Evidence	Location	Evidence found	Found in situ	Found out of situ	Evidence not found
E1 Dress	Base of C4	Yes	Yes	No	No
E2 Two pence coin	C5 (Fill 1)	No	-	-	Yes
E3 Lighter	C6 (Fill 2)	Yes	No	Yes	No
E4 Fake nail	C7 (Fill 3)	No	-	-	Yes
E5 ID card	C7 (Fill 3)	Yes	Yes	No	No
E6 Earring 2	C7 (Fill 3)	No	-	-	Yes
E7 Kirby grip	C8 (Fill 4)	No	-	-	Yes
E8 Earring 1	C8 (Fill 4)	No	-	-	Yes
E9 Cigarette papers	C9 (Fill 5)	Yes	Yes	No	No
	TOTAL:	4	3	1	5
	PERCENTAGE:	44.44%	75%	25%	55.56%

FIGURE 95: MATERIAL EVIDENCE IDENTIFIED BY ARCHAEOLOGIST 25 © Evis 2016.

Context number	Description	Context identified	Context not identified
C1	Natural	Yes	No
C2	Subsoil	Yes	No
C3	Topsoil and turf	Yes	No
C4	Feature cut	Yes	No
C5	Fill 1	Yes	No
C6	Fill 2	Yes	No
C7	Fill 3	Yes	No
C8	Fill 4	Yes	No
C9	Fill 5	Yes	No
C10	Replaced turf	Yes	No
	TOTAL:	10	0
	PERCENTAGE:	100%	0%

FIGURE 96: CONTEXTS IDENTIFIED BY ARCHAEOLOGIST 25 © Evis 2016.

Context 6 – Sand (north side). Contained: Lighter.

Context 5 – Sand (south side).

Blue and white striped dress placed on the bottom of the cut.

Cut – Grave cut through natural soil.

Earliest event

> **Archaeologist ID:** Archaeologist 26
>
> **Years of experience:** 28 years
>
> **Excavation approach:** Quadrant Excavation
>
> **Recording approach:** Standard Context Recording
>
> **Tools used to excavate the grave:** Trowel and hand shovel
>
> **Did the participant sieve the fill:** No
>
> **Weather conditions:** Overcast, damp and cold
>
> **Time taken:** 4 hours

Observations: The participant chose to use the Quadrant Excavation method and the Standard Context Recording method to excavate and record the grave. The participant started the excavation by recording the dimensions of the feature and dividing it into four quarters. The participant then began to excavate the first quarter. The participant excavated by removing small spits of soil approximately 1-2cm in depth across the quadrant until a change in fill was identified, the participant would then carefully delineate the boundaries of the underlying fill, and confirm its presence by inspecting the section faces.

Each time a new fill was identified a new context record would be completed, photographs taken, and notes made in the participant's notebook. As the participant uncovered material evidence more photographs were taken, and the notebook and context forms were updated. Additionally, as each new fill was uncovered the participant would take soil samples, in case such samples could provide more evidence for investigators. As the participant excavated the participant would also check the edges of the grave walls for the presence of any additional cuts intercutting the feature. Once the first quadrant had been excavated the participant recorded the section and long section faces. Again, the sand fills only appeared in the long section faces. The participant then excavated the opposing corner of the grave and followed the same procedure as outlined above.

Once the long section and section drawings had been completed after the excavation of the second quadrant, the participant sped up the excavation, as they were confident in how the grave was constructed. However, as other participants had found, the relatively recent filling of the grave meant that the fills had a tendency to start to collapse, therefore, the participant, again, strengthened the in situ quadrants by sloping them.

The participant did not sieve any of the spoil; instead the participant would inspect the spoil visually as it was being transferred to the tarpaulin. However, the participant justified this approach by stating that as the individual fills had been kept separately on the tarpaulins investigators could sieve the spoil if they wanted to look for additional evidence.

Archaeologist 26's narrative: The feature was rectangular in plan, oriented roughly east to west, 1.1m long, 0.4m wide, and 0.3m deep (orientation sometimes has a religious setting implication so this was noted). The feature was cut through woodland overburden down to the natural limestone bedrock. The excavator had then placed a blue and white dress, the bulk of which lay in the western quadrants. Then a light buff coloured building sand onto the base of the grave overlying the dress in which was found a 2p coin (useful for dating) and a cigarette lighter in the sand deposit on the eastern side.

The sand was sealed by a layer of loose brown silty clay (re-used overburden). Within this further artefacts were recorded – an ID card and an acrylic nail. A layer of imported compost sealed this deposit. Another

Evidence	Location	Evidence found	Found in situ	Found out of situ	Evidence not found
E1 Dress	Base of C4	Yes	Yes	No	No
E2 Two pence coin	C5 (Fill 1)	Yes	Yes	No	No
E3 Lighter	C6 (Fill 2)	Yes	Yes	No	No
E4 Fake nail	C7 (Fill 3)	Yes	No	Yes	No
E5 ID card	C7 (Fill 3)	Yes	Yes	No	No
E6 Earring 2	C7 (Fill 3)	No	-	-	Yes
E7 Kirby grip	C8 (Fill 4)	No	-	-	Yes
E8 Earring 1	C8 (Fill 4)	No	-	-	Yes
E9 Cigarette papers	C9 (Fill 5)	Yes	Yes	No	No
TOTAL:		6	5	1	3
PERCENTAGE:		66.67%	83.33%	16.67%	33.33%

FIGURE 97: MATERIAL EVIDENCE IDENTIFIED BY ARCHAEOLOGIST 26 © Evis 2016.

Context number	Description	Context identified	Context not identified
C1	Natural	Yes	No
C2	Subsoil	Yes	No
C3	Topsoil and turf	Yes	No
C4	Feature cut	Yes	No
C5	Fill 1	Yes	No
C6	Fill 2	Yes	No
C7	Fill 3	Yes	No
C8	Fill 4	Yes	No
C9	Fill 5	Yes	No
C10	Replaced turf	Yes	No
TOTAL:		10	0
PERCENTAGE:		100%	0%

FIGURE 98: CONTEXTS IDENTIFIED BY ARCHAEOLOGIST 26 © Evis 2016.

layer of re-deposited topsoil then sealed this compost layer. This layer contained further artefactual evidence – a packet of cigarette papers.

Throughout the excavation the deposition was fairly uniform, and there was no evidence of secondary internment. The formation of the grave can therefore be viewed as a single event. The grave cut was examined for other 'earlier' or 'later' cuts that may have been present within the immediate locality. None were evident. The feature did not truncate any other event.

Archaeologist ID: Archaeologist 27

Years of experience: 1 year

Excavation approach: Quadrant Excavation

Recording approach: Standard Context Recording

Tools used to excavate the grave: Trowel and hand shovel

Did the participant sieve the fill: No

Weather conditions: Clear and hot

Time taken: 3 1/2 hours

Observations: The participant chose to use the Quadrant Excavation method and the Standard Context Recording method to excavate and record the grave. The participant started the investigation by measuring the dimensions of the grave and dividing it into four quarters. The participant began to excavate the first quadrant of the feature by removing small spits of soil approximately 1-2cm in depth at a time, until the boundary of the next fill became apparent, after which the boundary was carefully followed.

As each new fill was uncovered the participant would complete a context form and take photographs. If material evidence was found it was placed in an evidence bag, photographed, and the context forms were updated. Once the participant reached the base of the first quadrant they recorded the long section and section. As with the other participants the sand fills were preserved in the long section but not the section. The participant also found that the quadrants did start to collapse and so supported them by angling the sections. Once the participant had recorded the first quadrant the second quadrant in the opposing corner was excavated following the same procedures, and the context sheets were updated with more information about the fills that had already been uncovered. Having excavated the second quadrant the participant completed the section drawing and the long section drawing. Then the participant excavated the remaining quadrants.

As with other participants, once the participant understood the formation sequence of the grave the participant sped up the excavation process. In terms of sieving, the participant chose not to. The participant did occasionally check for the presence of material on the spoil heaps of the separated individual fills, but did not put much effort into doing this.

Evidence	Location	Evidence found	Found in situ	Found out of situ	Evidence not found
E1 Dress	Base of C4	Yes	Yes	No	No
E2 Two pence coin	C5 (Fill 1)	No	-	-	Yes
E3 Lighter	C6 (Fill 2)	Yes	Yes	No	No
E4 Fake nail	C7 (Fill 3)	No	-	-	Yes
E5 ID card	C7 (Fill 3)	Yes	Yes	No	No
E6 Earring 2	C7 (Fill 3)	No	-	-	Yes
E7 Kirby grip	C8 (Fill 4)	No	-	-	Yes
E8 Earring 1	C8 (Fill 4)	No	-	-	Yes
E9 Cigarette papers	C9 (Fill 5)	Yes	Yes	No	No
TOTAL:		4	4	0	5
PERCENTAGE:		44.44%	100%	0%	55.56%

FIGURE 99: MATERIAL EVIDENCE IDENTIFIED BY ARCHAEOLOGIST 27 © Evis 2016.

Context number	Description	Context identified	Context not identified
C1	Natural	Yes	No
C2	Subsoil	Yes	No
C3	Topsoil and turf	Yes	No
C4	Feature cut	Yes	No
C5	Fill 1	Yes	No
C6	Fill 2	Yes	No
C7	Fill 3	Yes	No
C8	Fill 4	Yes	No
C9	Fill 5	Yes	No
C10	Replaced turf	Yes	No
	TOTAL:	10	0
	PERCENTAGE:	100%	0%

FIGURE 100: CONTEXTS IDENTIFIED BY ARCHAEOLOGIST 27 © Evis 2016.

Archaeologist 27's narrative: The feature was rectangular and measured approximately 110cm long, 40cm wide and 30cm deep. It contained five deposits each allocated a different context number. Four artefacts were located: a dress, a lighter, an ID card and cigarette papers.

The order in which the feature was formed is as follows:

The feature was cut (context 4).

A dress was placed along the base.

Two separate sand deposits were placed in the feature, context 5 covered the dress and context 6 contained a lighter.

A deposit of sandy light brown soil (context 7) overlaid the sand deposits and the dress. It contained an ID card.

A compost deposit (context 8) was then placed on top of context 7.

Context 8 was then covered over by some light brown soil (context 9), which comprised of a mix of topsoil and subsoil. It contained cigarette papers.

The turf was placed back over the feature.

Archaeologist ID: Archaeologist 28

Years of experience: 1 year

Excavation approach: Quadrant Excavation

Recording approach: Standard Context Recording

Tools used to excavate the grave: Trowel and hand shovel

Did the participant sieve the fill: No

Weather conditions: Clear and hot

Time taken: 3 1/4 hours

Observations: The participant chose to use the Quadrant Excavation method and the Standard Context Recording method to excavate and record the grave. The participant started the investigation by measuring the dimensions of the grave and dividing it into four quarters. The participant then excavated the first quadrant by excavating down in one area until a change in fill was noted, then the participant would work backwards from the known to the unknown so that each fill was defined correctly. As the participant identified new fills context forms would be filled in and photographs would be taken.

As the participant identified material evidence, additional photographs would be taken and the context forms updated. Each piece of evidence was stored in a labelled finds bag. As the participant finished excavating the first quadrant the long section and section drawings were completed. Then the participant went on to excavate the second opposing corner quadrant following the aforementioned approach. Once this quadrant was completed the individual completed the long section and section drawings. As more of a particular fill was uncovered the participant would update the records taken and take additional photographs.

Having recorded the formation process of the grave by record the participant proceeded to change excavation approach, and used a hand shovel to remove the remaining quadrants, during this process the participant would constantly check the section faces to ensure that the defined boundaries were being maintained, although this process undoubtedly chopped some of the fills surfaces by at least 1cm.

The participant decided not to sieve the fills or store them separately, rather, the participant dumped all of the extracted spoil onto a tarpaulin, as the participant claimed that it was unusual to sieve spoil in a commercial context.

Archaeologist 28's narrative: The grave contained no human remains, but did contain modern materials and imported soils, which could enable investigators to find the individual responsible for creating this bodiless grave.

The grave was approximately 110cm in length, 40cm in width and 30cm in depth. The grave was cut through the natural earth. After the grave had been cut a blue and white strapless dress was placed along the base of the grave. Two separate sand deposits were then dumped in. These deposits were not from the surrounding geology and clearly imported. One of these deposits contained a lighter. A deposit of mixed topsoil/subsoil from the local area was then deposited on top of these two earlier deposits. This deposit had an ID card and a fake fingernail within it. These might provide a clue as to the identity of the individual responsible for making the grave or the supposed victim. Another imported deposit composed of compost type soil then sealed this underlying deposit. Finally, another deposit of mixed topsoil/subsoil was placed on top, which contained a packet of cigarette papers. The grave was then sealed with the turf that had been removed to make the grave.

Evidence	Location	Evidence found	Found in situ	Found out of situ	Evidence not found
E1 Dress	Base of C4	Yes	Yes	No	No
E2 Two pence coin	C5 (Fill 1)	No	-	-	Yes
E3 Lighter	C6 (Fill 2)	Yes	Yes	No	No
E4 Fake nail	C7 (Fill 3)	Yes	Yes	No	No
E5 ID card	C7 (Fill 3)	Yes	Yes	No	No
E6 Earring 2	C7 (Fill 3)	No	-	-	Yes
E7 Kirby grip	C8 (Fill 4)	No	-	-	Yes
E8 Earring 1	C8 (Fill 4)	No	-	-	Yes
E9 Cigarette papers	C9 (Fill 5)	Yes	Yes	No	No
	TOTAL:	5	5	0	4
	PERCENTAGE:	55.56%	100%	0%	44.44%

FIGURE 101: MATERIAL EVIDENCE IDENTIFIED BY ARCHAEOLOGIST 28 © Evis 2016.

Context number	Description	Context identified	Context not identified
C1	Natural	Yes	No
C2	Subsoil	Yes	No
C3	Topsoil and turf	Yes	No
C4	Feature cut	Yes	No
C5	Fill 1	Yes	No
C6	Fill 2	Yes	No
C7	Fill 3	Yes	No
C8	Fill 4	Yes	No
C9	Fill 5	Yes	No
C10	Replaced turf	Yes	No
	TOTAL:	10	0
	PERCENTAGE:	100%	0%

FIGURE 102: CONTEXTS IDENTIFIED BY ARCHAEOLOGIST 28 © Evis 2016.

Archaeologist ID: Archaeologist 29

Years of experience: 1 year

Excavation approach: Quadrant Excavation

Recording approach: Standard Context Recording

Tools used to excavate the grave: Trowel and hand shovel

Did the participant sieve the fill: No

Weather conditions: Clear and hot

Time taken: 4 hours

Observations: The participant chose to use the Quadrant Excavation method and the Standard Context Recording method to excavate and record the grave. The participant started the investigation by measuring the dimensions of the grave and dividing it into four quarters. The participant then removed the turf and excavated the first quarter using miniature spits approximately 1-2cm in depth until a change in fill was identified. Then the participant would excavate the remaining fill being careful to maintain its boundaries.

Each time a new fill was identified the participant would photograph it and fill out a new context sheet. As material evidence was identified it was recorded on the relevant context form and was photographed and placed into a separate finds bag. This process continued until the participant reached the end of the first quadrant. After which, the participant completed the long section and section drawings. The participant then excavated the opposing corner quadrant using the same approach. Interestingly, the participant would use a new set of context sheets for each quadrant excavated, stating, if the participant was certain, that the current fill was the 'same as' a previous fill. Once the second quadrant was excavated the participant updated the section and long section drawing, and proceeded to excavate the remaining quadrants in the same manner.

As with the other participants the sand fills didn't appear in the section, but did appear in the long section drawing. The participant did not sieve the spoil extracted from the grave. However, the participant stored each fill from the different quadrants in different buckets, so that sampling and sieving could be conducted later if required.

Evidence	Location	Evidence found	Found in situ	Found out of situ	Evidence not found
E1 Dress	Base of C4	Yes	Yes	No	No
E2 Two pence coin	C5 (Fill 1)	Yes	Yes	No	No
E3 Lighter	C6 (Fill 2)	Yes	Yes	No	No
E4 Fake nail	C7 (Fill 3)	Yes	Yes	No	No
E5 ID card	C7 (Fill 3)	Yes	Yes	No	No
E6 Earring 2	C7 (Fill 3)	No	-	-	Yes
E7 Kirby grip	C8 (Fill 4)	No	-	-	Yes
E8 Earring 1	C8 (Fill 4)	No	-	-	Yes
E9 Cigarette papers	C9 (Fill 5)	Yes	Yes	No	No
TOTAL:		6	6	0	3
PERCENTAGE:		66.67%	100%	0%	33.33%

FIGURE 103: MATERIAL EVIDENCE IDENTIFIED BY ARCHAEOLOGIST 29 © Evis 2016.

Context number	Description	Context identified	Context not identified
C1	Natural	Yes	No
C2	Subsoil	Yes	No
C3	Topsoil and turf	Yes	No
C4	Feature cut	Yes	No
C5	Fill 1	Yes	No
C6	Fill 2	Yes	No
C7	Fill 3	Yes	No
C8	Fill 4	Yes	No
C9	Fill 5	Yes	No
C10	Replaced turf	Yes	No
TOTAL:		10	0
PERCENTAGE:		100%	0%

FIGURE 104: CONTEXTS IDENTIFIED BY ARCHAEOLOGIST 29 © Evis 2016.

Archaeologist 29's narrative: Following the excavation of the grave the following information was discerned:

The grave was cut (context 4) through the natural.

A blue and white striped dress was placed on the base of the cut on the eastern side.

This dress was then covered by a sand deposit (context 5). This deposit contained a coin dated to 1994.

On the western side another sand deposit (context 6) was placed into the grave and contained a lighter.

On top of both of these sand deposits was a re-deposited layer of topsoil/subsoil (context 7). This deposit contained an ID card and a false fingernail.

After this, a highly organic deposit (compost) (context 8) was added to the grave, sealing context 7.

The final deposit to enter the grave was again a re-deposited layer of topsoil/subsoil (context 9). This deposit contained a packet of cigarette papers.

The turf blocks that had been cut during the initial excavation of the grave were then placed back over the final deposit (context 10).

Archaeologist ID: Archaeologist 30

Years of experience: 30 years

Excavation approach: Quadrant Excavation

Recording approach: Standard Context Recording

Tools used to excavate the grave: Trowel and hand shovel

Did the participant sieve the fill: Yes

Weather conditions: Overcast and warm

Time taken: 6 hours

Observations: The participant chose to use the Quadrant Excavation method and the Standard Context Recording method to excavate and record the grave. The participant started the investigation by measuring the dimensions of the grave and dividing it into four quarters. The participant then conducted a fingertip search over the grave in order to identify if any surface evidence was present. The participant then trowelled the area adjacent to the grave in order to check that there were no later cuts intruding into the grave and to make the grave cut more distinct against the background in photographs.

Having completed these tasks the participant then removed the turf and excavated the first quarter using miniature spits approximately 1-2cm in depth until a change in fill was identified. Then the participant would excavate the remaining fill being careful to maintain its boundaries. The participant would verify the presence of a new fill by inspecting the section faces.

Once the participant was confident that a new fill had been identified a new context form would be completed, photographs would be taken, and the participant would add this data to an on site journal. The participant was very careful once a piece of material evidence was identified and would lift the object along with its surrounding soil matrix and store it in a specialised finds bag, in order to preserve any trace evidence the object might have on it. Any fills that were removed in the excavation process were sieved and stored in separate tubs and were labelled according to the quadrant and the fill from which they had originated. The participant also took additional soil samples from the fills and the undisturbed natural soil surrounding the grave in case further analysis would be required. Once the participant completed the excavation of the first quarter the section faces were both recorded. As with previous participants the sand fills were recorded in the long section and not in the section drawing. The participant then went on to excavate the second quadrant in the opposite quarter, and followed the same approach. Having excavated it, the participant completed the section and long section drawings and updated the journal, context forms and took additional photographs and soil samples. The participant then proceeded to excavate the remaining quadrants using the same techniques as described above.

Once the excavation had been completed, the participant inspected the edges of the cut to see if any other features had cut into the grave at a later date, once satisfied that this was not the case the participant ended the excavation.

Evidence	Location	Evidence found	Found in situ	Found out of situ	Evidence not found
E1 Dress	Base of C4	Yes	Yes	No	No
E2 Two pence coin	C5 (Fill 1)	Yes	Yes	No	No
E3 Lighter	C6 (Fill 2)	Yes	Yes	No	No
E4 Fake nail	C7 (Fill 3)	Yes	Yes	No	No
E5 ID card	C7 (Fill 3)	Yes	Yes	No	No
E6 Earring 2	C7 (Fill 3)	No	-	-	Yes
E7 Kirby grip	C8 (Fill 4)	Yes	Yes	No	No
E8 Earring 1	C8 (Fill 4)	Yes	Yes	No	No
E9 Cigarette papers	C9 (Fill 5)	Yes	Yes	No	No
TOTAL:		8	8	0	1
PERCENTAGE:		88.89%	100%	0%	11.11%

FIGURE 105: MATERIAL EVIDENCE IDENTIFIED BY ARCHAEOLOGIST 30 © Evis 2016.

Context number	Description	Context identified	Context not identified
C1	Natural	Yes	No
C2	Subsoil	Yes	No
C3	Topsoil and turf	Yes	No
C4	Feature cut	Yes	No
C5	Fill 1	Yes	No
C6	Fill 2	Yes	No
C7	Fill 3	Yes	No
C8	Fill 4	Yes	No
C9	Fill 5	Yes	No
C10	Replaced turf	Yes	No
TOTAL:		10	0
PERCENTAGE:		100%	0%

FIGURE 106: CONTEXTS IDENTIFIED BY ARCHAEOLOGIST 30 © Evis 2016.

Archaeologist 30's narrative: The excavation of the suspected grave uncovered the following artefacts: a dress, a two pence coin dated to 1994, a lighter, a false nail, an ID card, a metallic hairgrip, a gem earring and a packet of cigarette papers. Each of these items has been placed into labelled evidence bags along with its surrounding soil matrix in the hope that this may reveal more evidence regarding the victim and/or the perpetrator as no human remains were found.

The sequence of events starting from the initial cutting of the grave is as follows: After the grave had been cut (context 4), two sand deposits were placed into the grave (contexts 5 and 6). These deposits were found to be sloping towards the centre. This might indicate that these deposits were placed into the grave from the short side of the grave cut. Given the similar spatial placement of these deposits within the feature, they are most likely contemporaneous. Context 5 also contained a coin dated to 1994, which indicates that the time frame in which this grave was created (the year 1994 onwards) is of forensic interest. Context 6 contained a lighter. Above these sand contexts was another deposit (context 7), which was light brown in colour and most likely originated from the initial digging of the grave. This deposit sloped down into the middle of the cut following the boundaries of the two sand deposits. It contained a false nail and an ID card. This deposit was then entirely covered by a fourth (context 8), which comprised of what seemed to be potting soil, available in most garden centres. This deposit sloped upwards towards the eastern side. It contained a metallic hairgrip and an earring. This deposit was then entirely covered by a light brown soil (context 9), which again, most likely originated from the digging of the grave, and contained cigarette papers. This deposit also sloped upwards towards the eastern side. The grave was then sealed with turf blocks that were most likely from the initial excavation of the grave.

From knowledge of the local geology it is clear that the sand deposits (contexts 5 and 6) and the potting soil deposit (context 8) were imported soils. It is recommended that these soils be analysed by geoscientists, as they may be able to trace their origin.

Arbitrary Level Excavation method

Archaeologist ID: Archaeologist 31

Years of experience: 6 years

Excavation approach: Arbitrary Level Excavation

Recording approach: Unit Level Recording

Tools used to excavate the grave: Shovel, mattock, spade, trowel and hand shovel

Did the participant sieve the fill: No

Weather conditions: Overcast and warm

Time taken: 3 hours

Observations: The participant chose to use the Arbitrary Level Excavation method and the Unit Level Recording method to excavate and record the grave. The participant started the excavation by marking out an excavation unit of 1m by 1.5m around the area of the grave. The participant then drew a plan of the excavation unit. The participant then proceeded to de-turf the excavation unit, and remove the replaced turf from the grave structure. The participant then started to excavate the excavation unit by removing spits of soil 10cm at a time with a mattock, shovel and spade. When the participant reached the 10cm spit point the level was smoothed out using a trowel and a hand shovel. The participant would then plan the excavation unit. The participant would also annotate the plans and describe the composition of the 10cm spit that had just been excavated, complete a unit level recording form and note down any evidence that had been located.

When evidence was located, the participant would excavate around the item placing it on a pedestal until the spit had been completely excavated. Then the item was photographed, lifted, and placed in an evidence bag and the pedestal removed. This process continued until the participant reached the base of the grave. A total of three spits were excavated, each of which mixed the various fills contained within the grave, resulting in the participant only recognising two of the fills contained within the grave. The flat-bottomed spits also resulted in the dimensions of each fill being malformed, and therefore incorrectly recorded as flat topped and bottomed.

The participant stored each spit that was removed separately, on different tarpaulins. The participant, however, did not sieve any of the spits removed from the grave.

This process resulted in a number of the fills contained within the grave becoming intermixed, and the grave cut being destroyed. In addition, by not maintaining the boundaries of the grave cut, the participant contaminated the fills from within the cut with non-disturbed soils that had been removed using this method. This could have resulted in unrelated evidence from outside the grave cut becoming intermixed with the evidence from the actual grave, leading investigators to spend time and resources on extraneous evidence.

Evidence	Location	Evidence found	Found in situ	Found out of situ	Evidence not found
E1 Dress	Base of C4	Yes	Yes	No	No
E2 Two pence coin	C5 (Fill 1)	No	-	-	Yes
E3 Lighter	C6 (Fill 2)	No	-	-	Yes
E4 Fake nail	C7 (Fill 3)	No	-	-	Yes
E5 ID card	C7 (Fill 3)	Yes	No	Yes	No
E6 Earring 2	C7 (Fill 3)	No	-	-	Yes
E7 Kirby grip	C8 (Fill 4)	No	-	-	Yes
E8 Earring 1	C8 (Fill 4)	No	-	-	Yes
E9 Cigarette papers	C9 (Fill 5)	No	-	-	Yes
TOTAL:		2	1	1	7
PERCENTAGE:		22.22%	50%	50%	77.78%

FIGURE 107: MATERIAL EVIDENCE IDENTIFIED BY ARCHAEOLOGIST 31 © Evis 2016.

Context number	Description	Context identified	Context not identified
C1	Natural	Yes	No
C2	Subsoil	Yes	No
C3	Topsoil and turf	Yes	No
C4	Feature cut	No	Yes
C5	Fill 1	Yes	No
C6	Fill 2	No	Yes
C7	Fill 3	No	Yes
C8	Fill 4	No	Yes
C9	Fill 5	Yes	No
C10	Replaced turf	Yes	No
TOTAL:		6	4
PERCENTAGE:		60%	40%

FIGURE 108: CONTEXTS IDENTIFIED BY ARCHAEOLOGIST 31 © Evis 2016.

Archaeologist 31's narrative: Removed the turf (possibly recently replaced). Marked out an excavation unit perimeter.

Began to mattock the earth. Excavated the first spit to 10cm and drew a plan when reached. Then a period of more careful excavation ensued with the use of a tape measure and trowel. From the 10cm spits that were excavated and planned, there appeared to be changes in the colour of soil, but this wasn't apparent in the spits apart from the last spit in which distinctive sand appeared. Due to the method used layers didn't appear distinctive. About 30cm down a blue material dress began to become revealed, and following the spit method I excavated the area around the dress leaving it on a pedestal. When excavating around the dress, the soil was very compact and I noted that I must have reached the bottom of the feature. Once I removed the dress and placed it into an evidence bag I looked underneath the dress to search for other evidence. In terms of the formation process of the feature, it appears that the perpetrator dug a rectangular pit approximately 30cm deep, in which they placed a blue and white striped dress, laid out flat. After which sand was deposited. Then the excavated soil that was removed during the digging of the grave was put back in the grave and turf put on top to try and hide the grave. Later, when investigating the spoil heap an ID card was found. This was likely to have been deposited in the grave after the dress.

Archaeologist ID: Archaeologist 32

Years of experience: 1 year

Excavation approach: Arbitrary Level Excavation

Recording approach: Unit Level Recording

Tools used to excavate the grave: Trowel and hand shovel

Did the participant sieve the fill: No

Weather conditions: Clear and hot

Time taken: 2 1/4 hours

Observations: The participant chose to use the Arbitrary Level Excavation method and the Unit Level Recording method to excavate and record the grave. The participant started the excavation by recording a plan of the feature on a unit level recording form. After which the participant began to excavate the grave using 10cm spits within the grave cut using a trowel and a hand shovel. Three 10cm spits were excavated in total. Once each spit had been excavated the participant would complete a unit level recording form and would draw a plan of the spit.

When material evidence was identified the participant would circumscribe the item and keep it on a soil pedestal until the excavation of the spit was completed. Once the spit was recorded the item would be photographed and its spatial location recorded. It would then be transferred to an evidence bag. This excavation and recording processes continued until the base of the grave was reached.

By maintaining the grave cut the participant preserved any potential tool marks and spent less energy excavating the feature. However, as with the previous participant, through excavating 10cm spits at a time the multiple fills contained within the feature became intermixed, enabling the participant to only identify two fills. The dimensions of these fills were also inaccurately documented, as the methodological approach was unable to capture sloped fills. The participant also knocked several items off of pedestals. This meant that there was a risk, if this accident had gone unnoticed, that the evidence would become lost or contaminated.

The participant did not sieve the spoil. However, the spoil from each spit was kept in separate sections on a tarpaulin so if investigators found any evidence it could be associated with an approximate location i.e. spit 1.

Archaeologist 32's narrative: The grave was dug using spits of 10cm depth. After each spit was excavated a unit level form was completed and a plan was drawn. On the basis of the findings of this excavation the following conclusions can be drawn:

The grave was constructed with two deposits – one composed of soil removed in the initial excavation of the grave by the perpetrator, and the other composed of imported sand. These deposits were put into the grave after the perpetrator had laid the dress against the base of the grave.

Spit 1 (0-10cm) contained cigarette papers and an earring.

Spit 2 (10cm-20cm) no evidence was found.

Spit 3 (20cm-30cm) contained an ID card, a lighter, a two pence coin and a dress.

Evidence	Location	Evidence found	Found in situ	Found out of situ	Evidence not found
E1 Dress	Base of C4	Yes	Yes	No	No
E2 Two pence coin	C5 (Fill 1)	Yes	Yes	No	No
E3 Lighter	C6 (Fill 2)	Yes	Yes	No	No
E4 Fake nail	C7 (Fill 3)	No	-	-	Yes
E5 ID card	C7 (Fill 3)	Yes	Yes	No	No
E6 Earring 2	C7 (Fill 3)	No	-	-	Yes
E7 Kirby grip	C8 (Fill 4)	No	-	-	Yes
E8 Earring 1	C8 (Fill 4)	Yes	Yes	No	No
E9 Cigarette papers	C9 (Fill 5)	Yes	Yes	No	No
TOTAL:		6	6	0	3
PERCENTAGE:		66.67%	100%	0%	33.33%

FIGURE 109: MATERIAL EVIDENCE IDENTIFIED BY ARCHAEOLOGIST 32 © EVIS 2016.

Context number	Description	Context identified	Context not identified
C1	Natural	Yes	No
C2	Subsoil	Yes	No
C3	Topsoil and turf	Yes	No
C4	Feature cut	Yes	No
C5	Fill 1	Yes	No
C6	Fill 2	No	Yes
C7	Fill 3	No	Yes
C8	Fill 4	No	Yes
C9	Fill 5	Yes	No
C10	Replaced turf	Yes	No
TOTAL:		7	3
PERCENTAGE:		70%	30%

FIGURE 110: CONTEXTS IDENTIFIED BY ARCHAEOLOGIST 32 © EVIS 2016.

Archaeologist ID: Archaeologist 33

Years of experience: 1 year

Excavation approach: Arbitrary Level Excavation

Recording approach: Unit Level Recording

Tools used to excavate the grave: Trowel and hand shovel

Did the participant sieve the fill: No

Weather conditions: Clear and hot

Time taken: 2 3/4 hours

Observations: The participant chose to use the Arbitrary Level Excavation method and the Unit Level Recording method to excavate and record the grave. The participant started the excavation by recording a plan of the feature on a unit level recording form. After which the participant began to excavate the grave using 10cm spits within the grave cut, using a trowel and a hand shovel. Three 10cm spits were excavated in total. After each spit was excavated a plan would be drawn, photographs taken, and a unit level recording form completed. The participant also made additional notes in a notebook.

When material evidence was located it was kept on a soil pedestal until the excavation of the spit was completed. As with the previous participant, this led to some evidence being knocked off of pedestals as the excavation progressed. Additionally, as was found with the other participants using this technique, the process of excavating arbitrarily defined spits led to the different fills becoming mixed and the participant only identifying two of the fills within the grave. The dimensions of these fills were distorted by the excavation approach as the method failed to preserve the integrity of the sloped fills.

The participant did not sieve the extracted spoil, but did store the spoil from each spit separately.

Evidence	Location	Evidence found	Found in situ	Found out of situ	Evidence not found
E1 Dress	Base of C4	Yes	Yes	No	No
E2 Two pence coin	C5 (Fill 1)	No	-	-	Yes
E3 Lighter	C6 (Fill 2)	Yes	Yes	No	No
E4 Fake nail	C7 (Fill 3)	Yes	Yes	No	No
E5 ID card	C7 (Fill 3)	Yes	Yes	No	No
E6 Earring 2	C7 (Fill 3)	No	-	-	Yes
E7 Kirby grip	C8 (Fill 4)	No	-	-	Yes
E8 Earring 1	C8 (Fill 4)	Yes	Yes	No	No
E9 Cigarette papers	C9 (Fill 5)	Yes	Yes	No	No
TOTAL:		6	6	0	3
PERCENTAGE:		66.67%	100%	0%	33.33%

FIGURE 111: MATERIAL EVIDENCE IDENTIFIED BY ARCHAEOLOGIST 33 © Evis 2016.

Context number	Description	Context identified	Context not identified
C1	Natural	Yes	No
C2	Subsoil	Yes	No
C3	Topsoil and turf	Yes	No
C4	Feature cut	Yes	No
C5	Fill 1	Yes	No
C6	Fill 2	No	Yes
C7	Fill 3	No	Yes
C8	Fill 4	No	Yes
C9	Fill 5	Yes	No
C10	Replaced turf	Yes	No
TOTAL:		7	3
PERCENTAGE:		70%	30%

FIGURE 112: CONTEXTS IDENTIFIED BY ARCHAEOLOGIST 33 © Evis 2016.

Archaeologist 33's narrative: A rectangular feature was dug (110cm long, 40cm wide, 30cm deep).

A blue and white striped dress was placed on the bottom of the feature.

Sand was then added to the feature, on top of the dress, into which a lighter, fake nail and ID card was thrown.

Spoil from the original digging of the feature was then put back into the feature, into which an earring and packet of cigarette papers were thrown.

Turf was then put back over the grave in an attempt to conceal it.

Archaeologist ID: Archaeologist 34

Years of experience: 1 year

Excavation approach: Arbitrary Level Excavation

Recording approach: Unit Level Recording

Tools used to excavate the grave: Trowel and hand shovel

Did the participant sieve the fill: No

Weather conditions: Clear and hot

Time taken: 3 1/2 hours

Observations: The participant chose to use the Arbitrary Level Excavation method and the Unit Level Recording method to excavate and record the grave. The participant started the excavation by recording a plan of the feature on a unit level recording form. After which the participant began to excavate the grave using 10cm spits within the grave cut, using a trowel and a hand shovel. Three 10cm spits were excavated in total. As with previous users of this method, once a spit was excavated, the participant would draw a plan, take photographs, and complete a unit level recording form.

Once material evidence was identified it would be kept on a soil pedestal until the excavation of the spit was completed, after which, the participant would record its location on the plans and unit level recording form and remove it. This approach again, resulted in the participant only identifying two fills, as the fills all became intermingled as the excavation progressed. Therefore, the dimensions recorded for each of the fills recognised were incorrect and evidence items were defined as originating from the wrong fill.

The participant did not sieve the spoil extracted from the grave. Instead, the participant would inspect the spoil by hand as it was being transferred to the tarpaulin.

Evidence	Location	Evidence found	Found in situ	Found out of situ	Evidence not found
E1 Dress	Base of C4	Yes	Yes	No	No
E2 Two pence coin	C5 (Fill 1)	No	-	-	Yes
E3 Lighter	C6 (Fill 2)	Yes	Yes	No	No
E4 Fake nail	C7 (Fill 3)	No	-	-	Yes
E5 ID card	C7 (Fill 3)	Yes	Yes	No	No
E6 Earring 2	C7 (Fill 3)	Yes	Yes	No	No
E7 Kirby grip	C8 (Fill 4)	Yes	Yes	No	No
E8 Earring 1	C8 (Fill 4)	Yes	Yes	No	No
E9 Cigarette papers	C9 (Fill 5)	Yes	Yes	No	No
TOTAL:		7	7	0	2
PERCENTAGE:		77.78%	100%	0%	22.22%

FIGURE 113: MATERIAL EVIDENCE IDENTIFIED BY ARCHAEOLOGIST 34 © Evis 2016.

Context number	Description	Context identified	Context not identified
C1	Natural	Yes	No
C2	Subsoil	Yes	No
C3	Topsoil and turf	Yes	No
C4	Feature cut	Yes	No
C5	Fill 1	Yes	No
C6	Fill 2	No	Yes
C7	Fill 3	No	Yes
C8	Fill 4	No	Yes
C9	Fill 5	Yes	No
C10	Replaced turf	Yes	No
TOTAL:		7	3
PERCENTAGE:		70%	30%

FIGURE 114: CONTEXTS IDENTIFIED BY ARCHAEOLOGIST 34 © Evis 2016.

Archaeologist 34's narrative: The grave was constructed using either a spade or a shovel, as there was evidence of tool marks in the turf clods removed from the grave. The grave itself had straight sides and a flat bottom; this would have taken the culprit some time to do. The grave was rectangular and measured 110cm x 40cm x 30cm.

After the culprit had dug the grave, a dress was placed on the bottom; this dress was from Primark. Then, (imported) sand was put into the grave. This sand deposit contained a lighter and an ID card. The ID card might belong to the victim or the perpetrator.

Then, a light brown sandy soil with high organic content was put over the sand deposit. This deposit was present over two of the spits 0-20cm. Spit 2, contained an earring, and spit 1, contained another matching earring, a hair grip and a packet of cigarette papers. The earring found in spit 2 would have been placed in the grave before the artefacts found in spit 1 as it was found deeper in the grave.

After the culprit had finished filling in the grave the turf clods were put back over it.

Archaeologist ID: Archaeologist 35

Years of experience: 1 year

Excavation approach: Arbitrary Level Excavation

Recording approach: Unit Level Recording

Tools used to excavate the grave: Trowel and hand shovel

Did the participant sieve the fill: No

Weather conditions: Clear and hot

Time taken: 3 1/4 hours

Observations: The participant chose to use the Arbitrary Level Excavation method and the Unit Level Recording method to excavate and record the grave. The participant started the excavation by recording a plan of the feature on a unit level recording form. After which, the participant began to excavate the grave using 10cm spits within the grave cut, using a trowel and a hand shovel. Three 10cm spits were excavated in total. When the participant had finished excavating a spit a unit level recording form would be completed, photographs taken, and a plan drawn. This plan would document the location of any material evidence found.

As evidence was found it was kept on a soil pedestal until the excavation of the spit was completed. This process continued until the participant reached the base of the grave. As with the previous participants, the method resulted in only two fills being identified. This again, meant that material evidence was recorded as originating from a spit, and an incorrect fill.

The participant did not sieve the spoil removed from the grave. Instead, the participant would occasionally look through the extracted spoil whilst it was being moved onto the tarpaulin.

Evidence	Location	Evidence found	Found in situ	Found out of situ	Evidence not found
E1 Dress	Base of C4	Yes	Yes	No	No
E2 Two pence coin	C5 (Fill 1)	No	-	-	Yes
E3 Lighter	C6 (Fill 2)	Yes	Yes	No	No
E4 Fake nail	C7 (Fill 3)	Yes	Yes	No	No
E5 ID card	C7 (Fill 3)	Yes	Yes	No	No
E6 Earring 2	C7 (Fill 3)	No	-	-	Yes
E7 Kirby grip	C8 (Fill 4)	No	-	-	Yes
E8 Earring 1	C8 (Fill 4)	Yes	Yes	No	No
E9 Cigarette papers	C9 (Fill 5)	Yes	Yes	No	No
TOTAL:		6	6	0	3
PERCENTAGE:		66.67%	100%	0%	33.33%

FIGURE 115: MATERIAL EVIDENCE IDENTIFIED BY ARCHAEOLOGIST 35 © Evis 2016.

Context number	Description	Context identified	Context not identified
C1	Natural	Yes	No
C2	Subsoil	Yes	No
C3	Topsoil and turf	Yes	No
C4	Feature cut	Yes	No
C5	Fill 1	Yes	No
C6	Fill 2	No	Yes
C7	Fill 3	No	Yes
C8	Fill 4	No	Yes
C9	Fill 5	Yes	No
C10	Replaced turf	Yes	No
TOTAL:		7	3
PERCENTAGE:		70%	30%

FIGURE 116: CONTEXTS IDENTIFIED BY ARCHAEOLOGIST 35 © Evis 2016.

Archaeologist 35's narrative: The feature was recently dug. This is based on the fact that the feature contained modern artefacts and the soil within it was relatively loose. The feature contained no human remains and so is probably not of forensic interest.

The feature was 110cm long, 40cm wide and 30cm deep with straight edges and a flat base. Spit 1 (0-10cm) contained a packet of cigarette papers and an earring. Spit 2 (10-20cm) contained no artefacts. Spit 3 (20-30cm) contained an ID card, an acrylic nail, a lighter and a dress. The artefacts found in spit 1 were placed into the grave later than the artefacts found in spit 3.

Archaeologist ID: Archaeologist 36

Years of experience: 5 years

Excavation approach: Arbitrary Level Excavation

Recording approach: Unit Level Recording

Tools used to excavate the grave: Trowel and hand shovel

Did the participant sieve the fill: No

Weather conditions: Overcast and mild

Time taken: 2 hours

Observations: The participant chose to use the Arbitrary Level Excavation method and the Unit Level Recording method to excavate and record the grave. The participant started the excavation by recording a plan of the feature on a unit level recording form. After which, the participant began to excavate the grave using 10cm spits within the grave cut, using a trowel and a hand shovel. After the completion of a spit the participant would record a plan of the feature, and write notes in a journal regarding what was found, as well as complete the unit level recording form.

146

As material evidence was found the participant would not leave it on a pedestal but would record its position on the plan and in the journal and the unit level recording form. Then the participant would place the item into an evidence bag. This process was repeated as each spit was removed until the excavation was completed. Like the previous participants, this method resulted in only two fills being identified.

The participant also failed to sieve the spoil removed from the grave. However, the spoil from each spit was stored on a separate tarpaulin.

Archaeologist 36's narrative: A feature was dug to the rough depth of 300mm. A small body was placed in the feature and covered with a layer of sand. The rest of the feature was refilled with a mix of subsoil and topsoil and then some turf to hide the excavated feature.

Various artefacts relating to the individual were extracted on excavation.

The size of the grave suggests an adolescent but the artefacts relating to smoking suggest an older individual. The smoking paraphernalia could be related to the individual who excavated the feature as they were within the subsoil/topsoil.

Evidence	Location	Evidence found	Found In situ	Found out of situ	Evidence not found
E1 Dress	Base of C4	Yes	Yes	No	No
E2 Two pence coin	C5 (Fill 1)	Yes	Yes	No	No
E3 Lighter	C6 (Fill 2)	No	-	-	Yes
E4 Fake nail	C7 (Fill 3)	No	-	-	Yes
E5 ID card	C7 (Fill 3)	No	-	-	Yes
E6 Earring 2	C7 (Fill 3)	No	-	-	Yes
E7 Kirby grip	C8 (Fill 4)	No	-	-	Yes
E8 Earring 1	C8 (Fill 4)	No	-	-	Yes
E9 Cigarette papers	C9 (Fill 5)	Yes	Yes	No	No
TOTAL:		3	3	0	6
PERCENTAGE:		33.33%	100%	0%	66.67%

FIGURE 117: MATERIAL EVIDENCE IDENTIFIED BY ARCHAEOLOGIST 36 © Evis 2016.

Context number	Description	Context identified	Context not identified
C1	Natural	Yes	No
C2	Subsoil	Yes	No
C3	Topsoil and turf	Yes	No
C4	Feature cut	Yes	No
C5	Fill 1	Yes	No
C6	Fill 2	No	Yes
C7	Fill 3	No	Yes
C8	Fill 4	No	Yes
C9	Fill 5	Yes	No
C10	Replaced turf	Yes	No
TOTAL:		7	3
PERCENTAGE:		70%	30%

FIGURE 118: CONTEXTS IDENTIFIED BY ARCHAEOLOGIST 36 © Evis 2016.

Archaeologist ID: Archaeologist 37

Years of experience: 13 years

Excavation approach: Arbitrary Level Excavation

Recording approach: Unit Level Recording

Tools used to excavate the grave: Trowel and hand shovel

Did the participant sieve the fill: No

Weather conditions: Overcast and mild

Time taken: 2 hours

Observations: The participant chose to use the Arbitrary Level Excavation method and the Unit Level Recording method to excavate and record the grave. The participant started the excavation by recording a plan of the feature on a unit level recording form. After which, the participant began to excavate the grave using 10cm spits within the grave cut, using a trowel and a hand shovel. After the completion of a spit the participant would record a plan of the feature, complete the unit level recording form, and take photographs.

Evidence	Location	Evidence found	Found in situ	Found out of situ	Evidence not found
E1 Dress	Base of C4	Yes	Yes	No	No
E2 Two pence coin	C5 (Fill 1)	No	-	-	Yes
E3 Lighter	C6 (Fill 2)	Yes	Yes	No	No
E4 Fake nail	C7 (Fill 3)	Yes	Yes	No	No
E5 ID card	C7 (Fill 3)	Yes	Yes	No	No
E6 Earring 2	C7 (Fill 3)	No	-	-	Yes
E7 Kirby grip	C8 (Fill 4)	No	-	-	Yes
E8 Earring 1	C8 (Fill 4)	No	-	-	Yes
E9 Cigarette papers	C9 (Fill 5)	Yes	Yes	No	No
TOTAL:		5	5	0	4
PERCENTAGE:		55.56%	100%	0%	44.44%

FIGURE 119: MATERIAL EVIDENCE IDENTIFIED BY ARCHAEOLOGIST 37 © Evis 2016.

Context number	Description	Context identified	Context not identified
C1	Natural	Yes	No
C2	Subsoil	Yes	No
C3	Topsoil and turf	Yes	No
C4	Feature cut	Yes	No
C5	Fill 1	Yes	No
C6	Fill 2	No	Yes
C7	Fill 3	No	Yes
C8	Fill 4	No	Yes
C9	Fill 5	Yes	No
C10	Replaced turf	Yes	No
TOTAL:		7	3
PERCENTAGE:		70%	30%

FIGURE 120: CONTEXTS IDENTIFIED BY ARCHAEOLOGIST 37 © Evis 2016.

As material evidence was found the participant would record its position on the plan and on the unit level recording form. Then the participant would place the item into an evidence bag. This process was repeated as each spit was removed until the excavation was completed. Like the previous participants, this method resulted in only two fills being identified.

As with previous participants, the participant did not sieve the spoil from the grave. Although, the participant did store the spoil from each spit in a different storage container, so that samples could be taken if necessary.

Archaeologist 37's narrative: Steep-sided square edged cut approximately 110cm in length, 40cm in width and 30cm in depth. The termination of roots and rootlets from the surrounding soil suggests recent excavation. The feature was filled initially with fine, lightly dark beige white sand, containing cloth, cigarette lighter, false nail and an ID card. A mixed topsoil containing cigarette papers covered the layer of sand.

Archaeologist ID: Archaeologist 38

Years of experience: 12 years

Excavation approach: Arbitrary Level Excavation

Recording approach: Unit Level Recording

Tools used to excavate the grave: Trowel and hand shovel

Did the participant sieve the fill: No

Weather conditions: Overcast and mild

Time taken: 2 1/2 hours

Observations: The participant chose to use the Arbitrary Level Excavation method and the Unit Level Recording method to excavate and record the grave. The participant started the excavation by recording a plan of the feature on a unit level recording form. After which, the participant began to excavate the grave using 10cm spits within the grave cut, using a trowel and a hand shovel. After the completion of a spit the participant would record a plan of the feature, complete the unit level recording form, and take photographs.

As material evidence was found the participant would record its position on the plan and on the unit level recording form. Then the participant would place the item into an evidence bag. This process was repeated as each spit was removed until the excavation was completed. Like the previous participants, this method resulted in only two fills being identified.

The participant did not sieve the spoil from the grave. However, the participant would check the spoil as it was being moved to the tarpaulin by scraping it with a trowel.

Archaeologist 38's narrative: A rectangular cut with straight edges. Beneath the turf, is a uniform loose sediment mid-brown in colour contains: a packet of cigarette papers towards the north side. In the same layer towards the south side a single white metal earring. Below this layer was a distinctive sandy layer with a sharp boundary, which contained an ID card in the middle of the layer, and a lighter towards the south side. Towards the north of the feature on the base of the feature is an item of blue clothing. Immediately beneath the sandy layer there was a highly compacted dark sediment which appeared to be undisturbed. The sides of the cut were sharply cut from the surface into the base, which was flat. Roots were encountered throughout and occasional stones.

Evidence	Location	Evidence found	Found in situ	Found out of situ	Evidence not found
E1 Dress	Base of C4	Yes	Yes	No	No
E2 Two pence coin	C5 (Fill 1)	No	-	-	Yes
E3 Lighter	C6 (Fill 2)	Yes	No	Yes	No
E4 Fake nail	C7 (Fill 3)	No	-	-	Yes
E5 ID card	C7 (Fill 3)	Yes	Yes	No	No
E6 Earring 2	C7 (Fill 3)	Yes	Yes	No	No
E7 Kirby grip	C8 (Fill 4)	No	-	-	Yes
E8 Earring 1	C8 (Fill 4)	No	-	-	Yes
E9 Cigarette papers	C9 (Fill 5)	Yes	Yes	No	No
TOTAL:		5	4	1	4
PERCENTAGE:		55.56%	80%	20%	44.44%

FIGURE 121: MATERIAL EVIDENCE IDENTIFIED BY ARCHAEOLOGIST 38 © Evis 2016.

Context number	Description	Context identified	Context not identified
C1	Natural	Yes	No
C2	Subsoil	Yes	No
C3	Topsoil and turf	Yes	No
C4	Feature cut	Yes	No
C5	Fill 1	Yes	No
C6	Fill 2	No	Yes
C7	Fill 3	No	Yes
C8	Fill 4	No	Yes
C9	Fill 5	Yes	No
C10	Replaced turf	Yes	No
TOTAL:		7	3
PERCENTAGE:		70%	30%

FIGURE 122: CONTEXTS IDENTIFIED BY ARCHAEOLOGIST 38 © Evis 2016.

Archaeologist ID: Archaeologist 39

Years of experience: 30 years

Excavation approach: Arbitrary Level Excavation

Recording approach: Unit Level Recording

Tools used to excavate the grave: Trowel and hand shovel

Did the participant sieve the fill: No

Weather conditions: Overcast and mild

Time taken: 2 hours

Observations: The participant chose to use the Arbitrary Level Excavation method and the Unit Level Recording method to excavate and record the grave. The participant started the excavation by recording a plan of the feature on a unit level recording form. After which, the participant began to excavate the

150

grave using 10cm spits within the grave cut, using a trowel and a hand shovel. After the completion of a spit the participant would record a plan of the feature, complete the unit level recording form, and take photographs.

As material evidence was found the participant would record its position on the plan and on the unit level recording form. Then the participant would place the item into an evidence bag. This process was repeated as each spit was removed until the excavation was completed. Like the previous participants, this method resulted in only two fills being identified.

The participant did not sieve the spoil from the grave. All spoil that was extracted from the grave was placed onto a tarpaulin in one large pile.

Evidence	Location	Evidence found	Found in situ	Found out of situ	Evidence not found
E1 Dress	Base of C4	Yes	Yes	No	No
E2 Two pence coin	C5 (Fill 1)	No	-	-	Yes
E3 Lighter	C6 (Fill 2)	No	-	-	Yes
E4 Fake nail	C7 (Fill 3)	No	-	-	Yes
E5 ID card	C7 (Fill 3)	Yes	Yes	No	No
E6 Earring 2	C7 (Fill 3)	No	-	-	Yes
E7 Kirby grip	C8 (Fill 4)	No	-	-	Yes
E8 Earring 1	C8 (Fill 4)	No	-	-	Yes
E9 Cigarette papers	C9 (Fill 5)	Yes	Yes	No	No
TOTAL:		3	3	0	6
PERCENTAGE:		33.33%	100%	0%	66.67%

FIGURE 123: MATERIAL EVIDENCE IDENTIFIED BY ARCHAEOLOGIST 39 © Evis 2016.

Context number	Description	Context identified	Context not identified
C1	Natural	Yes	No
C2	Subsoil	Yes	No
C3	Topsoil and turf	Yes	No
C4	Feature cut	Yes	No
C5	Fill 1	Yes	No
C6	Fill 2	No	Yes
C7	Fill 3	No	Yes
C8	Fill 4	No	Yes
C9	Fill 5	Yes	No
C10	Replaced turf	Yes	No
TOTAL:		7	3
PERCENTAGE:		70%	30%

FIGURE 124: CONTEXTS IDENTIFIED BY ARCHAEOLOGIST 39 © Evis 2016.

Archaeologist 39's narrative: The police called me out to investigate a suspicious small rectangular plot of earth 110cm x 40cm in area. The police had contacted the coroner and called me in for archaeological advice. The following was observed archaeologically.

The upper fill of the feature comprised of mid-brown silt with moderate fragments of modern brick and pebbles. Within this fill there was a packet of cigarette papers.

The underlying fill covered the whole feature and was loose and brown-buff coloured sand. Clearly brought in from elsewhere and dumped in the cut. This contained an ID card.

Underneath this second fill was a modern blue and white striped dress.

The cut feature was rectangular with vertical sides and a flat base.

The feature is clearly of really modern date; it is not a grave, as there were no human remains. Nothing of archaeological interest was found.

Archaeologist ID: Archaeologist 40

Years of experience: 4 years

Excavation approach: Arbitrary Level Excavation

Recording approach: Unit Level Recording

Tools used to excavate the grave: Trowel and hand shovel

Did the participant sieve the fill: No

Weather conditions: Overcast and mild

Time taken: 2 hours

Observations: The participant chose to use the Arbitrary Level Excavation method and the Unit Level Recording method to excavate and record the grave. The participant started the excavation by recording a plan of the feature on a unit level recording form. After which, the participant began to excavate the grave using 10cm spits within the grave cut, using a trowel and a hand shovel. After the completion of a spit the participant would record a plan of the feature, complete the unit level recording form, and take photographs.

As material evidence was found the participant would record its position on the plan and on the unit level recording form. Then the participant would place the item into an evidence bag. This process was repeated as each spit was removed until the excavation was completed. This method resulted in only two fills being identified.

The participant did not sieve the spoil from the grave. However, the participant would attempt to inspect the spoil by spreading out the spoil from each spit over a wide area on the tarpaulin in the hope that any additional evidence would become apparent.

Archaeologist 40's narrative: Rectangular cut feature 40cm x 110cm x 30cm.

This feature has been dug recently. The uppermost fill comprised of very loose material clearly from the initial digging of the feature.

Below this was a layer of clean building sand, which sealed a piece of blue cotton fabric/garment.

The garment was located at the eastern end of the feature.

The base of the cut feature was very compact and I believe it represents undisturbed sediment.

This feature, I believe has been dug recently to discard the garment. The deposit of sand suggests to me that this was used to seal in the garment to facilitate later recovery, although the feature was untouched.

Evidence	Location	Evidence found	Found in situ	Found out of situ	Evidence not found
E1 Dress	Base of C4	Yes	Yes	No	No
E2 Two pence coin	C5 (Fill 1)	No	-	-	Yes
E3 Lighter	C6 (Fill 2)	No	-	-	Yes
E4 Fake nail	C7 (Fill 3)	No	-	-	Yes
E5 ID card	C7 (Fill 3)	Yes	Yes	No	No
E6 Earring 2	C7 (Fill 3)	No	-	-	Yes
E7 Kirby grip	C8 (Fill 4)	No	-	-	Yes
E8 Earring 1	C8 (Fill 4)	Yes	Yes	No	No
E9 Cigarette papers	C9 (Fill 5)	No	-	-	Yes
TOTAL:		3	3	0	6
PERCENTAGE:		33.33%	100%	0%	66.67%

FIGURE 125: MATERIAL EVIDENCE IDENTIFIED BY ARCHAEOLOGIST 40 © Evis 2016.

Context number	Description	Context identified	Context not identified
C1	Natural	Yes	No
C2	Subsoil	Yes	No
C3	Topsoil and turf	Yes	No
C4	Feature cut	Yes	No
C5	Fill 1	Yes	No
C6	Fill 2	No	Yes
C7	Fill 3	No	Yes
C8	Fill 4	No	Yes
C9	Fill 5	Yes	No
C10	Replaced turf	Yes	No
TOTAL:		7	3
PERCENTAGE:		70%	30%

FIGURE 126: CONTEXTS IDENTIFIED BY ARCHAEOLOGIST 40 © Evis 2016.

Control participants

Control ID: Control 01

Years of experience: No archaeological experience

Tools used to excavate the grave: Trowel, hand shovel and shovel

Did the participant sieve the fill: Yes

Weather conditions: Flurries of snow and cold

Time taken: 5 hours

Observations: This participant had no archaeological experience whatsoever. The conditions in which the participant excavated the feature were dreadful, with snow continuing to fall throughout the experiment. Prior to starting the excavation, the participant recorded the dimensions of the grave as a whole, and then proceeded to measure the dimensions of each of the turf clumps that had been placed back over the grave. The participant then removed the overlying turf with a shovel, and placed a peg in the top left corner of the grave in order to be able to use it as a base point from which to measure the placement of evidence and fills within the grave.

The participant excavated the fills of the grave using a hand shovel, removing a shovelful of spoil at a time, working from one end of the grave to the other until the base of the grave was reached. As each shovelful was removed, the participant would sieve it in order to see if any evidence could be found. There was however, a problem, as the participant failed to define the edges of the grave correctly, leaving approximately 10cm of grave fill in situ without realising it. This meant that the cigarette papers and the two pence coin failed to be retrieved. In terms of the structure of the grave the participant identified two fills.

When the participant identified a piece of evidence in the grave its place within the grave structure would be measured using the base point peg, and then it would be transferred to an evidence bag.

In terms of recording, the participant kept a notebook throughout the entire process and would write all of the findings (measurements, finds, soil layers) in it. The participant also took photographs to document the placement of evidence within the grave. The participant did not take pictures of the different fills within the grave.

Evidence	Location	Evidence found	Found in situ	Found out of situ	Evidence not found
E1 Dress	Base of C4	Yes	Yes	No	No
E2 Two pence coin	C5 (Fill 1)	No	-	-	Yes
E3 Lighter	C6 (Fill 2)	Yes	Yes	No	No
E4 Fake nail	C7 (Fill 3)	Yes	No	Yes	No
E5 ID card	C7 (Fill 3)	Yes	Yes	No	No
E6 Earring 2	C7 (Fill 3)	Yes	No	Yes	No
E7 Kirby grip	C8 (Fill 4)	Yes	No	Yes	No
E8 Earring 1	C8 (Fill 4)	Yes	No	Yes	No
E9 Cigarette papers	C9 (Fill 5)	No	-	-	Yes
TOTAL:		7	3	4	2
PERCENTAGE:		77.78%	42.86%	57.14%	22.22%

FIGURE 127: MATERIAL EVIDENCE IDENTIFIED BY CONTROL 01 © Evis 2016.

Context number	Description	Context identified	Context not identified
C1	Natural	Yes	No
C2	Subsoil	Yes	No
C3	Topsoil and turf	Yes	No
C4	Feature cut	Yes	No
C5	Fill 1	No	Yes
C6	Fill 2	Yes	No
C7	Fill 3	No	Yes
C8	Fill 4	No	Yes
C9	Fill 5	Yes	No
C10	Replaced turf	Yes	No
TOTAL:		7	3
PERCENTAGE:		70%	30%

FIGURE 128: CONTEXTS IDENTIFIED BY CONTROL 01 © Evis 2016.

Control 01's narrative: The items recovered suggest that they belong to a female aged between 16 and 28. This is due to the size and style of the dress found. I believe it would be difficult for the majority of male transsexuals or cross-dressers to be able to fit into such a small dress. The fact that the earrings still retained their 'butterfly' backs suggest that they were not forcibly removed. However, the false nail could indicate some struggle. The cigarette lighter suggests that the owner of the items was a smoker. The ID card could indicate that the owner has some disposable income however; it is not clear whether this card has been used. The fact that the dress was from Primark and the lighter was disposable could imply that money is tight. It is my belief that the owner of the items has brown hair – this is due to the colour of the single kirby grip recovered.

The feature itself indicates that the person or persons digging were not rushed. They had time to section the turf before removing it piece by piece. The geometric shape of the hole could also indicate that some time had been taken during digging. In addition, the dress and cigarette lighter had been placed at the bottom of the hole and then covered with a layer of sand. This sand was not mixed with the upper soil.

I do not believe that the owner of the items was responsible for digging the feature.

Control ID: Control 02

Years of experience: No archaeological experience

Tools used to excavate the grave: Hand shovel, trowel and shovel

Did the participant sieve the fill: No

Weather conditions: Overcast and warm

Time taken: 2 hours

Observations: This participant had no archaeological experience whatsoever. The participant started the process by measuring the dimensions of the grave. The participant then removed the turf overlying the grave with a shovel. By using a shovel to remove the turf clumps the participant accidentally went too deep, and removed the majority of the fifth fill (context 9) too. The participant then noticed the presence of cigarette papers and altered the excavation approach. The participant then used a trowel and a hand shovel to excavate the grave. The participant would dig across the length of the grave excavating in one direction and then would excavate back in the opposite direction until a change was noticed in the soil type. Whilst using this technique the participant noticed changes in soil texture and colouration and so would maintain the dimensions of the fills as the excavation progressed. The participant would document these changes in a notebook, and would measure the dimensions of each of the fills, as they were uncovered. However, as the participant reached the base of the grave and uncovered the top of the dress, the participant pulled it out from its original position and ended up disrupting the sand fill (context 5), resulting in it being strewn all over the base of the grave. This led to the participant incorrectly recording the two separate sand fills as being one, but unlike the first control, this participant did note that the fill sloped in the middle and was not a uniform flat layer.

When the participant found an item of evidence the item would be photographed and transferred into an evidence bag. The participant would measure the placement of an item using the grave walls as a reference point. The participant would write this information in a notebook.

The participant did not sieve as the experiment progressed, but did inspect each shovelful of soil as it was being transferred to the tarpaulin.

Evidence	Location	Evidence found	Found in situ	Found out of situ	Evidence not found
E1 Dress	Base of C4	Yes	Yes	No	No
E2 Two pence coin	C5 (Fill 1)	Yes	Yes	No	No
E3 Lighter	C6 (Fill 2)	Yes	Yes	No	No
E4 Fake nail	C7 (Fill 3)	Yes	No	Yes	No
E5 ID card	C7 (Fill 3)	Yes	Yes	No	No
E6 Earring 2	C7 (Fill 3)	No	-	-	Yes
E7 Kirby grip	C8 (Fill 4)	No	-	-	Yes
E8 Earring 1	C8 (Fill 4)	Yes	No	Yes	No
E9 Cigarette papers	C9 (Fill 5)	Yes	Yes	No	No
TOTAL:		7	5	2	2
PERCENTAGE:		77.78%	71.43%	28.57%	22.22%

FIGURE 129: MATERIAL EVIDENCE IDENTIFIED BY CONTROL 02 © Evis 2016.

Context number	Description	Context identified	Context not identified
C1	Natural	Yes	No
C2	Subsoil	Yes	No
C3	Topsoil and turf	Yes	No
C4	Feature cut	Yes	No
C5	Fill 1	No	Yes
C6	Fill 2	Yes	No
C7	Fill 3	Yes	No
C8	Fill 4	Yes	No
C9	Fill 5	No	Yes
C10	Replaced turf	Yes	No
TOTAL:		8	2
PERCENTAGE:		80%	20%

FIGURE 130: CONTEXTS IDENTIFIED BY CONTROL 02 © Evis 2016.

Control 02's narrative: Grave dug recently. The top layer of turf still divided into pieces where it was removed using a spade. The ground still disturbed around the area. It would appear as though an 110cm long pit was dug to a depth of approximately 30cm, with the original material of soil and stone supplemented with commercially available topsoil and sand.

It terms of formation it would appear as though a pit was dug, the material was removed and placed next to the pit. After which sand was carefully placed into the bottom of the pit. Into the sand some items were placed. Namely, a green cigarette lighter, a two pence coin and a blue and white stripy dress. After the sand some of the original material was returned coupled with a false fingernail and an ID. The following material returned to the pit was a mixture of topsoil (shop bought) and the original material. In which it can be assumed a silver earring was placed, although this was found later in the spoil heap, and a packet of cigarette papers.

In terms of a human narrative, it would appear as though the feature was constructed recently, probably using a spade by a few people or indeed one person, under fairly clement environmental conditions. The feature contained several pieces of "evidence" suggesting the remains of a young woman. Obviously, no human remains were found. However, the conditions under which the evidence was discovered suggests an attempt to hide evidence from an illegal activity or event, presumably, murder, rape or abduction.

> **Control ID:** Control 03
>
> **Years of experience:** No archaeological experience
>
> **Tools used to excavate the grave:** Hand shovel and shovel
>
> **Did the participant sieve the fill:** No
>
> **Weather conditions:** Overcast and damp
>
> **Time taken:** 1 1/2 hours

Observations: This participant had no archaeological experience whatsoever. The participant started the experiment by measuring the dimensions of the grave, after which they removed the replaced turf. The participant then proceeded to excavate the fills of the grave using a mix of a shovel and a hand shovel, varying between the two tools as the excavation proceeded. The participant would use these tools to remove longitudinal sweeps of the grave fills. This resulted in the participant digging through the fourth fill and the third fill without realising that there was a change in fill composition. Furthermore, as the participant used the shovel to remove the sand fills at the base of the grave, the participant was unable to determine that these were two separate fills, as the participant had mixed up the interface between them. Moreover, as the participant uncovered the dress at the base of the grave it was pulled out from its original position, further mixing up the sand fills at the base of the grave. This approach also resulted in the participant failing to define the boundaries of the feature, leaving approximately 10cm of the fills intact at either end of the grave. The participant only realised that this had occurred when a piece of material evidence was found to be sticking out of the fill. The participant then used the shovel to dig straight down to access this item and further disrupted the fills in the grave.

When pieces of evidence were found the participant would note down the location in which it was found in a notebook and transfer each item into an evidence bag. The participant also used the notebook to document the change in soil colouration between the sand and the overlying fill. The participant also took photographs throughout the experiment, to document the procedure that was being used and the location in which evidence was recovered. The participant did not take photographs of the soil, as the participant was primarily focused on finding material evidence.

The participant failed to sieve the spoil extracted from the grave. However, occasionally, the participant would inspect the spoil contained within the hand shovel or shovel as it was being transferred to the tarpaulin, but this was rare.

Control 03's narrative: Excavation began at 11:42. I began by removing the loose layer off grass with a shovel. The total area of the site was 110cm x 40cm. I then moved onto using a small hand shovel and shovel to careful remove small amounts of soil and place them onto the tarpaulin provided and I had a quick scan for finds.

The first item I found was an ID card located in the centre of the grave. I continued to dig around the edges of the grave and discovered a second item – cigarette papers at the top left corner. When I reached towards the base of the feature there was a layer of fine sand across the whole area. I moved to using the hand shovel to remove it from the grave. I then came across a ladies blue stripy dress. I continued to dig around the edges of the grave and found a blue lighter. All items were removed and placed into labelled bags with the depths and the time at which the objects were found written on them. Photos were taken throughout.

Evidence	Location	Evidence found	Found in situ	Found out of situ	Evidence not found
E1 Dress	Base of C4	Yes	Yes	No	No
E2 Two pence coin	C5 (Fill 1)	No	-	-	Yes
E3 Lighter	C6 (Fill 2)	Yes	Yes	No	No
E4 Fake nail	C7 (Fill 3)	No	-	-	Yes
E5 ID card	C7 (Fill 3)	Yes	Yes	No	No
E6 Earring 2	C7 (Fill 3)	No	-	-	Yes
E7 Kirby grip	C8 (Fill 4)	No	-	-	Yes
E8 Earring 1	C8 (Fill 4)	No	-	-	Yes
E9 Cigarette papers	C9 (Fill 5)	Yes	Yes	No	No
TOTAL:		4	4	0	5
PERCENTAGE:		44.44%	100%	0%	55.56%

FIGURE 131: MATERIAL EVIDENCE IDENTIFIED BY CONTROL 03 © Evis 2016.

Context number	Description	Context identified	Context not identified
C1	Natural	Yes	No
C2	Subsoil	Yes	No
C3	Topsoil and turf	Yes	No
C4	Feature cut	Yes	No
C5	Fill 1	Yes	No
C6	Fill 2	No	Yes
C7	Fill 3	No	Yes
C8	Fill 4	No	Yes
C9	Fill 5	Yes	No
C10	Replaced turf	Yes	No
TOTAL:		7	3
PERCENTAGE:		70%	30%

FIGURE 132: CONTEXTS IDENTIFIED BY CONTROL 03 © Evis 2016.

Control ID: Control 04

Years of experience: No archaeological experience

Tools used to excavate the grave: Trowel and hand shovel

Did the participant sieve the fill: Yes

Weather conditions: Clear and warm

Time taken: 2 hours

Observations: This participant had no archaeological experience whatsoever. The participant started the experiment by measuring the dimensions of the grave and removing the turf that was overlying the grave by hand. The participant then began to excavate the grave using a hand shovel. The participant would remove fills by dragging the hand shovel across the length of the feature and placing the spoil onto a tarpaulin. The depth of each shovelful of soil varied throughout the entire process as the participant did not follow a strict protocol. Due to using this approach the participant failed to identify the fourth and third fills.

As the excavation progressed the participant identified the presence of sand at either end of the feature. The participant then began to use a trowel to carefully reveal the sand fills. This resulted in the participant successfully uncovering the two separate sand fills, allowing the participant to note the fact that these fills sloped down towards the middle of the grave but did not join. Unlike the previous control participants, when the participant noticed the presence of the dress the participant did not pull it out but excavated the fill until the dress was fully uncovered.

When material evidence was identified the participant would record its position using the grave walls as a reference point and noted down these details in a notebook. The participant would also take photographs of the objects and then transfer them into an evidence bag. The participant also documented the change in fills via taking photographs and writing notes in the notebook to aid with interpreting the feature.

The participant also sieved all of the spoil that was removed from the grave in order to search for more material evidence. The participant would, however, attempt to sieve large volumes of spoil at a time, and became lethargic as the process went on. This might explain why the participant failed to retrieve all of the pieces of material evidence whilst sieving.

Control 04's narrative: Turf was placed onto the grave and soil placed into the grave. Cigarette papers were put into the grave, then an earring, then an ID card. Then separate deposits of sand were placed either side of the grave, one contained a lighter and the other contained a two pence coin. Under which there was a dress.

Evidence	Location	Evidence found	Found in situ	Found out of situ	Evidence not found
E1 Dress	Base of C4	Yes	Yes	No	No
E2 Two pence coin	C5 (Fill 1)	Yes	Yes	No	No
E3 Lighter	C6 (Fill 2)	Yes	Yes	No	No
E4 Fake nail	C7 (Fill 3)	No	-	-	Yes
E5 ID card	C7 (Fill 3)	Yes	Yes	No	No
E6 Earring 2	C7 (Fill 3)	No	-	-	Yes
E7 Kirby grip	C8 (Fill 4)	No	-	-	Yes
E8 Earring 1	C8 (Fill 4)	Yes	No	Yes	No
E9 Cigarette papers	C9 (Fill 5)	Yes	Yes	No	No
TOTAL:		6	5	1	3
PERCENTAGE:		66.67%	83.33%	16.67%	33.33%

FIGURE 133: MATERIAL EVIDENCE IDENTIFIED BY CONTROL 04 © Evis 2016.

Context number	Description	Context identified	Context not identified
C1	Natural	Yes	No
C2	Subsoil	Yes	No
C3	Topsoil and turf	Yes	No
C4	Feature cut	Yes	No
C5	Fill 1	Yes	No
C6	Fill 2	Yes	No
C7	Fill 3	No	Yes
C8	Fill 4	No	Yes
C9	Fill 5	Yes	No
C10	Replaced turf	Yes	No
TOTAL:		8	2
PERCENTAGE:		80%	20%

FIGURE 134: CONTEXTS IDENTIFIED BY CONTROL 04 © Evis 2016.

Control ID: Control 05

Years of experience: No archaeological experience

Tools used to excavate the grave: Trowel and hand shovel

Did the participant sieve the fill: No

Weather conditions: Clear and warm

Time taken: 2 hours

Observations: This participant had no archaeological experience whatsoever. The participant started the experiment by measuring the dimensions of the grave. The participant then removed the turfs that had been placed on top of the grave by hand. The participant then began to excavate the grave using a hand shovel. The participant would start digging down at one end of the grave and then would drag the hand shovel across the longitudinal axis of the grave until the shovel needed to be emptied. This resulted in the participant failing to identify the two middle fills of the grave (fill 3 and fill 4). As with the previous participant, once the participant noticed the presence of sand at either end of the grave the participant adapted their approach and began to use the trowel to carefully uncover the boundary between the sand fills and the fill that lay over them. This resulted in the participant being able to determine that the two sand fills were separate fills rather than one. Furthermore, the participant did not pull the dress from under the sand fill, but excavated this fill until the dress was uncovered. This prevented the participant from mixing the sand fills accidentally and again, assisted the participant in determining that there were in fact two separate sand fills within the grave.

As material evidence was identified the participant would write down its location in a notebook, take photographs and transfer it into an evidence bag. The participant also took progress photos throughout

Evidence	Location	Evidence found	Found in situ	Found out of situ	Evidence not found
E1 Dress	Base of C4	Yes	Yes	No	No
E2 Two pence coin	C5 (Fill 1)	No	-	-	Yes
E3 Lighter	C6 (Fill 2)	Yes	Yes	No	No
E4 Fake nail	C7 (Fill 3)	No	-	-	Yes
E5 ID card	C7 (Fill 3)	Yes	Yes	No	No
E6 Earring 2	C7 (Fill 3)	No	-	-	Yes
E7 Kirby grip	C8 (Fill 4)	No	-	-	Yes
E8 Earring 1	C8 (Fill 4)	No	-	-	Yes
E9 Cigarette papers	C9 (Fill 5)	Yes	Yes	No	No
TOTAL:		4	4	0	5
PERCENTAGE:		44.44%	100%	0%	55.56%

FIGURE 135: MATERIAL EVIDENCE IDENTIFIED BY CONTROL 05 © Evis 2016.

Context number	Description	Context identified	Context not identified
C1	Natural	Yes	No
C2	Subsoil	Yes	No
C3	Topsoil and turf	Yes	No
C4	Feature cut	Yes	No
C5	Fill 1	Yes	No
C6	Fill 2	Yes	No
C7	Fill 3	No	Yes
C8	Fill 4	No	Yes
C9	Fill 5	Yes	No
C10	Replaced turf	Yes	No
TOTAL:		8	2
PERCENTAGE:		80%	20%

FIGURE 136: CONTEXTS IDENTIFIED BY CONTROL 05 © Evis 2016.

the experiment and documented any thoughts about the finds and the different fills within the grave in the notebook. The participant also measured each of the fills and wrote down these measurements in the notebook.

The participant did not sieve any of the spoil extracted from the grave.

Control 05's narrative: A rectangular hole was dug, dimensions: 110cm x 40cm x 30cm deep.

At the bottom of the grave, the blue and white shapeless dress was laid. The dress was placed on the topside of the grave and reached the middle of the grave.

On top of the dress, sand was deposited. The sand was deeper at the top of the grave than it was in the middle. Up to 20cm of sand covered the top of the dress, whereas the lower part of the dress was covered with 2cm of sand.

Sand was also placed in the lower half of the grave. There was more sand at the bottom of the grave, up to 20cm deep.

Whilst the sand in the lower part of the grave was being placed a lighter fell in the sand.

Darker topsoil was then placed on top of the sand. An ID card was found in the centre of the grave. More soil was then placed on top of the ID card.

Cigarette papers were found in the top right side of the grave. More soil was on top of the cigarette papers.

Summary: Hole dug, dress, sand, lighter, sand, dark topsoil, ID card, dark topsoil, cigarette paper, dark topsoil, turf.

Control ID: Control 06

Years of experience: No archaeological experience

Tools used to excavate the grave: Trowel and hand shovel

Did the participant sieve the fill: No

Weather conditions: Overcast and warm

Time taken: 2 1/4 hours

Observations: This participant had no archaeological experience whatsoever. The participant started the process by measuring the dimensions of the grave. The participant then divided the grave into three sections across its width using test pegs and string. The participant allocated each of these sections a letter 'A', 'B' and 'C'. The participant then removed the turf off of the first section (section A) and then proceeded to remove the fills in this section with a hand shovel. Once the participant reached the base of section A, sections B and C were excavated in the same manner. Unfortunately, through using a hand shovel to remove the majority of the fills the participant was unable to identify the presence of fills 4 and 3, and as the participant failed to check or realise that the different fills would be present in the section face that had been created, these fills were not documented. Furthermore, as the participant removed each section individually, and did not document the position of the sand fills at the bottom, the participant mistakenly believed there to be only one sand fill at the base of the grave that sloped at either end.

As items of material evidence were uncovered the participant would document their positions in a notebook. The participant did not deem it relevant to document the dimensions of the different fills identified, but briefly described them in a sketch in the notebook. Material evidence was, however, photographed in situ.

The participant did not sieve any of the spoil removed from the grave.

Control 06's narrative: The dimensions of the grave were 110cm x 40cm x 30cm. The grave contained two different layers, one was made from brown topsoil and the other was yellow sand. The brown topsoil layer had cigarette papers, a pair of earrings, an ID card and a fake nail in. The sand had a two pence coin (1994) and a lighter in. This sand layer was on top of a blue and white stripy Primark dress. No bones or body parts were found.

Evidence	Location	Evidence found	Found in situ	Found out of situ	Evidence not found
E1 Dress	Base of C4	Yes	Yes	No	No
E2 Two pence coin	C5 (Fill 1)	Yes	No	Yes	No
E3 Lighter	C6 (Fill 2)	Yes	Yes	No	No
E4 Fake nail	C7 (Fill 3)	Yes	Yes	No	No
E5 ID card	C7 (Fill 3)	Yes	Yes	No	No
E6 Earring 2	C7 (Fill 3)	Yes	No	Yes	No
E7 Kirby grip	C8 (Fill 4)	No	-	-	Yes
E8 Earring 1	C8 (Fill 4)	Yes	No	Yes	No
E9 Cigarette papers	C9 (Fill 5)	Yes	Yes	No	No
TOTAL:		8	5	3	1
PERCENTAGE:		88.89%	62.50%	37.50%	11.11%

FIGURE 137: MATERIAL EVIDENCE IDENTIFIED BY CONTROL 06 © EVIS 2016.

Context number	Description	Context identified	Context not identified
C1	Natural	Yes	No
C2	Subsoil	Yes	No
C3	Topsoil and turf	Yes	No
C4	Feature cut	Yes	No
C5	Fill 1	Yes	No
C6	Fill 2	No	Yes
C7	Fill 3	No	Yes
C8	Fill 4	No	Yes
C9	Fill 5	Yes	No
C10	Replaced turf	Yes	No
TOTAL:		7	3
PERCENTAGE:		70%	30%

FIGURE 138: CONTEXTS IDENTIFIED BY CONTROL 06 © Evis 2016.

Control ID: Control 07

Years of experience: No archaeological experience

Tools used to excavate the grave: Trowel and hand shovel

Did the participant sieve the fill: Yes

Weather conditions: Overcast and warm

Time taken: 2 1/2 hours

Observations: This participant had no archaeological experience whatsoever. The participant started the excavation process by measuring the boundaries of the grave. Then the participant removed the turf that had been placed over the grave and began to excavate the grave fills. The participant excavated the grave using a hand shovel. The participant would remove shovelfuls of soil at a time, with no structured approach. This meant that the participant failed to maintain the boundaries of the individual fills in the grave. This also meant that the participant failed to recognise the presence of fills three and four, as the participant dug straight through them. When the participant reached the sand fills the approach was changed and the participant began to use a trowel rather than a hand shovel. This change in approach is probably due to the fact that these fills were obviously distinct. The participant began to follow the boundaries of these sand fills and realised that they were sloping. However, when the participant noticed the dress on the base it was pulled from its position, resulting in the participant contaminating the area between the two sand fills with sand, leading the participant to think that the sand at the base of the grave was one fill that sloped in the middle.

As finds were identified the participant would record their position, photograph them and then transfer them into an evidence bag. Notes regarding finds were written in a notebook. In terms of recording fills, the participant did draw sketch plans of them, but did not add any detail in regards to the dimensions of the different fills.

The participant did sieve all of the excavated spoil. This led to a high evidence recovery rate as the participant spent a long time inspecting the sieve for the presence of material evidence.

Evidence	Location	Evidence found	Found in situ	Found out of situ	Evidence not found
E1 Dress	Base of C4	Yes	Yes	No	No
E2 Two pence coin	C5 (Fill 1)	Yes	No	Yes	No
E3 Lighter	C6 (Fill 2)	Yes	Yes	No	No
E4 Fake nail	C7 (Fill 3)	Yes	Yes	No	No
E5 ID card	C7 (Fill 3)	Yes	Yes	No	No
E6 Earring 2	C7 (Fill 3)	Yes	No	Yes	No
E7 Kirby grip	C8 (Fill 4)	Yes	No	Yes	No
E8 Earring 1	C8 (Fill 4)	Yes	Yes	No	No
E9 Cigarette papers	C9 (Fill 5)	Yes	Yes	No	No
TOTAL:		9	6	3	0
PERCENTAGE:		100%	66.67%	33.33%	0%

FIGURE 139: MATERIAL EVIDENCE IDENTIFIED BY CONTROL 07 © Evis 2016.

Context number	Description	Context identified	Context not identified
C1	Natural	Yes	No
C2	Subsoil	Yes	No
C3	Topsoil and turf	Yes	No
C4	Feature cut	Yes	No
C5	Fill 1	Yes	No
C6	Fill 2	No	Yes
C7	Fill 3	No	Yes
C8	Fill 4	No	Yes
C9	Fill 5	Yes	No
C10	Replaced turf	Yes	No
TOTAL:		7	3
PERCENTAGE:		70%	30%

FIGURE 140: CONTEXTS IDENTIFIED BY CONTROL 07 © Evis 2016.

Control 07's narrative: A grave would have been dug of 110cm long by 40cm wide and 30cm deep.

Then the dress was laid out in the western end of the grave with the top at the western most end and lying towards the east. Sand was covering this layer/item. In the eastern end a pink lighter was placed in the sand as well.

Soil was piled over this and a fake nail dropped close to the middle. More soil was placed on this and an ID card was placed in the middle of the grave perfectly at right angles to the rectangle of the grave's outline. The two pence coin was found on the soil near the position of the fake nail. One gold/green earring was in the north side or the eastern end of the grave. Still with the rubber stopper attached. A hair clip was placed in the soil near the western end of the grave. A second earring gold/green, was found in the south side of the western edge of the grave close to parallel with the line the other earring was on. It also still had the rubber stopper attached. Near the top of the grave a cigarette paper packet had been placed close to the western end of the grave, perpendicular to the grave's outline.

Both the cigarette paper packet and the ID card could be read by a person stood at the western end of the grave looking in per the orientation they were laid.

The turf that had been removed when the grave was dug was placed back over the grave.

Control ID: Control 08

Years of experience: No archaeological experience

Tools used to excavate the grave: Shovel and hand shovel

Did the participant sieve the fill: No

Weather conditions: Clear and hot

Time taken: 3 hours

Observations: This participant had no archaeological training whatsoever. The participant started the excavation by measuring the dimensions of the grave. The participant then removed the overlying turfs by hand and then proceeded to excavate the grave using a shovel. The participant would dig down into the grave with the shovel and then transfer the spoil to the tarpaulin, moving across the length of the grave as each shovelful was removed. This resulted in the participant digging through fills three and four. When the participant noticed that there was sand in the base of the grave they continued to use the shovel for the first few centimetres but then altered their approach and used a hand shovel. By this stage, however, the dimensions of the sand fills at the base of the grave had been disrupted and the participant had transferred sand across the base of the feature, leading the participant to interpret that the two sand fills were one flat, level fill. This issue was exacerbated by the fact that when the participant uncovered the dress it was pulled from its original position, further blurring the definition between the two sand fills.

As material evidence was found photographs were taken, and notes written in the participant's notebook regarding its location. The participant also took photographs of the sand layer.

The participant chose not to sieve the spoil and made no attempt to inspect the spoil for any material evidence.

Evidence	Location	Evidence found	Found in situ	Found out of situ	Evidence not found
E1 Dress	Base of C4	Yes	Yes	No	No
E2 Two pence coin	C5 (Fill 1)	No	-	-	Yes
E3 Lighter	C6 (Fill 2)	Yes	Yes	No	No
E4 Fake nail	C7 (Fill 3)	No	-	-	Yes
E5 ID card	C7 (Fill 3)	Yes	Yes	No	No
E6 Earring 2	C7 (Fill 3)	No	-	-	Yes
E7 Kirby grip	C8 (Fill 4)	No	-	-	Yes
E8 Earring 1	C8 (Fill 4)	No	-	-	Yes
E9 Cigarette papers	C9 (Fill 5)	Yes	Yes	No	No
TOTAL:		**4**	**4**	**0**	**5**
PERCENTAGE:		**44.44%**	**100%**	**0%**	**55.56%**

FIGURE 141: MATERIAL EVIDENCE IDENTIFIED BY CONTROL 08 © Evis 2016.

Control 08's narrative: The grave was dug, and then a dress was placed on the bottom. Then sand was put in the grave. In this sand there was a lighter and an ID card. Some brown soil was then put on top of the sand and this soil had a packet of cigarette papers in. Then grass was put back over the grave. There didn't appear to be any bones in the grave. The grave was approximately 110cm long, 40cm wide and 30cm deep.

Context number	Description	Context identified	Context not identified
C1	Natural	Yes	No
C2	Subsoil	Yes	No
C3	Topsoil and turf	Yes	No
C4	Feature cut	Yes	No
C5	Fill 1	Yes	No
C6	Fill 2	No	Yes
C7	Fill 3	No	Yes
C8	Fill 4	No	Yes
C9	Fill 5	Yes	No
C10	Replaced turf	Yes	No
TOTAL:		7	3
PERCENTAGE:		70%	30%

FIGURE 142: CONTEXTS IDENTIFIED BY CONTROL 08 © Evis 2016.

Control ID: Control 09

Years of experience: No archaeological experience

Tools used to excavate the grave: Trowel and hand shovel

Did the participant sieve the fill: Yes

Weather conditions: Overcast and warm

Time taken: 2 1/2 hours

Observations: This participant had no archaeological experience whatsoever. The participant started the excavation by recording the dimensions of the grave. The participant then began to excavate the grave using a hand shovel. The participant would drag the hand shovel across the length of the grave and then sieve the extracted spoil. This resulted in the participant digging straight through the third and the fourth fills. When the participant got towards the end of the grave, the participant noticed the presence of the sand fills. The participant then used a combination of a trowel and a hand shovel to remove these fills. This resulted in the participant realising that the two sand fills sloped down towards the centre of the grave. However, by using the hand shovel to remove some of the sand the participant dragged sand into the middle of the grave, where the boundary between these split sand fills was, which resulted in the participant misinterpreting these two different fills as one.

When the participant identified pieces of material evidence their locations would be noted down in the participant's notebook and photographs would be taken. The participant also took photographs of the two different fills and measured their dimensions.

The participant sieved all of the extracted spoil. However, the participant did not pay full attention during this process and several items of material evidence were thrown onto the tarpaulin without being noticed.

Evidence	Location	Evidence found	Found in situ	Found out of situ	Evidence not found
E1 Dress	Base of C4	Yes	Yes	No	No
E2 Two pence coin	C5 (Fill 1)	Yes	No	Yes	No
E3 Lighter	C6 (Fill 2)	Yes	Yes	No	No
E4 Fake nail	C7 (Fill 3)	No	-	-	Yes
E5 ID card	C7 (Fill 3)	Yes	Yes	No	No
E6 Earring 2	C7 (Fill 3)	No	-	-	Yes
E7 Kirby grip	C8 (Fill 4)	No	-	-	Yes
E8 Earring 1	C8 (Fill 4)	No	-	-	Yes
E9 Cigarette papers	C9 (Fill 5)	Yes	Yes	No	No
	TOTAL:	5	4	1	4
	PERCENTAGE:	55.56%	80%	20%	44.44%

FIGURE 143: MATERIAL EVIDENCE IDENTIFIED BY CONTROL 09 © EVIS 2016.

Context number	Description	Context identified	Context not identified
C1	Natural	Yes	No
C2	Subsoil	Yes	No
C3	Topsoil and turf	Yes	No
C4	Feature cut	Yes	No
C5	Fill 1	Yes	No
C6	Fill 2	No	Yes
C7	Fill 3	No	Yes
C8	Fill 4	No	Yes
C9	Fill 5	Yes	No
C10	Replaced turf	Yes	No
	TOTAL:	7	3
	PERCENTAGE:	70%	30%

FIGURE 144: CONTEXTS IDENTIFIED BY CONTROL 09 © EVIS 2016.

Control 09's narrative: The grave was composed of two layers. One layer was made from sand; this layer was at the bottom of the grave. The other layer was made from a light brown soil with organic matter in, which had been placed on top of the sand. Turf had then been placed back on top of the grave.

The sand layer looked like building sand and so was probably brought in from elsewhere and contained: a blue and white Primark dress, a plastic lighter and a two pence coin.

The light brown soil contained: an ID card and a packet of cigarette papers.

There was no evidence of human remains.

Control ID: Control 10

Years of experience: No archaeological experience

Tools used to excavate the grave: Trowel and hand shovel

Did the participant sieve the fill: Yes

Weather conditions: Overcast and warm

Time taken: 2 1/4 hours

Observations: This participant had no archaeological experience whatsoever. The participant started the excavation by recording the dimensions of the grave. The participant then removed the turf by hand and began to dig out the fills with a hand shovel. The participant would dig down approximately 5-10cm in one area and would then put the hand shovel in the gap that had been created to remove the rest of the in situ fill. This resulted in the participant digging through fills four and three without noticing the change in composition. However, as with the other participants, once the sand fills were noticed the participant altered their approach. The participant began to use the trowel and carefully removed the fill overlaying the sand fills, ensuring that the boundaries of the sand fills were maintained. This enabled the participant to determine that there were two distinct sand fills in the grave rather than one. The participant also resisted the urge to pull the dress from its position and waited until the overlaying sand had been removed before lifting it.

As items of material evidence were uncovered the participant was careful to document where each item was found in their notebook. The participant also took photographs of each item in the place in which it was found and in the evidence bag it was transferred to. The participant also took photographs of the different fills that they had identified, measured their dimensions and noted them down in their notebook.

The participant also ensured that the two sand fills were sieved separately from each other and from the other fill that the participant had identified, in order for the participant to be able to state where a piece of evidence had originated.

Evidence	Location	Evidence found	Found in situ	Found out of situ	Evidence not found
E1 Dress	Base of C4	Yes	Yes	No	No
E2 Two pence coin	C5 (Fill 1)	Yes	No	Yes	No
E3 Lighter	C6 (Fill 2)	Yes	No	Yes	No
E4 Fake nail	C7 (Fill 3)	Yes	No	Yes	No
E5 ID card	C7 (Fill 3)	Yes	Yes	No	No
E6 Earring 2	C7 (Fill 3)	No	-	-	Yes
E7 Kirby grip	C8 (Fill 4)	No	-	-	Yes
E8 Earring 1	C8 (Fill 4)	No	-	-	Yes
E9 Cigarette papers	C9 (Fill 5)	Yes	Yes	No	No
TOTAL:		6	3	3	3
PERCENTAGE:		66.67%	50%	50%	33.33%

FIGURE 145: MATERIAL EVIDENCE IDENTIFIED BY CONTROL 10 © Evis 2016.

Context number	Description	Context identified	Context not identified
C1	Natural	Yes	No
C2	Subsoil	Yes	No
C3	Topsoil and turf	Yes	No
C4	Feature cut	Yes	No
C5	Fill 1	Yes	No
C6	Fill 2	Yes	No
C7	Fill 3	No	Yes
C8	Fill 4	No	Yes
C9	Fill 5	Yes	No
C10	Replaced turf	Yes	No
TOTAL:		8	2
PERCENTAGE:		80%	20%

FIGURE 146: CONTEXTS IDENTIFIED BY CONTROL 10 © Evis 2016.

Control 10's narrative: The grave was dug by the perpetrator(s) to the following dimensions: 110cm x 40cm x 30cm.

The perpetrator(s) then put two different sand layers in, both of which sloped down towards the middle but didn't join.

Sand layer 1 covered a blue and white striped dress and also had a two pence coin in dated to 1994.

Sand layer 2 had a green plastic lighter in.

Above these two sand layers a layer of brown topsoil was put in the grave and filled the grave to the brim. This layer had a fake fingernail, an ID card and a packet of cigarette papers in.

Then the perpetrator(s) attempted to hide the grave by putting turf over it. This turf was likely to have been the turf that was removed when the perpetrator(s) dug the grave.

Evidence	Location	Evidence found	Found in situ	Found out of situ	Evidence not found
E1 Dress	Base of C4	10	10	0	0
E2 Two pence coin	C5 (Fill 1)	8	3	5	2
E3 Lighter	C6 (Fill 2)	10	9	1	0
E4 Fake nail	C7 (Fill 3)	7	5	2	3
E5 ID card	C7 (Fill 3)	10	10	0	0
E6 Earring 2	C7 (Fill 3)	4	2	2	6
E7 Kirby grip	C8 (Fill 4)	1	0	1	9
E8 Earring 1	C8 (Fill 4)	4	3	1	6
E9 Cigarette papers	C9 (Fill 5)	10	10	0	0
TOTAL:		64	52	12	26
PERCENTAGE:		71.11%	81.25%	18.75%	28.89%

FIGURE 147: STRATIGRAPHIC EXCAVATION MATERIAL EVIDENCE IDENTIFICATION AVERAGES © EVIS 2016.

Evidence	Location	Evidence found	Found in situ	Found out of situ	Evidence not found
E1 Dress	Base of C4	10	10	0	0
E2 Two pence coin	C5 (Fill 1)	7	5	2	3
E3 Lighter	C6 (Fill 2)	10	9	1	0
E4 Fake nail	C7 (Fill 3)	9	8	1	1
E5 ID card	C7 (Fill 3)	10	10	0	0
E6 Earring 2	C7 (Fill 3)	5	3	2	5
E7 Kirby grip	C8 (Fill 4)	0	0	0	10
E8 Earring 1	C8 (Fill 4)	6	4	2	4
E9 Cigarette papers	C9 (Fill 5)	9	9	0	1
TOTAL:		66	58	8	24
PERCENTAGE:		73.33%	87.88%	12.12%	26.67%

FIGURE 148: DEMIRANT EXCAVATION MATERIAL EVIDENCE IDENTIFICATION AVERAGES © EVIS 2016.

Evidence	Location	Evidence found	Found in situ	Found out of situ	Evidence not found
E1 Dress	Base of C4	10	10	0	0
E2 Two pence coin	C5 (Fill 1)	7	6	1	3
E3 Lighter	C6 (Fill 2)	10	9	1	0
E4 Fake nail	C7 (Fill 3)	8	7	1	2
E5 ID card	C7 (Fill 3)	10	10	0	0
E6 Earring 2	C7 (Fill 3)	3	3	0	7
E7 Kirby grip	C8 (Fill 4)	1	1	0	9
E8 Earring 1	C8 (Fill 4)	5	4	1	5
E9 Cigarette papers	C9 (Fill 5)	10	10	0	0
TOTAL:		64	60	4	26
PERCENTAGE:		71.11%	93.75%	6.25%	28.89%

FIGURE 149: QUADRANT EXCAVATION MATERIAL EVIDENCE IDENTIFICATION AVERAGES © EVIS 2016.

Evidence	Location	Evidence found	Found in situ	Found out of situ	Evidence not found
E1 Dress	Base of C4	10	10	0	0
E2 Two pence coin	C5 (Fill 1)	2	2	0	8
E3 Lighter	C6 (Fill 2)	6	5	1	4
E4 Fake nail	C7 (Fill 3)	3	3	0	7
E5 ID card	C7 (Fill 3)	9	8	1	1
E6 Earring 2	C7 (Fill 3)	2	2	0	8
E7 Kirby grip	C8 (Fill 4)	1	1	0	9
E8 Earring 1	C8 (Fill 4)	5	5	0	5
E9 Cigarette papers	C9 (Fill 5)	8	8	0	2
TOTAL		46	44	2	44
PERCENTAGE:		51.11%	95.65%	4.35%	48.89%

FIGURE 150: ARBITRARY LEVEL EXCAVATION MATERIAL EVIDENCE IDENTIFICATION AVERAGES © EVIS 2016.

Evidence	Location	Evidence found	Found in situ	Found out of situ	Evidence not found
E1 Dress	Base of C4	10	10	0	0
E2 Two pence coin	C5 (Fill 1)	6	2	4	4
E3 Lighter	C6 (Fill 2)	10	9	1	0
E4 Fake nail	C7 (Fill 3)	5	2	3	5
E5 ID card	C7 (Fill 3)	10	10	0	0
E6 Earring 2	C7 (Fill 3)	3	0	3	7
E7 Kirby grip	C8 (Fill 4)	2	0	2	8
E8 Earring 1	C8 (Fill 4)	5	1	4	5
E9 Cigarette papers	C9 (Fill 5)	9	9	0	1
TOTAL:		60	43	17	30
PERCENTAGE:		66.67%	71.67%	28.33%	33.33%

FIGURE 151: CONTROL EXCAVATION MATERIAL EVIDENCE IDENTIFICATION AVERAGES © EVIS 2016.

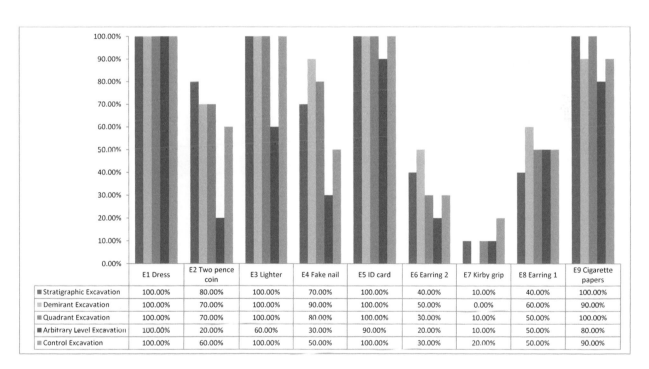

	E1 Dress	E2 Two pence coin	E3 Lighter	E4 Fake nail	E5 ID card	E6 Earring 2	E7 Kirby grip	E8 Earring 1	E9 Cigarette papers
Stratigraphic Excavation	100.00%	80.00%	100.00%	70.00%	100.00%	40.00%	10.00%	40.00%	100.00%
Demirant Excavation	100.00%	70.00%	100.00%	90.00%	100.00%	50.00%	0.00%	60.00%	90.00%
Quadrant Excavation	100.00%	70.00%	100.00%	80.00%	100.00%	30.00%	10.00%	50.00%	100.00%
Arbitrary Level Excavation	100.00%	20.00%	60.00%	30.00%	90.00%	20.00%	10.00%	50.00%	80.00%
Control Excavation	100.00%	60.00%	100.00%	50.00%	100.00%	30.00%	20.00%	50.00%	90.00%

FIGURE 152: RECOVERY RATES OF MATERIAL EVIDENCE ITEMS FOR EACH EXCAVATION METHOD © EVIS 2016.

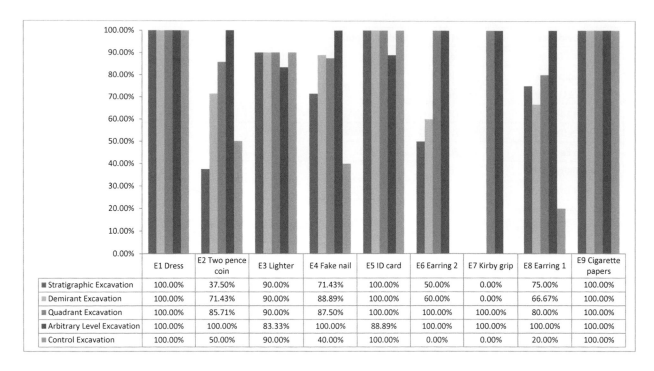

	E1 Dress	E2 Two pence coin	E3 Lighter	E4 Fake nail	E5 ID card	E6 Earring 2	E7 Kirby grip	E8 Earring 1	E9 Cigarette papers
■ Stratigraphic Excavation	100.00%	37.50%	90.00%	71.43%	100.00%	50.00%	0.00%	75.00%	100.00%
▨ Demirant Excavation	100.00%	71.43%	90.00%	88.89%	100.00%	60.00%	0.00%	66.67%	100.00%
▨ Quadrant Excavation	100.00%	85.71%	90.00%	87.50%	100.00%	100.00%	100.00%	80.00%	100.00%
■ Arbitrary Level Excavation	100.00%	100.00%	83.33%	100.00%	88.89%	100.00%	100.00%	100.00%	100.00%
▨ Control Excavation	100.00%	50.00%	90.00%	40.00%	100.00%	0.00%	0.00%	20.00%	100.00%

FIGURE 153: IN SITU RECOVERY RATES OF MATERIAL EVIDENCE ITEMS FOR EACH EXCAVATION METHOD © EVIS 2016.

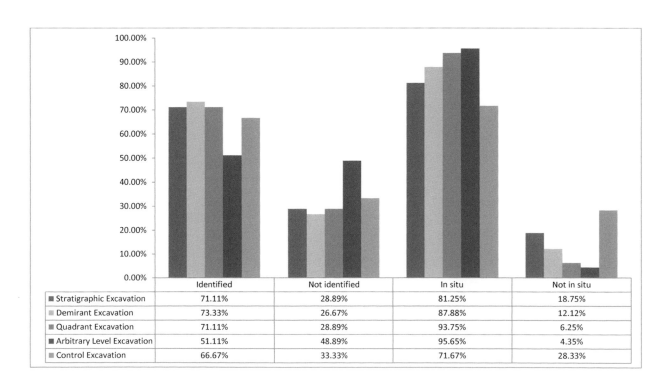

	Identified	Not identified	In situ	Not in situ
■ Stratigraphic Excavation	71.11%	28.89%	81.25%	18.75%
▨ Demirant Excavation	73.33%	26.67%	87.88%	12.12%
▨ Quadrant Excavation	71.11%	28.89%	93.75%	6.25%
■ Arbitrary Level Excavation	51.11%	48.89%	95.65%	4.35%
▨ Control Excavation	66.67%	33.33%	71.67%	28.33%

FIGURE 154: OVERALL RECOVERY RATES OF MATERIAL EVIDENCE FOR EACH EXCAVATION METHOD © EVIS 2016.

Context number	Description	Context identified	Context not identified
C1	Natural	10	0
C2	Subsoil	10	0
C3	Topsoil and turf	10	0
C4	Feature cut	10	0
C5	Fill 1	9	1
C6	Fill 2	9	1
C7	Fill 3	10	0
C8	Fill 4	7	3
C9	Fill 5	10	0
C10	Replaced turf	10	0
	TOTAL	**5**	**5**
	PERCENTAGE:	95%	5%

FIGURE 155: STRATIGRAPHIC EXCAVATION CONTEXT IDENTIFICATION AVERAGES © EVIS 2016.

Context number	Description	Context identified	Context not identified
C1	Natural	10	0
C2	Subsoil	10	0
C3	Topsoil and turf	10	0
C4	Feature cut	10	0
C5	Fill 1	9	1
C6	Fill 2	9	1
C7	Fill 3	10	0
C8	Fill 4	10	0
C9	Fill 5	10	0
C10	Replaced turf	10	0
	TOTAL:	**98**	**2**
	PERCENTAGE:	98%	2%

FIGURE 156: DEMIRANT EXCAVATION CONTEXT IDENTIFICATION AVERAGES © EVIS 2016.

Context number	Description	Context identified	Context not identified
C1	Natural	10	0
C2	Subsoil	10	0
C3	Topsoil and turf	10	0
C4	Feature cut	10	0
C5	Fill 1	10	0
C6	Fill 2	10	0
C7	Fill 3	9	1
C8	Fill 4	9	1
C9	Fill 5	10	0
C10	Replaced turf	10	0
	TOTAL:	**98**	**2**
	PERCENTAGE:	98%	2%

FIGURE 157: QUADRANT EXCAVATION CONTEXT IDENTIFICATION AVERAGES © EVIS 2016.

Context number	Description	Context identified	Context not identified
C1	Natural	10	0
C2	Subsoil	10	0
C3	Topsoil and turf	10	0
C4	Feature cut	9	1
C5	Fill 1	10	0
C6	Fill 2	0	10
C7	Fill 3	0	10
C8	Fill 4	0	10
C9	Fill 5	10	0
C10	Replaced turf	10	0
	TOTAL:	69	31
	PERCENTAGE:	69%	31%

FIGURE 158: ARBITRARY LEVEL CONTEXT IDENTIFICATION AVERAGES © EVIS 2016.

Context number	Description	Context identified	Context not identified
C1	Natural	10	0
C2	Subsoil	10	0
C3	Topsoil and turf	10	0
C4	Feature cut	10	0
C5	Fill 1	8	2
C6	Fill 2	5	5
C7	Fill 3	1	9
C8	Fill 4	1	9
C9	Fill 5	9	1
C10	Replaced turf	10	0
	TOTAL:	74	26
	PERCENTAGE:	74%	26%

FIGURE 159: CONTROL EXCAVATION CONTEXT IDENTIFICATION AVERAGES © EVIS 2016.

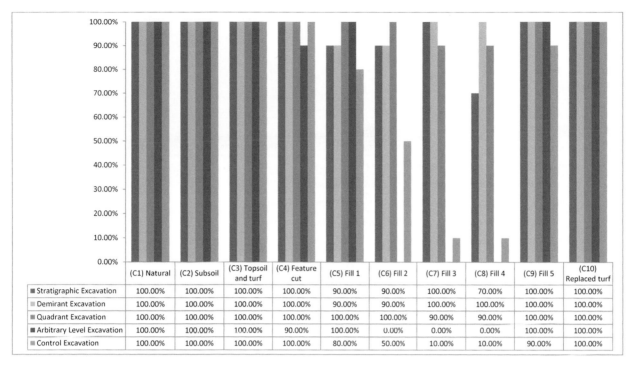

	(C1) Natural	(C2) Subsoil	(C3) Topsoil and turf	(C4) Feature cut	(C5) Fill 1	(C6) Fill 2	(C7) Fill 3	(C8) Fill 4	(C9) Fill 5	(C10) Replaced turf
■ Stratigraphic Excavation	100.00%	100.00%	100.00%	100.00%	90.00%	90.00%	100.00%	70.00%	100.00%	100.00%
▨ Demirant Excavation	100.00%	100.00%	100.00%	100.00%	90.00%	90.00%	100.00%	100.00%	100.00%	100.00%
▨ Quadrant Excavation	100.00%	100.00%	100.00%	100.00%	100.00%	100.00%	90.00%	90.00%	100.00%	100.00%
■ Arbitrary Level Excavation	100.00%	100.00%	100.00%	90.00%	100.00%	0.00%	0.00%	0.00%	100.00%	100.00%
▨ Control Excavation	100.00%	100.00%	100.00%	100.00%	80.00%	50.00%	10.00%	10.00%	90.00%	100.00%

FIGURE 160: IDENTIFICATION OF INDIVIDUAL CONTEXTS FOR EACH EXCAVATION METHOD © EVIS 2016.

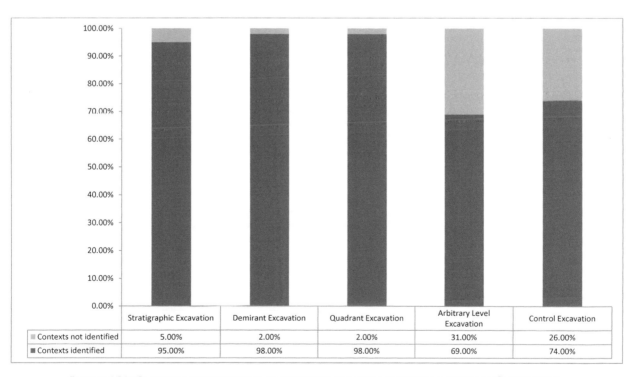

	Stratigraphic Excavation	Demirant Excavation	Quadrant Excavation	Arbitrary Level Excavation	Control Excavation
▨ Contexts not identified	5.00%	2.00%	2.00%	31.00%	26.00%
■ Contexts identified	95.00%	98.00%	98.00%	69.00%	74.00%

FIGURE 161: OVERALL IDENTIFICATION OF CONTEXTS FOR EACH EXCAVATION METHOD © EVIS 2016.

Stage	Description	Identified	Not identified
1	The feature was cut (C4) through the top soil and turf (C3), subsoil (C2) and natural (C1)	10	0
2	The dress (E1) was placed along the base of the cut feature (C4)	10	0
3	Fill 1 (C5) started to be added to the feature overlaying the dress (E1)	9	1
4	A two pence coin (E2) was added to fill 1 (C5) at 10 cm L, 35 cm W, 25 cm D	8	2
5	The rest of fill 1 (C5) was added to the feature, covering the two pence coin (E2)	8	2
6	Fill 2 (C6) started to be added to the feature	9	1
7	A lighter (E3) was added to fill 2 (C6) at 105 cm L, 2 cm W, 28 cm D	10	0
8	The rest of fill 2 (C6) was added to the feature, covering the lighter (E3)	9	1
9	Fill 3 (C7) started to be added to the feature overlaying fill 1 (C5) and fill 2 (C6)	10	0
10	A fake nail (E4) was added to fill 3 (C7) at 70 cm L, 10 cm W, 25 cm D	7	3
11	More of fill 3 (C7) was added to the feature, covering the fake nail (E4)	7	3
12	An ID card (E5) was added to fill 3 (C7) at 50 cm L, 20 cm W, 20 cm D	10	0
13	More of fill 3 (C7) was added to the feature, covering the ID card (E5)	10	0
14	Earring 2 (E6) was added to fill 3 (C7) at 90 cm L, 20 cm W, 15 cm D	4	6
15	The rest of fill 3 (C7) was added to the feature, covering earring 2 (E6)	4	6
16	Fill 4 (C8) started to be added to the feature overlaying fill 3 (C7)	7	3
17	A kirby grip (E7) was added to fill 4 (C8) at 30 cm L, 15 cm W, 10 cm D	1	9
18	More of fill 4 (C8) was added to the feature, covering the kirby grip (E7)	1	9
19	Earring 1 (E8) was added to fill 4 (C8) at 90 cm L, 35 cm W, 5 cm D	4	6
20	The rest of fill 4 (C8) was added to the feature, covering earring 1 (E8)	4	6
21	Fill 5 (C9) started to be added to the feature overlaying fill 4 (C8)	10	0
22	A packet of cigarette papers (E9) was added to fill 5 (C9) at 10 cm L, 10 cm W, 4 cm D	10	0
23	The rest of fill 5 (C9) was added to the feature, covering the cigarette papers (E9)	10	0
24	The turf (C10) that had been removed during stage 1 was placed back over the feature, overlaying fill 5 (C9)	10	0
	TOTAL:	182	58
	PERCENTAGE:	75.83%	24.17%

FIGURE 162: STRATIGRAPHIC EXCAVATION FORMATION SEQUENCE AVERAGES © EVIS 2016.

Stage	Description	Identified	Not identified
1	The feature was cut (C4) through the top soil and turf (C3), subsoil (C2) and natural (C1)	10	0
2	The dress (E1) was placed along the base of the cut feature (C4)	10	0
3	Fill 1 (C5) started to be added to the feature overlaying the dress (E1)	9	1
4	A two pence coin (E2) was added to fill 1 (C5) at 10 cm L, 35 cm W, 25 cm D	7	3
5	The rest of fill 1 (C5) was added to the feature, covering the two pence coin (E2)	7	3
6	Fill 2 (C6) started to be added to the feature	9	1
7	A lighter (E3) was added to fill 2 (C6) at 105 cm L, 2 cm W, 28 cm D	10	0
8	The rest of fill 2 (C6) was added to the feature, covering the lighter (E3)	9	1
9	Fill 3 (C7) started to be added to the feature overlaying fill 1 (C5) and fill 2 (C6)	10	0
10	A fake nail (E4) was added to fill 3 (C7) at 70 cm L, 10 cm W, 25 cm D	9	1
11	More of fill 3 (C7) was added to the feature, covering the fake nail (E4)	9	1
12	An ID card (E5) was added to fill 3 (C7) at 50 cm L, 20 cm W, 20 cm D	10	0
13	More of fill 3 (C7) was added to the feature, covering the ID card (E5)	10	0
14	Earring 2 (E6) was added to fill 3 (C7) at 90 cm L, 20 cm W, 15 cm D	5	5
15	The rest of fill 3 (C7) was added to the feature, covering earring 2 (E6)	5	5
16	Fill 4 (C8) started to be added to the feature overlaying fill 3 (C7)	10	0
17	A kirby grip (E7) was added to fill 4 (C8) at 30 cm L, 15 cm W, 10 cm D	0	10
18	More of fill 4 (C8) was added to the feature, covering the kirby grip (E7)	0	10
19	Earring 1 (E8) was added to fill 4 (C8) at 90 cm L, 35 cm W, 5 cm D	6	4
20	The rest of fill 4 (C8) was added to the feature, covering earring 1 (E8)	6	4
21	Fill 5 (C9) started to be added to the feature overlaying fill 4 (C8)	10	0
22	A packet of cigarette papers (E9) was added to fill 5 (C9) at 10 cm L, 10 cm W, 4 cm D	9	1
23	The rest of fill 5 (C9) was added to the feature, covering the cigarette papers (E9)	9	1
24	The turf (C10) that had been removed during stage 1 was placed back over the feature, overlaying fill 5 (C9)	10	0
	TOTAL:	189	51
	PERCENTAGE:	78.75%	21.25%

FIGURE 163: DEMIRANT EXCAVATION FORMATION SEQUENCE AVERAGES © EVIS 2016.

Stage	Description	Identified	Not identified
1	The feature was cut (C4) through the top soil and turf (C3), subsoil (C2) and natural (C1)	10	0
2	The dress (E1) was placed along the base of the cut feature (C4)	10	0
3	Fill 1 (C5) started to be added to the feature overlaying the dress (E1)	10	0
4	A two pence coin (E2) was added to fill 1 (C5) at 10 cm L, 35 cm W, 25 cm D	7	3
5	The rest of fill 1 (C5) was added to the feature, covering the two pence coin (E2)	7	3
6	Fill 2 (C6) started to be added to the feature	10	0
7	A lighter (E3) was added to fill 2 (C6) at 105 cm L, 2 cm W, 28 cm D	10	0
8	The rest of fill 2 (C6) was added to the feature, covering the lighter (E3)	10	0
9	Fill 3 (C7) started to be added to the feature overlaying fill 1 (C5) and fill 2 (C6)	9	1
10	A fake nail (E4) was added to fill 3 (C7) at 70 cm L, 10 cm W, 25 cm D	8	2
11	More of fill 3 (C7) was added to the feature, covering the fake nail (E4)	7	3
12	An ID card (E5) was added to fill 3 (C7) at 50 cm L, 20 cm W, 20 cm D	10	0
13	More of fill 3 (C7) was added to the feature, covering the ID card (E5)	9	1
14	Earring 2 (E6) was added to fill 3 (C7) at 90 cm L, 20 cm W, 15 cm D	3	7
15	The rest of fill 3 (C7) was added to the feature, covering earring 2 (E6)	2	8
16	Fill 4 (C8) started to be added to the feature overlaying fill 3 (C7)	9	1
17	A kirby grip (E7) was added to fill 4 (C8) at 30 cm L, 15 cm W, 10 cm D	1	9
18	More of fill 4 (C8) was added to the feature, covering the kirby grip (E7)	1	9
19	Earring 1 (E8) was added to fill 4 (C8) at 90 cm L, 35 cm W, 5 cm D	5	5
20	The rest of fill 4 (C8) was added to the feature, covering earring 1 (E8)	4	6
21	Fill 5 (C9) started to be added to the feature overlaying fill 4 (C8)	10	0
22	A packet of cigarette papers (E9) was added to fill 5 (C9) at 10 cm L, 10 cm W, 4 cm D	10	0
23	The rest of fill 5 (C9) was added to the feature, covering the cigarette papers (E9)	10	0
24	The turf (C10) that had been removed during stage 1 was placed back over the feature, overlaying fill 5 (C9)	10	0
	TOTAL:	182	58
	PERCENTAGE:	75.83%	24.17%

FIGURE 164: QUADRANT EXCAVATION FORMATION SEQUENCE AVERAGES © EVIS 2016.

Stage	Description	Identified	Not identified
1	The feature was cut (C4) through the top soil and turf (C3), subsoil (C2) and natural (C1)	9	1
2	The dress (E1) was placed along the base of the cut feature (C4)	10	0
3	Fill 1 (C5) started to be added to the feature overlaying the dress (E1)	10	0
4	A two pence coin (E2) was added to fill 1 (C5) at 10 cm L, 35 cm W, 25 cm D	2	8
5	The rest of fill 1 (C5) was added to the feature, covering the two pence coin (E2)	2	8
6	Fill 2 (C6) started to be added to the feature	0	10
7	A lighter (E3) was added to fill 2 (C6) at 105 cm L, 2 cm W, 28 cm D	6	4
8	The rest of fill 2 (C6) was added to the feature, covering the lighter (E3)	0	10
9	Fill 3 (C7) started to be added to the feature overlaying fill 1 (C5) and fill 2 (C6)	0	10
10	A fake nail (E4) was added to fill 3 (C7) at 70 cm L, 10 cm W, 25 cm D	3	7
11	More of fill 3 (C7) was added to the feature, covering the fake nail (E4)	0	10
12	An ID card (E5) was added to fill 3 (C7) at 50 cm L, 20 cm W, 20 cm D	9	1
13	More of fill 3 (C7) was added to the feature, covering the ID card (E5)	0	10
14	Earring 2 (E6) was added to fill 3 (C7) at 90 cm L, 20 cm W, 15 cm D	2	8
15	The rest of fill 3 (C7) was added to the feature, covering earring 2 (E6)	0	10
16	Fill 4 (C8) started to be added to the feature overlaying fill 3 (C7)	0	10
17	A kirby grip (E7) was added to fill 4 (C8) at 30 cm L, 15 cm W, 10 cm D	1	9
18	More of fill 4 (C8) was added to the feature, covering the kirby grip (E7)	0	10
19	Earring 1 (E8) was added to fill 4 (C8) at 90 cm L, 35 cm W, 5 cm D	5	5
20	The rest of fill 4 (C8) was added to the feature, covering earring 1 (E8)	0	10
21	Fill 5 (C9) started to be added to the feature overlaying fill 4 (C8)	10	0
22	A packet of cigarette papers (E9) was added to fill 5 (C9) at 10 cm L, 10 cm W, 4 cm D	8	2
23	The rest of fill 5 (C9) was added to the feature, covering the cigarette papers (E9)	8	2
24	The turf (C10) that had been removed during stage 1 was placed back over the feature, overlaying fill 5 (C9)	10	0
	TOTAL:	95	145
	PERCENTAGE:	39.58%	60.42%

FIGURE 165: ARBITRARY LEVEL EXCAVATION FORMATION SEQUENCE AVERAGES © EVIS 2016.

Stage	Description	Identified	Not identified
1	The feature was cut (C4) through the top soil and turf (C3), subsoil (C2) and natural (C1)	10	0
2	The dress (E1) was placed along the base of the cut feature (C4)	10	0
3	Fill 1 (C5) started to be added to the feature overlaying the dress (E1)	8	2
4	A two pence coin (E2) was added to fill 1 (C5) at 10 cm L, 35 cm W, 25 cm D	6	4
5	The rest of fill 1 (C5) was added to the feature, covering the two pence coin (E2)	5	5
6	Fill 2 (C6) started to be added to the feature	5	5
7	A lighter (E3) was added to fill 2 (C6) at 105 cm L, 2 cm W, 28 cm D	10	0
8	The rest of fill 2 (C6) was added to the feature, covering the lighter (E3)	5	5
9	Fill 3 (C7) started to be added to the feature overlaying fill 1 (C5) and fill 2 (C6)	1	9
10	A fake nail (E4) was added to fill 3 (C7) at 70 cm L, 10 cm W, 25 cm D	5	5
11	More of fill 3 (C7) was added to the feature, covering the fake nail (E4)	1	9
12	An ID card (E5) was added to fill 3 (C7) at 50 cm L, 20 cm W, 20 cm D	10	0
13	More of fill 3 (C7) was added to the feature, covering the ID card (E5)	1	9
14	Earring 2 (E6) was added to fill 3 (C7) at 90 cm L, 20 cm W, 15 cm D	3	7
15	The rest of fill 3 (C7) was added to the feature, covering earring 2 (E6)	1	9
16	Fill 4 (C8) started to be added to the feature overlaying fill 3 (C7)	1	9
17	A kirby grip (E7) was added to fill 4 (C8) at 30 cm L, 15 cm W, 10 cm D	2	8
18	More of fill 4 (C8) was added to the feature, covering the kirby grip (E7)	0	10
19	Earring 1 (E8) was added to fill 4 (C8) at 90 cm L, 35 cm W, 5 cm D	5	5
20	The rest of fill 4 (C8) was added to the feature, covering earring 1 (E8)	1	9
21	Fill 5 (C9) started to be added to the feature overlaying fill 4 (C8)	9	1
22	A packet of cigarette papers (E9) was added to fill 5 (C9) at 10 cm L, 10 cm W, 4 cm D	9	1
23	The rest of fill 5 (C9) was added to the feature, covering the cigarette papers (E9)	8	2
24	The turf (C10) that had been removed during stage 1 was placed back over the feature, overlaying fill 5 (C9)	10	0
	TOTAL:	126	114
	PERCENTAGE:	52.50%	47.50%

FIGURE 166: CONTROL EXCAVATION FORMATION SEQUENCE AVERAGES © EVIS 2016.

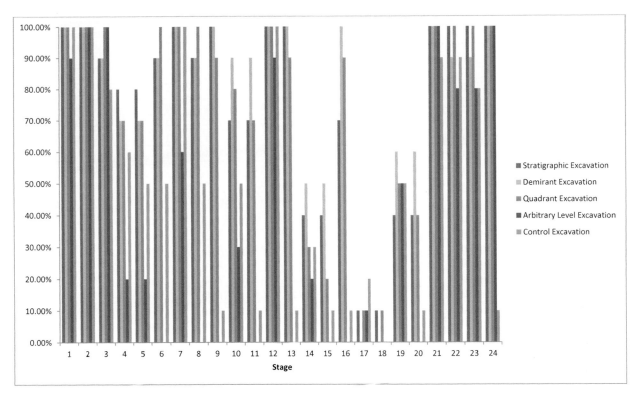

FIGURE 167: OVERALL IDENTIFICATION OF EACH STAGE IN THE FEATURE'S FORMATION PROCESS FOR EACH EXCAVATION METHOD © EVIS 2016.

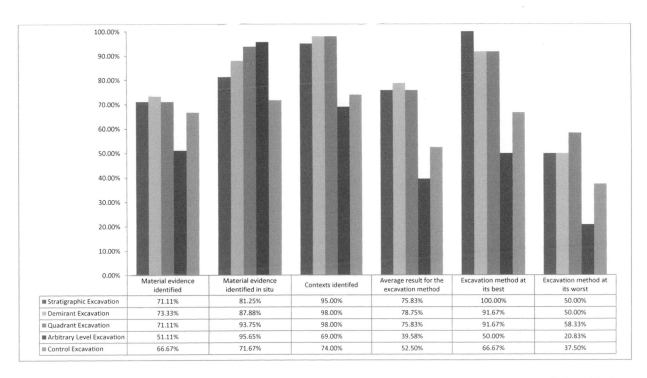

	Material evidence identified	Material evidence identified in situ	Contexts identifed	Average result for the excavation method	Excavation method at its best	Excavation method at its worst
■ Stratigraphic Excavation	71.11%	81.25%	95.00%	75.83%	100.00%	50.00%
■ Demirant Excavation	73.33%	87.88%	98.00%	78.75%	91.67%	50.00%
■ Quadrant Excavation	71.11%	93.75%	98.00%	75.83%	91.67%	58.33%
■ Arbitrary Level Excavation	51.11%	95.65%	69.00%	39.58%	50.00%	20.83%
■ Control Excavation	66.67%	71.67%	74.00%	52.50%	66.67%	37.50%

FIGURE 168: OVERALL PERFORMANCE OF EACH EXCAVATION METHOD AGAINST ALL ANALYTICAL CRITERIA © EVIS 2016.

MATERIAL EVIDENCE FOUND	Average		Standard deviation		T-test *P*-value
	1	2	1	2	
(1) Stratigraphic Excavation versus (2) Control Excavation	6.4	6	1.71	1.76	0.61316
(1) Demirant Excavation versus (2) Control Excavation	6.6	6	1.71	1.76	0.45028
(1) Quadrant Excavation versus (2) Control Excavation	6.4	6	1.65	1.76	0.60652
(1) Arbitrary Level Excavation versus (2) Control Excavation	4.6	6	1.71	1.76	0.08852
(1) Stratigraphic Excavation versus (2) Demirant Excavation	6.4	6.6	1.71	1.71	0.79697
(1) Stratigraphic Excavation versus (2) Quadrant Excavation	6.4	6.4	1.71	1.65	1.00000
(1) Stratigraphic Excavation versus (2) Arbitrary Level Excavation	6.4	4.6	1.71	1.71	0.03038
(1) Demirant Excavation versus (2) Quadrant Excavation	6.6	6.4	1.71	1.65	0.79311
(1) Demirant Excavation versus (2) Arbitrary Level Excavation	6.6	4.6	1.71	1.71	0.01768
(1) Quadrant Excavation versus (2) Arbitrary Level Excavation	6.4	4.6	1.65	1.71	0.02766
MATERIAL EVIDENCE FOUND IN SITU	**Average**		**Standard deviation**		**T-test *P*-value**
	1	2	1	2	
(1) Stratigraphic Excavation versus (2) Control Excavation	5.2	4.3	1.48	0.95	0.12213
(1) Demirant Excavation versus (2) Control Excavation	5.8	4.3	1.55	0.95	0.01768
(1) Quadrant Excavation versus (2) Control Excavation	5.9	4.3	1.66	0.95	0.01656
(1) Arbitrary Level Excavation versus (2) Control Excavation	4.4	4.3	1.90	0.95	0.88315
(1) Stratigraphic Excavation versus (2) Demirant Excavation	5.2	5.8	1.48	1.55	0.38688
(1) Stratigraphic Excavation versus (2) Quadrant Excavation	5.2	5.9	1.48	1.66	0.33269
(1) Stratigraphic Excavation versus (2) Arbitrary Level Excavation	5.2	4.4	1.48	1.90	0.30651
(1) Demirant Excavation versus (2) Quadrant Excavation	5.8	5.9	1.55	1.66	0.89090
(1) Demirant Excavation versus (2) Arbitrary Level Excavation	5.8	4.4	1.55	1.90	0.08744
(1) Quadrant Excavation versus (2) Arbitrary Level Excavation	5.9	4.4	1.66	1.90	0.07641
CONTEXTS IDENTIFIED	**Average**		**Standard deviation**		**T-test *P*-value**
	1	2	1	2	
(1) Stratigraphic Excavation versus (2) Control Excavation	9.5	7.4	0.85	0.52	0.00000289
(1) Demirant Excavation versus (2) Control Excavation	9.8	7.4	0.42	0.52	0.000000001
(1) Quadrant Excavation versus (2) Control Excavation	9.8	7.4	0.63	0.52	0.00000003
(1) Arbitrary Level Excavation versus (2) Control Excavation	6.9	7.4	0.32	0.52	0.01768
(1) Stratigraphic Excavation versus (2) Demirant Excavation	9.5	9.8	0.85	0.42	0.33056
(1) Stratigraphic Excavation versus (2) Quadrant Excavation	9.5	9.8	0.85	0.63	0.38232
(1) Stratigraphic Excavation versus (2) Arbitrary Level Excavation	9.5	6.9	0.85	0.32	0.00000004
(1) Demirant Excavation versus (2) Quadrant Excavation	9.8	9.8	0.42	0.63	1.00000
(1) Demirant Excavation versus (2) Arbitrary Level Excavation	9.8	6.9	0.42	0.32	0.000000000001
(1) Quadrant Excavation versus (2) Arbitrary Level Excavation	9.8	6.9	0.63	0.32	0.0000000001
FORMATION STAGES IDENTIFIED (OVERALL PERFORMANCE)	**Average**		**Standard deviation**		**T-test *P*-value**
	1	2	1	2	
(1) Stratigraphic Excavation versus (2) Control Excavation	18.2	12.5	4.13	2.46	0.00147
(1) Demirant Excavation versus (2) Control Excavation	18.9	12.5	3.90	2.46	0.00035
(1) Quadrant Excavation versus (2) Control Excavation	18.2	12.5	3.19	2.46	0.00029
(1) Arbitrary Level Excavation versus (2) Control Excavation	9.5	12.5	2.27	2.46	0.01105
(1) Stratigraphic Excavation versus (2) Demirant Excavation	18.2	18.9	4.13	3.90	0.70139
(1) Stratigraphic Excavation versus (2) Quadrant Excavation	18.2	18.2	4.13	3.19	1.00000
(1) Stratigraphic Excavation versus (2) Arbitrary Level Excavation	18.2	9.5	4.13	2.27	0.0000158
(1) Demirant Excavation versus (2) Quadrant Excavation	18.9	18.2	3.90	3.19	0.66566
(1) Demirant Excavation versus (2) Arbitrary Level Excavation	18.9	9.5	3.90	2.27	0.000003
(1) Quadrant Excavation versus (2) Arbitrary Level Excavation	18.2	9.5	3.19	2.27	0.00000148

FIGURE 169: OVERALL PERFORMANCE OF EACH EXCAVATION METHOD AGAINST ALL ANALYTICAL CRITERIA (STATISTICAL ANALYSIS).

Stratigraphic Excavation (hours)		Demirant Excavation (hours)		Quadrant Excavation (hours)		Arbitrary Level Excavation (hours)		Control Excavation (hours)	
Archaeologist 01	4	Archaeologist 11	14	Archaeologist 21	3.5	Archaeologist 31	3	Control 01	5
Archaeologist 02	14	Archaeologist 12	4	Archaeologist 22	4	Archaeologist 32	2.25	Control 02	2
Archaeologist 03	14	Archaeologist 13	4	Archaeologist 23	7.5	Archaeologist 33	2.75	Control 03	1.5
Archaeologist 04	2	Archaeologist 14	4	Archaeologist 24	6	Archaeologist 34	3.5	Control 04	2
Archaeologist 05	4	Archaeologist 15	4	Archaeologist 25	3	Archaeologist 35	3.25	Control 05	2
Archaeologist 06	2	Archaeologist 16	3.5	Archaeologist 26	4	Archaeologist 36	2	Control 06	2.25
Archaeologist 07	2.5	Archaeologist 17	3.5	Archaeologist 27	3.5	Archaeologist 37	2	Control 07	2.5
Archaeologist 08	3.5	Archaeologist 18	5.5	Archaeologist 28	3.25	Archaeologist 38	2.5	Control 08	3
Archaeologist 09	3.5	Archaeologist 19	4	Archaeologist 29	4	Archaeologist 39	2	Control 09	2.5
Archaeologist 10	4	Archaeologist 20	4	Archaeologist 30	6	Archaeologist 40	2	Control 10	2.25
AVERAGE HOURS:	5.35	AVERAGE HOURS:	5.05	AVERAGE HOURS:	4.47	AVERAGE HOURS:	2.52	AVERAGE HOURS:	2.5

FIGURE 170: INDIVIDUAL TIME TAKEN AND AVERAGE TIME TAKEN TO INVESTIGATE THE GRAVE SIMULATION © EVIS 2016.

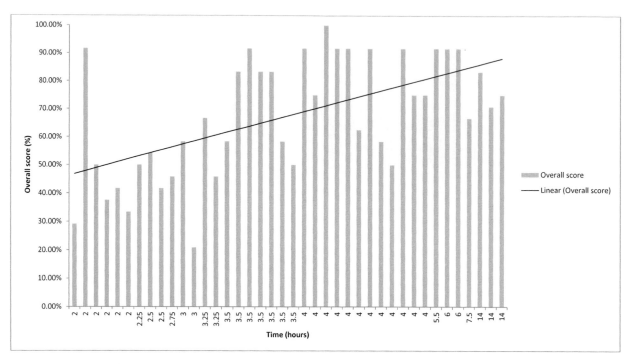

FIGURE 171: TIME SPENT INVESTIGATING THE GRAVE SIMULATION AGAINST OVERALL PERFORMANCE © EVIS 2016.

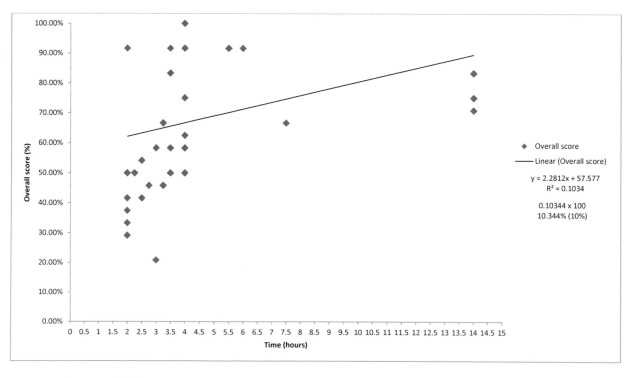

FIGURE 172: TIME SPENT INVESTIGATING THE GRAVE SIMULATION AGAINST OVERALL PERFORMANCE
(LINEAR REGRESSION ANALYSIS) © EVIS 2016.

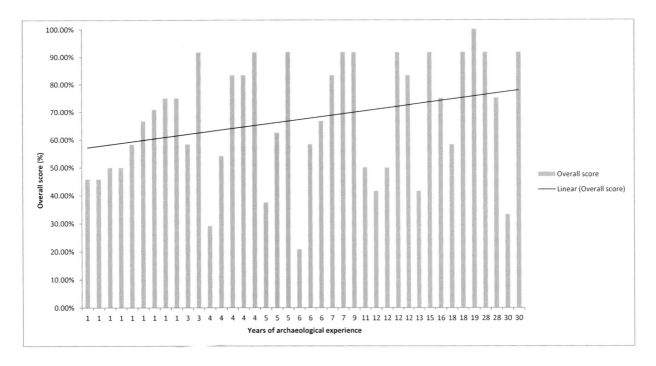

FIGURE 173: YEARS OF ARCHAEOLOGICAL EXPERIENCE AGAINST OVERALL PERFORMANCE © EVIS 2016.

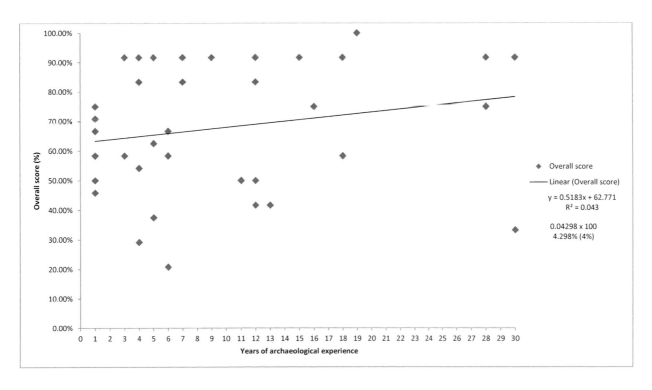

FIGURE 174: YEARS OF ARCHAEOLOGICAL EXPERIENCE AGAINST OVERALL PERFORMANCE (LINEAR REGRESSION ANALYSIS) © EVIS 2016.

The selection of archaeological excavation methods and recording systems for forensic investigations

Material evidence found using the Stratigraphic Excavation method

An average of 71.11% of material evidence was recovered using the Stratigraphic method of excavation (Figure 147; Figure 154), although the total material evidence retrieval rate varied between 44.44% and 100% amongst each of the participants (Figure 57; Figure 61; Figure 65).

Each of the participants identified material evidence both in and out of situ. Material evidence that was identified out of situ was recovered either by the participants when sieving the extracted spoil, or whilst they were transferring spoil to tarpaulins. An average of 18.75% of material evidence was found out of situ, although, the percentage of material evidence found out of situ varied from 0% and 44.44% (Figure 51; Figure 57; Figure 61; Figure 63; Figure 65; Figure 154). Despite finding material evidence out of situ, due to the participants using the Stratigraphic Excavation approach, the participants were able to reassociate the material evidence recovered with the individual contexts they had defined, and were thus able to place these items within a stratigraphic sequence and determine their relative depositional chronology. However, as Archaeologist 03, Archaeologist 06, and Archaeologist 07 failed to define all ten contexts present within the grave structure, they subsequently associated some of the recovered material evidence with incorrect contexts, making their reconstruction of the stratigraphic sequence and overall interpretation of the material evidence deposition sequence incorrect (Figure 155). However, the extent to which their reconstructions were incorrect varied in accordance to the number of contexts each participant identified. Therefore, Archaeologist 03 who identified nine contexts correctly was able to define the material evidences' relative depositional sequence the most accurately, then Archaeologist 06 and Archaeologist 07 both of whom identified eight contexts correctly (Figure 52; Figure 58; Figure 60).

Some material evidence failed to be recovered whilst utilising the Stratigraphic Excavation method. The percentage of evidence not recovered averaged at 28.89%, however, between participants the percentage of evidence not recovered varied from 0% and 55.56% (Figure 57; Figure 61; Figure 65; Figure 147; Figure 154). The failure of Archaeologist 06 and Archaeologist 08 to identify the two pence coin, and Archaeologist 06, Archaeologist 07 and Archaeologist 08 to identify the fake nail can be explained by the fact that these participants chose not to sieve during the excavation experiment. However, 90% of the participants failed to identify the kirby grip and 60% of participants failed to identify the earrings placed into the grave (Figure 147; Figure 152). The reason why the majority of participants did not find these items can be attributed to the size of these items, which made them difficult to identify whilst excavating and sieving the soil. Although, as all of the graves contained exactly the same evidence and contexts, each participant had an equal chance of retrieving the material evidence present, particularly the six participants that chose to sieve as they excavated, as Archaeologist 10 demonstrated by recovering all of the material evidence placed into the grave. Therefore, the failure of these participants to retrieve all of the material evidence present cannot be attributed to the method, as all of the contexts removed from the grave were sieved by these participants, therefore, any evidence not retrieved in situ, should have been retrieved whilst they were sieving. Consequently, the failure of these participants to retrieve material evidence must be attributed to the individual archaeologist and the care and attention they paid whilst excavating and sieving.

Material evidence found using the Demirant Excavation method

An average of 73.33% of material evidence was recovered using the Demirant method of excavation (Figure 148; Figure 154), although the total material evidence retrieval rate varied between 44.44% and 88.89% amongst each of the participants (Figure 69; Figure 71; Figure 75; Figure 77; Figure 81; Figure 83; Figure 85).

As found in the Stratigraphic Excavation experiments, material evidence was recovered both in and out of situ whilst participants were using the Demirant Excavation method (Figure 148; Figure 153; Figure 154). As with the Stratigraphic Excavation experiments, material evidence that was found out of situ was found either when the participants were sieving the extracted spoil, or whilst they were transferring spoil to tarpaulins. An average of 12.12% of material evidence was found out of situ, however, the percentage of material evidence found out of situ varied from 0% and 25% (Figure 148; Figure 69; Figure 71; Figure 73; Figure 75; Figure 77; Figure 81; Figure 83). As with the Stratigraphic Excavation method, despite participants finding evidence out of situ, because the Demirant Excavation method ensures that individual contexts are excavated and recorded separately, the participants were able to reassociate the material evidence that was recovered with the contexts from which they had originated. However, Archaeologist 14 and Archaeologist 20 failed to identify all ten contexts correctly, each missing one (Figure 74; Figure 86; Figure 156). Therefore, as with the participants who missed contexts in the Stratigraphic Excavation experiments, these individuals' reconstructions of the stratigraphic sequence and material evidence deposition sequence were incorrect (Figure 74; Figure 86).

As was found with the Stratigraphic Excavation experiments, some material evidence failed to be recovered during the Demirant Excavation experiments (Figure 148; Figure 154). An average of 26.67% of material evidence was not identified during the experiments, although the amount of material evidence not recovered varied between 11.11% and 55.56% (Figure 148; Figure 152; Figure 154; Figure 69; Figure 71; Figure 75; Figure 77; Figure 81; Figure 83; Figure 85). As was found in the Stratigraphic Excavation experiments, the failure of Archaeologist 14, Archaeologist 19, and Archaeologist 20 to find items such as the two pence coin, the fake nail and the packet of cigarette papers is due to the fact that these participants did not attempt to sieve the spoil during the excavation process (Figure 73; Figure 83; Figure 85). Again, however, there was a trend amongst participants, even amongst those who sieved, for smaller material evidence items such as the earrings and the kirby grip not to be located (Figure 148; Figure 152). This is likely to be due to the care and attention that individual participants paid whilst excavating and sieving.

Material evidence found using the Quadrant Excavation method

An average of 71.11% of material evidence was recovered using the Quadrant Excavation method (Figure 149; Figure 154). Although, the recovery rate of material evidence varied between 44.44% and 88.89% (Figure 89; Figure 91; Figure 93; Figure 95; Figure 99; Figure 105).

As with both the Stratigraphic Excavation and Demirant Excavation experiments, participants identified material evidence both in and out of situ, and such evidence was found whist the participants were sieving or when they were transferring spoil to the tarpaulins. An average of 6.25% of evidence was found out of situ, but the percentage of evidence found out of situ varied between 0% and 25% (Figure 87; Figure 89; Figure 91; Figure 93; Figure 95; Figure 99; Figure 101; Figure 103; Figure 105; Figure 149; Figure 153; Figure 154). As with both the Stratigraphic and Demirant methods of excavation, because this methodological approach ensures that each context is defined and recorded separately, the fact that material evidence was found out of situ was irrelevant, as the participants were able to reassociate these items with the contexts from which they had originated and determine the material evidence deposition sequence. However, as with the Stratigraphic and Demirant Excavation experiments, some contexts were not identified. Archaeologist 23 only identified eight out of the ten contexts present in the grave, and therefore, this participant's reconstruction of the stratigraphic and material evidence deposition sequence was incorrect (Figure 92; Figure 157).

As with the Stratigraphic and Demirant Excavation experiments, some participants failed to identify items of material evidence during the Quadrant Excavation experiments (Figure 149; Figure 154). On average, 28.89% of material evidence failed to be found (Figure 149; Figure 154). The amount of material evidence that was not found varied between 11.11% and 55.56% (Figure 89; Figure 91; Figure 93; Figure 95; Figure 99; Figure 105). The failure of Archaeologist 25, Archaeologist 27 and Archaeologist 28 to recover items such as the two pence coin and the fake nail can be attributed to the fact that these participants failed to sieve the spoil (Figure 95; Figure 99; Figure 101). However, seven out of the ten participants did not sieve the spoil, and yet, the majority managed to locate these items, so the fact that these participants failed to locate these items is probably due to these participants not being particularly observant. Moreover, as was found with the Stratigraphic and Demirant Excavation experiments, the material evidence items that were not located tended to be small in size, such as the earrings and the kirby grip, however, as some of the participants did manage to find these items even when they did not sieve, one must again attribute the participants' failure to locate such items to the participants' lack of attention when excavating and sieving the soil (Figure 149; Figure 152).

Material evidence found using the Arbitrary Level Excavation method

Using the Arbitrary Level Excavation method an average of 51.11% of material evidence was recovered (Figure 150; Figure 154). The total material evidence retrieval rate varied between 22.22% and 77.78% (Figure 107; Figure 113).

As with the previous excavation experiments, material evidence was recovered both in and out of situ whilst the participants were using the Arbitrary Level Excavation method. As not one of the participants using the Arbitrary Level Excavation method sieved the spoil, material evidence items that were found out of situ were identified as the participants were transferring the spoil to the tarpaulin. On average, 4.35% of material evidence was found out of situ, however, the amount of evidence found out of situ varied between 0% and 50% (Figure 150; Figure 153; Figure 154; Figure 107; Figure 109; Figure 111; Figure 113; Figure 115; Figure 117; Figure 119; Figure 123; Figure 125). Although, unlike in the previous excavation experiments, all of the participants who used the Arbitrary Level Excavation method failed to identify all of the ten contexts present in the grave, and therefore, reassociated some of the recovered material evidence with incorrect contexts, making all of the Arbitrary Level Excavation participants' reconstructions of both the stratigraphic sequence and material evidence deposition sequence incorrect (Figure 158; Figure 108; Figure 110; Figure 112; Figure 114; Figure 116; Figure 118; Figure 120; Figure 122; Figure 124; Figure 126).

As with the previous excavation experiments, some material evidence failed to be recovered by participants using the Arbitrary Level Excavation method (Figure 150; Figure 152; Figure 154). An average of 48.89% of material evidence failed to be identified, although the percentage varied between 22.22% and 77.78% (Figure 107; Figure 113; Figure 150). The type of evidence that tended to be lost was again small in size, such as the earrings and the kirby grip (Figure 150). Although, participants using this method also failed to recover larger items, such as the cigarette papers, two pence coin, fake nail and ID card (Figure 150). The failure to recover both the smaller and larger items can be attributed to the fact that not one of the participants attempted to sieve the spoil. Furthermore, the participants using this method rapidly removed the fills contained within the grave, removing hand shovelfuls of spoil at a time, and made barely any attempts to inspect the spoil for the presence of material evidence, and so this too may account for the failure of these archaeologists to recover as much material evidence as the other methodological approaches did.

Material evidence found by the Control participants

An average of 66.67% of material evidence was recovered by the Control participants (Figure 151). Although, this varied between 44.44% and 100% (Figure 131; Figure 154; Figure 135; Figure 139; Figure 141).

As with all of the previous excavation experiments, participants found evidence in and out of situ. An average of 28.33% of evidence was found out of situ, although this varied between 0% and 57.14% (Figure 151; Figure 153; Figure 154; Figure 127; Figure 131; Figure 135; Figure 141). As was found in the Arbitrary Level Excavation experiments, all of the Control participants failed to identify all of the ten contexts present in the grave, and therefore, reassociated some of the recovered material evidence with the wrong contexts, resulting in all of the Control participants' stratigraphic sequences and material evidence deposition sequences being incorrect (Figure 159; Figure 128; Figure 130; Figure 132; Figure 134; Figure 136; Figure 138; Figure 140; Figure 142; Figure 144; Figure 146).

During the experiment, some material evidence failed to be located. On average 33.33% of the material evidence was not found, although, this varied between 0% and 55.56% (Figure 151; Figure 131; Figure 135; Figure 139; Figure 141; Figure 152; Figure 154). As was found in all of the previous excavation experiments, it was smaller items such as the earrings and the kirby grip that were not found (Figure 151). Although, as with the Arbitrary Level Excavation experiments, some participants failed to find larger items such as the cigarette papers, two pence coin and the fake nail (Figure 151). In part, this can be attributed to the fact that not all of the Control participants sieved the spoil, but also again, like with the Arbitrary Level Excavation experiments, many participants used hand shovels, spades and shovels to remove the fills of the grave and did not attempt to inspect the spoil for the presence of material evidence, resulting in them failing to find many of the items. Nevertheless, as with the Stratigraphic, Demirant, and Quadrant Excavation experiments some of the participants who sieved the spoil still failed to recover all of the material evidence items, therefore, their failure to find some of the material evidence placed into the grave must be due to the fact that they did not pay enough attention when excavating and sieving (Figure 151).

By comparing the results of each of the excavation methods it is clear that the Demirant Excavation method recovered the most material evidence, with an average result of 73.33% of the material evidence being found. However, the Stratigraphic and Quadrant Excavation methods were close behind both recovering an average of 71.11% of the material evidence. Statistically, the differences between the Demirant Excavation method, Quadrant Excavation method and Stratigraphic Excavation method material evidence recovery rates were not significant (Figure 169). As expected, the Control participants recovered less material evidence than these three methods, having found an average of 66.67% of the material evidence, however, this difference was found not to be statistically significant (Figure 169). The descriptive statistical difference between the results of the Control participants and the three other excavation methods results can be explained by the fact that these participants had no archaeological training or archaeological experience. However, what is surprising is that the Arbitrary Level Excavation approach recovered the least amount of material evidence, finding an average of 51.11% of the material evidence present, these results were also found to exhibit statistically significant differences from those obtained by participants using the Stratigraphic Excavation method, Demirant Excavation method and Quadrant Excavation method (Figure 169). The differences exhibited between the Control participants and the participants using the Arbitrary Level Excavation method were found not to be statistically significant (Figure 169).

When comparing the range of variation in the overall amount of material evidence recovered by participants using each of the excavation methods, it is apparent that the Demirant and the Quadrant Excavation methods had the smallest amount of variability in their results, both having a range of 44.45%. In comparison, the Stratigraphic Excavation method, the Arbitrary Level Excavation method, and the Control participants had the largest variation range, totalling 55.56%.

In terms of in situ recovery of material evidence, the Arbitrary Level Excavation method had the highest in situ recovery rate, with 95.65% of material evidence being found in situ. The Quadrant Excavation method then followed, with an average of 93.75% of material evidence being found in situ. The Demirant Excavation method was third, with an average of 87.88% of material evidence being found in situ. The Stratigraphic Excavation method found the least items of material evidence in situ out of the archaeological methods tested, recovering an average of 81.25% of material evidence in situ. Finally, the Control participants found an average of 71.67% of material evidence in situ, which again, is not unexpected as these participants had no archaeological training or archaeological knowledge. However, the differences between the Stratigraphic Excavation method, Demirant Excavation method, Quadrant Excavation method and Arbitrary Level Excavation method were found to have no statistical significance (Figure 169). The only statistically significant results found when in situ material evidence recovery was examined were between the results of the participants using the Demirant Excavation method and the Control participants, and the participants using the Quadrant Excavation method and the Control participants (Figure 169). This is due to the fact that participants using the Demirant Excavation method and Quadrant Excavation method recovered a large number of material evidence items and found them in situ, in comparison to the other methodological approaches tested.

However, despite the Arbitrary Level Excavation method recovering the highest percentage of material evidence in situ, all participants using this method failed to correctly identify the complete set of contexts present in the grave. This means that the participants using this method, despite finding the highest percentage of material evidence in situ, associated such material evidence items with incorrect contexts, and therefore, incorrectly reconstructed both the stratigraphic sequence and the deposition sequence of the material evidence. This criticism is also relevant to Archaeologist 03, Archaeologist 06 and Archaeologist 07 using the Stratigraphic Excavation method, Archaeologist 14 and Archaeologist 20 using the Demirant Excavation method, Archaeologist 23 using the Quadrant Excavation method and all of the Control participants. However, for all of the Stratigraphic Excavation method, Demirant Excavation method, and Quadrant Excavation method participants who did correctly identify all of the contexts present in the grave, the fact that they failed to recover all of the material evidence in situ is irrelevant, as they identified, defined, excavated, and recorded each of the separate contexts present in the grave individually, and therefore, any material evidence that was found out of situ whilst excavating any one of these contexts could be put back into its place within both the stratigraphic sequence and material evidence deposition sequence of the grave.

When these results are paired with those of Evis (2009), Pelling (2008) and Tuller and Đurić (2006), each of whom tested two of the excavation methods evaluated in this research project – the Stratigraphic Excavation method and the Arbitrary Level Excavation method, their findings correlate with those of this study. They found that the Stratigraphic Excavation method recovered more material evidence than the Arbitrary Level Excavation method, and that the Stratigraphic method of excavation enabled the archaeologists to reconstruct the deposition sequence of material evidence more accurately than the Arbitrary Level Excavation method (Evis 2009; Pelling 2008; Tuller and Đurić 2006). However, not one of these aforementioned studies tested the Demirant Excavation method or the Quadrant Excavation method. The results obtained for material evidence recovery using these two previously untested methods indicate that the Demirant method had the highest material evidence recovery rate of all of the methods tested, although, the Quadrant Excavation method did achieve the same recovery rate as the Stratigraphic Excavation method. Additionally, both the Demirant and the Quadrant Excavation methods had the most consistent rate of evidence recovery amongst all of the excavation methods. Furthermore, as discussed earlier, as these two methods ensure that the boundaries of contexts are defined, maintained, and recorded throughout the excavation process, they, as with the Stratigraphic Excavation method, ensure that the deposition sequence of the material evidence is more accurately recorded than with the Arbitrary Level Excavation method. The Quadrant Excavation method also recovered the second highest percentage of material evidence in situ, followed by the Demirant Excavation method, whereas the Stratigraphic Excavation method came last out of the excavation methods tested. Therefore, on the basis of the data gained in this research project, the Demirant and Quadrant Excavation methods proved to be more productive, in terms of material evidence recovery, than both the Stratigraphic and Arbitrary Level Excavation methods.

Contexts identified using the Stratigraphic Excavation method

When using the Stratigraphic Excavation approach, each of the participants proceeded to remove individual fills, defined by differences in texture, composition, volume, compactness and colouration in the reverse order in which they were deposited, from the latest to the earliest. Each participant would define and remove individual fills in their entirety, and would then complete context forms and draw plans of each of the fills that they had identified. Through maintaining the boundaries of the individual fills within the grave, the participants also preserved the grave cut, and as a result, at the end of the excavation, were able to plan and photograph its dimensions. Through following this approach, it ensured that the participants were able to define the different stages of the grave's formation process.

An average of 95% of the contexts present in the grave were correctly identified by the participants using the Stratigraphic Excavation method (Figure 155; Figure 160; Figure 161). However, the number of contexts that were correctly identified varied from eight (80%) to ten (100%) (Figure 48; Figure 50; Figure 54; Figure 56; Figure 58; Figure 60; Figure 62; Figure 64; Figure 66). Although, seven out of ten participants successfully identified all of the contexts present in the grave. The three participants who failed to identify all of the contexts present were Archaeologist 03 who identified 9 (90%) contexts correctly, Archaeologist 06 who identified 8 (80%) contexts correctly, and Archaeologist 07 who again identified 8 (80%) contexts correctly (Figure 52; Figure 58; Figure 60).

The reason why Archaeologist 03 failed to identify context 8 (fill 4) was due to the rapidity with which context 9 (fill 5) was removed. The participant presumed, based on previous experience of excavating graves, that the grave would lack complex stratigraphy and therefore proceeded to excavate context 9 (fill 5) rapidly, using both a hand shovel and trowel, this resulted in the participant intermixing context 8 (fill 4) and context 9 (fill 5) whilst excavating, resulting in the participant failing to identify or record context 8 (fill 4).

The reason why Archaeologist 06 failed to identify context 8 (fill 4) and context 5 (fill 5) can again be attributed to the way in which this archaeologist chose to excavate the grave. Having started the excavation of context 9 (fill 5) using a trowel, the participant adapted their approach, approximately half way through context 9s (fill 5) excavation, and used a hand shovel and spade to remove the fills within the grave. This resulted in the participant digging straight through context 8 (fill 4). However, having noticed the presence of context 7 (fill 3), the participant went back to using a trowel, and was then able to successfully document context 7 (fill 3) and context 6 (fill 2). However, when excavating context 6 (fill 2), which was constructed of sand, the participant went back to using a spade to excavate the fill, this resulted in the participant failing to identify that, despite context 6 (fill 2) and context 5 (fill 1) being constructed of the same type of fill (sand), these two contexts were stratigraphically distinct, and were not joined in any way. Therefore, by using the spade to excavate context 6 (fill 2), the space between these two contexts became contaminated with sand, thus making the stratigraphic distinctness of these two contexts difficult to discern, and resulted in the participant classifying context 6 (fill 2) and context 5 (fill 1) as being one and the same.

The reason why Archaeologist 07 failed to identify context 8 (fill 4) and context 6 (fill 2) can again be attributed to the manner in which this participant excavated the grave. This participant decided to use a mattock and a shovel to remove the fills of the grave, stating that this was the approach that the participant had been taught to use when excavating cut features in commercial contexts. Through using a mattock to excavate context 9 (fill 5) the participant dug straight through context 8 (fill 4). However, as with Archaeologist 06, the participant noticed that there was a change in fill composition and so began to use a trowel and therefore successfully identified context 7 (fill 3). The participant then identified context 5 (fill 1) and having uncovered a portion of context 6 (fill 2) after the removal of context 7 (fill 3) presumed, like

Archaeologist 07 had done, that these two separate contexts were one and the same. The participant then decided to revert back to using a shovel and a mattock to remove these fills and therefore failed to define the space between these two contexts, as they had become intermixed and their boundaries conjoined, leading to the participant recording these two contexts as one.

Therefore, the failure of some participants to successfully identify all of the contexts present in the grave using the Stratigraphic Excavation method should not be attributed to the methodology itself, as 70% of the participants managed to identify all of the contexts successfully. Consequently, the failure of some participants to identify the contexts present in the grave must be attributed to individual archaeologists and their ability to choose the right equipment for the job, their rushed approach, their lack of attention to detail and visual skills.

Contexts identified using the Demirant Excavation method

When using the Demirant Excavation approach, each of the participants defined the boundaries of the grave, set up a section line across the middle of the grave, and then proceeded to excavate the first half of the grave, then record the half section and complete context sheets, and then remove the remaining half of the grave. Individual fills were defined by differences in texture, composition, volume, compactness and colouration, and were excavated in the reverse order in which they were deposited, from the latest to the earliest. However, as the majority of the participants divided the grave across its width, the contexts in the base of the grave context 5 (fill 1) and context 6 (fill 2) which sloped downwards, often did not reach the point at which the participant had set up their half section, resulting in the participants having to plan and complete context forms for these contexts, as these contexts did not appear in the half section. Despite this, by defining, maintaining, and recording individual contexts during the excavation of the first half of the feature, and then confirming their presence in the half section, and then again, whilst excavating the second half of the grave, participants were able to preserve the boundaries of the grave cut and were able to define the different stages of the grave's formation process.

An average of 98% of the contexts present in the grave were correctly identified by the participants using the Demirant Excavation method (Figure 156; Figure 160; Figure 161). However, the number of contexts that were correctly identified varied from nine (90%) to ten (100%) (Figure 68; Figure 70; Figure 72; Figure 74; Figure 76; Figure 78; Figure 80; Figure 82; Figure 84; Figure 86). Although eight out of ten participants successfully identified all of the contexts present in the grave. The two participants who failed to identify all of the contexts present were Archaeologist 14 who identified 9 (90%) contexts correctly and Archaeologist 20 who also identified 9 (90%) contexts correctly (Figure 74; Figure 86).

The reason why Archaeologist 14 failed to identify context 5 (fill 1) was due to the fact that when the participant was excavating the first half of the feature, context 6 (fill 2) didn't reach completely into the half section, sloping down to millimetre thickness at the half section point. Therefore, when the participant was excavating the second half of the grave, and came upon context 5 (fill 1), composed of the same sand as context 6 (fill 2), the participant mistakenly thought that they had accidentally dug through the sand that had connected these two deposits, as the participant could not recall the exact point to which context 6 (fill 2) had reached. This resulted in the participant misclassifying these two different contexts as one.

The reason why Archaeologist 20 failed to identify context 6 (fill 2) was due to the fact that when this participant excavated the grave they chose to divide the grave in half across its length. Although this meant that context 5 (fill 1) and context 6 (fill 2) were preserved in the half section, when the participant was removing these two contexts in the first half of the grave they moved the sand, of which both of these contexts were composed, across the space between these two contexts and across the half section. As the participant failed to clean the half section properly, this meant that these two contexts appeared to be connected, and therefore the participant misclassified these two separate contexts as being one context.

Therefore, the failure of these two participants to successfully identify all of the contexts present in the grave using the Demirant Excavation approach should in part be attributed to the method, as through dividing the grave in half it meant that all of the contexts present in the grave could not be seen or analysed in their entirety individually, increasing the possibility for participants to miss certain contexts during the excavation. However, as 80% of the participants successfully identified all of the contexts present in the grave, the major factors leading to the two participants missing the contexts was their failure to record all of the contexts in sufficient detail to determine their boundaries, and also, their failure to follow procedure, and clean the half section face prior to recording.

Contexts identified using the Quadrant Excavation method

When using the Quadrant Excavation approach, each of the participants defined the boundaries of the grave, set up a section line across the length and width of the grave, and then proceeded to excavate the first quadrant of the grave. They then recorded the long section and half section of the grave and completed context sheets. The participants would then excavate the opposing quadrant and updated and completed the long section and half section drawings and their context sheets. Each participant would then excavate the remaining two quadrants and add any new data to their context sheets. Individual fills were defined by differences in texture, composition, volume, compactness and colouration, and were excavated in the reverse order in which they were deposited, from latest to earliest. One problem noted by participants using this technique, was the tendency for the quadrants to crumble if they were left standing with a sharp vertical edge. However, this problem was resolved by slightly angling the edges of the quadrants, which provided them with more stability. Despite this issue, through the participants being careful to define, record, and maintain individual contexts during the excavation, they were able to maintain the grave cut and accurately determine the formation sequence of the grave. By completing both the long section and half section drawings the archaeologists also had a visual record of the grave's formation sequence.

An average of 98% of the contexts contained within the grave were identified by the participants using the Quadrant Excavation method (Figure 157; Figure 160; Figure 161). However, the number of contexts that were correctly identified varied from eight (80%) to ten (100%) (Figure 88; Figure 90; Figure 92; Figure 94; Figure 96; Figure 98; Figure 100; Figure 102; Figure 104; Figure 106). Although, nine out of ten participants successfully identified all of the contexts present in the grave. The participant who failed to identify all of the contexts was Archaeologist 23 who identified 8 (80%) contexts correctly (Figure 92).

The reason why Archaeologist 23 failed to identify context 7 (fill 3) and context 8 (fill 4) is due to the fact that the participant chose to use a hand shovel to remove context 9 (fill 5). This resulted in the participant excavating straight through context 8 (fill 4) and context 7 (fill 3), as the participant believed that these contexts represented a mixed singular backfill. Even when the participant was excavating the three other quadrants, they still continued to use the hand shovel resulting in the removal of large volumes of soil and the continual failure to define these two additional contexts. It was only when the participant reached context 5 (fill 1) and context 6 (fill 2) that they chose to use a trowel, and as a result, successfully determined that these two contexts were stratigraphically distinct.

Therefore, the failure of this one participant to identify all of the contexts contained within the grave cannot be attributed to the excavation method, as all of the other participants successfully identified and recorded all of the fills contained within the grave. Instead, this participant's failure to identify all of the contexts within the grave is due to the fact that they chose to use the wrong equipment to excavate the grave.

Contexts identified using the Arbitrary Level Excavation method

When using the Arbitrary Level Excavation approach all but one of the participants maintained the boundaries of the grave. One participant, Archaeologist 31 chose to create an excavation unit instead, which meant that both the grave and the 1m x 1.5m excavation unit surrounding it were excavated. Each of the archaeologists excavated the grave by removing 10cm spits at a time, after which, they would draw a plan of the grave and complete a unit level recording form. This process was then repeated until the participants reached the base of the grave.

Through excavating the grave in this manner, the participants identified an average of 69% of contexts correctly (Figure 158; Figure 160; Figure 161). However, the number of contexts that were correctly identified varied from six (60%) to seven (70%) (Figure 108; Figure 110; Figure 112; Figure 114; Figure 116; Figure 118; Figure 120; Figure 122; Figure 124; Figure 126). All participants failed to identify all of the contexts within the grave correctly. With nine out of ten of the participants only identifying 7 of the contexts (Figure 110; Figure 112; Figure 114; Figure 116; Figure 118; Figure 120; Figure 122; Figure 124; Figure 126). The participant who failed to identify the additional context, the grave cut, was Archaeologist 31, the participant who chose to create and excavate an entire excavation unit, and as a result destroyed and failed to define the grave cut (context 4) (Figure 108).

Due to the fact that not one of the participants using the Arbitrary Level Excavation approach managed to identify all of the contexts present in the grave, and that there was little variation in the number of contexts that were successfully identified using this method, it would appear that the inability of the participants to define all of the contexts within the grave is attributable to the methodology itself. This is due to the fact that by excavating in 10cm spits, the participants unwittingly introduced artificial divisions in the formation sequence of the grave. This, in turn, meant that the participants were only able to record three of the fills within the grave, as only three spits had to be excavated to reach the base of the grave. The fills that were recorded were those that corresponded to the base of the spit that had been excavated. By excavating in this manner the participants intermixed the various contexts within the grave, and in turn, also failed to accurately define the angled dimensions of each of the fills within the grave. Consequently, this led to all of the participants failing to accurately excavate and interpret the formation sequence of the grave.

Contexts identified by the Control participants

Although the Control participants had no archaeological training or archaeological knowledge, they successfully identified an average of 74% of the contexts within the grave (Figure 159; Figure 160; Figure 161). However, the number of contexts that were correctly identified varied from seven (70%) to eight (80%) (Figure 128; Figure 130; Figure 132; Figure 134; Figure 136; Figure 138; Figure 140; Figure 142; Figure 144; Figure 146).

Similarly to the Arbitrary Level Excavation participants, not one of the Control participants managed to correctly identify all of the contexts present in the grave. This is not unsurprising as not one of the Control participants had any archaeological training or archaeological knowledge. Therefore, they had not been trained to detect and define different contexts. What is perhaps most surprising is that Control 04, Control 05, and Control 10 all correctly established that context 5 (fill 1) and context 6 (fill 2) were separate contexts and classified them as such, which several of the trained archaeologists failed to do (Figure 134; Figure 136; Figure 146). Another explanation for why the Control participants failed to detect all of the contexts within the grave was due to the equipment that they chose to use, with the majority using hand shovels, spades and shovels to remove the fills of the grave. This resulted in large volumes of soil being removed from the grave at a time, making all but the boundaries of the most obvious fills - context 5 (fill 1) and context 6 (fill 2), the sand deposits, hard to differentiate.

When comparing the results of each of the excavation methods against one another, it is apparent that the Quadrant Excavation method was the most successful at identifying contexts, as on average it correctly identified the most contexts, with only one participant failing to recover all of the contexts within the grave. However, on average, the Demirant Excavation method recovered the same number of contexts as the Quadrant Excavation method, but due to the fact that two participants failed to accurately define the stratigraphic sequence, rather than one, it would suggest that this method has an increased potential for archaeologists to miss contexts whilst excavating. The Stratigraphic Excavation method proved to be the third most successful technique at recovering contexts within the grave structure, with three participants failing to identify all of the contexts. This result would suggest that this technique is, again, more susceptible to errors whilst excavating when compared with the two aforementioned techniques. However, as discussed earlier, if the right methodological approach is taken when using these three techniques it is possible for archaeologists to recover and document all of the contexts present successfully. This is supported by the statistical analyses of these three techniques against each other, which found that there was not a statistically significant difference between the numbers of contexts identified (Figure 169). The least successful technique for identifying contexts was the Arbitrary Level Excavation method, and considering that even the Control participants results exceeded those of this method, it would suggest that this method is highly unreliable, and leaves the archaeologist unable to accurately reconstruct the stratigraphic sequence of an archaeological feature. This finding is supported by the statistical analyses of the results, which found that the participants who used the Arbitrary Level Excavation method produced results that were significantly different, and in this case, poorer than those participants using the other methodological approaches (Figure 169). Furthermore, the statistical analyses of the number of contexts that were identified also indicated that the Stratigraphic Excavation method, Demirant Excavation method, Quadrant Excavation method and Arbitrary Level Excavation method recovery rates were significantly different to those of the Control participants (Figure 169). This indicates that the Stratigraphic Excavation method, Demirant Excavation method and Quadrant Excavation method were significantly more successful at identifying contexts than both the Control participants and participants using the Arbitrary Level Excavation method (Figure 169). Moreover, it indicates that the Control participants results were significantly different, and in this case, better than participants using the Arbitrary Level Excavation method (Figure 169).

Such findings correlate with the results of Evis (2009), Pelling (2008) and Tuller and Đurić (2006) who found that the Arbitrary Level Excavation method resulted in less contexts being identified than the Stratigraphic Excavation method.

Identification of the formation sequence of the grave

The formation sequence of the grave was broken down into 24 different stages. Each stage represented an activity that occurred when creating the grave, for example, the placement of an artefact would represent one stage, and then the deposition of a fill on top of that artefact would represent another stage. By breaking the formation sequence of the grave down into these different stages it was possible to determine the overall accuracy or score for each of the excavation methods tested in the research project.

The Stratigraphic Excavation method had an overall accuracy rate of 75.83%, however, the overall scores of participants varied between 50% and 100% (Figure 162; Figure 167; Figure 168). In comparison, the Demirant Excavation method had an overall accuracy rate of 78.75%, although, the overall scores of participants varied between 50% and 91.67% (Figure 163; Figure 167; Figure 168). The Quadrant Excavation method had an overall accuracy rate of 75.83%, however, the overall scores of participants varied between 58.33% and 91.67% (Figure 164; Figure 167; Figure 168). Whereas, the Arbitrary Level Excavation method had an overall accuracy rate of 39.58%, but the overall scores of participants varied from 20.83% to 50% (Figure 165; Figure 167; Figure 168). Finally, the Control participants had an overall accuracy rate of 52.50%, although, the overall scores of participants varied between 37.5% and 66.67% (Figure 166; Figure 167; Figure 168).

By comparing the overall accuracy scores for each of the excavation methods, it is evident that the Demirant Excavation method had the highest overall score. The Stratigraphic and Quadrant Excavation methods came next, both achieving the same overall score. The Control participants came in fourth, and then, finally, the Arbitrary Level Excavation method came in last achieving the lowest overall score. Although, when these results were subjected to statistical testing, the overall scores of the Stratigraphic Excavation method, Demirant Excavation method and Quadrant Excavation method were found to not be significantly different from one another (Figure 169). There were, however, significant differences in the results of participants using the Stratigraphic Excavation method, Demirant Excavation method, Quadrant Excavation method, Control participants and those of the participants who used the Arbitrary Level Excavation method (Figure 169). This finding indicates that those participants who used the Arbitrary Level Excavation method during experimental testing produced significantly poorer overall scores than those participants using the other methodological approaches (Figure 169). Furthermore, the statistical analyses of the overall scores indicated that the Stratigraphic Excavation method, Demirant Excavation method, Quadrant Excavation method and Arbitrary Level Excavation method overall scores were significantly different to those of the Control participants (Figure 169). This indicates that the Stratigraphic Excavation method, Demirant Excavation method and Quadrant Excavation method were significantly more successful techniques than both the Control participants and participants using the Arbitrary Level Excavation method (Figure 169). Moreover, it indicates that the Control participants overall results were significantly different, and in this case, better than participants using the Arbitrary Level Excavation method (Figure 169).

However, as the overall scores of the Stratigraphic, Demirant and Quadrant Excavation methods were relatively close – 75.83% and 78.75%, and showed no statistically significant differences, it is important to compare the consistency of these three methods by reviewing the variation in the overall scores achieved using these three methods. By analysing these three methods in this manner, the Quadrant Excavation method proved to be the most consistent with a variation rate of 33.34%. The Demirant Excavation method was the next most consistent with a variation rate of 41.67%, and finally, the Stratigraphic Excavation method was the least consistent with a variation rate of 50%. These findings are supported by the results shown in Figure 169. This data suggests that the Quadrant Excavation method was the most consistent of these three methods, and therefore, the most reliable in terms of recovering the most accurate record of the formation sequence of the grave simulation.

The Single Context Recording system

Whilst excavating using the Stratigraphic Excavation method, the participants used a recording system known as the Single Context Recording system to document their findings. When using this recording system, as each new context was identified and defined a participant would allocate this context a context number, plan it, and then excavate it. Both during and after the excavation of an individual context, the participant would also complete a context form. This form provided each of the participants with prompts to describe the context, note down any material evidence that was located, and take note of any photographs or samples that were taken. Participants tended to take photographs at the start and end of the excavation of an individual context and also when a piece of material evidence was found. The participants would also complete a Harris Matrix on each of these context forms. This matrix documented the stratigraphic relationships that the context that was being documented had with those contexts that stratigraphically preceded it and succeeded it.

In addition, some participants also documented their findings in a journal. The participants would use this journal to note down their theories relating to how the grave was created and what evidence was found. They would then use the notes that were written down in this journal to form their interpretations and construct their narrative of the grave's formation process.

Some participants also created long section drawings of the contexts within the grave by either measuring the dimensions of the different contexts that had adhered to the grave walls, or, by measuring the dimensions of the individual contexts as they were excavated. This resulted in the participants not only having plans of the individual contexts, but also an overall diagram of how the contexts related to one another in the grave.

Overall, the Single Context Recording system provided a comprehensive set of records for each of the contexts that were identified. Although, one disadvantage of this method was the amount of time that was spent on planning each context as it was excavated. Furthermore, for the participants who only planned the individual contexts and did not create long section drawings of the contexts, it could be argued that it was difficult to interpret how the grave was constructed by merely relying on the plan drawings, particularly for those individuals with no archaeological training, as such plan drawings merely depicted a rectangular box with hashers and level measurements on it, making it hard to visualise how the grave was constructed in its entirety. Therefore, this recording system would require participants to spend extra time, in comparison to the other methods, on reconstructing the grave structure after the excavation, by manually or digitally, superimposing each plan drawing on top of one another to create a depiction of the grave in its entirety.

The Standard Context Recording system

Whilst excavating using the Demirant Excavation method and the Quadrant Excavation method, the participants used a recording system known as the Standard Context Recording system to document their findings, although its application varied according to which of the two methodological approaches was being used.

When the participants were using the Demirant Excavation method to excavate the grave simulation, the participants divided the grave in half, and would then excavate one half. Whilst excavating this half they would allocate each new context that was identified a context number and then complete a context form. This context form is the same form as the one used in the Single Context Recording system, and contains the same prompts. However, due to the fact that only one half of the grave is removed at a time, the context form is not completed until both halves and the entirety of each context has been excavated. Some participants would use a different approach, however, and would record separate context forms for each half of the grave that was excavated, and would record the two halves of a singular context using different context numbers, but would record them as the 'same as' the first half of the context that was excavated during the removal of the first half of the grave.

Some participants also decided to use a journal in order to document their findings in a more casual manner. Such journals helped participants keep track of which contexts individual pieces of material evidence were found in across the two halves of the feature, and prevented participants from incorrectly associating material evidence with incorrect contexts. In addition, the journal provided a space in which the participants could jot down theories about how the grave was constructed as they were excavating, which in turn, helped the participants create their narrative of the grave's formation sequence.

In addition to completing the context forms, participants would also record a half section. This half section was drawn once each participant had completed the excavation of one half of the grave and would display the formation sequence of the grave at the point at which they chose to set up their section point. This half section provided a good visual aid to interpreting the stratigraphic sequence of the grave, as it showed how each context related to one another in a single drawing. However, in the case of this experiment, due to the fact that context 5 and context 6 sloped down and did not reach the mid-point of the grave at which the majority of the participants set up their half section, participants using the Demirant Excavation method found that, these contexts did not extend into the half section face, and consequently, did not appear in the half section drawing, meaning that the participants had to record a separate plan drawing of these contexts in order to document their presence. In addition, as the half section drawing only displays the stratigraphic sequence of the grave at one given point in the grave, this type of recording, on its own, fails to document the shape and dimensions of any contexts that slope or vary in shape across the grave structure. Therefore, any individuals wishing to accurately reconstruct and visually depict the formation sequence of the grave must refer to all of the records completed during the excavation, and as with the Single Context Recording system, reconstruct the structure of the grave, either manually or digitally, after the excavation using the drawings and the measurements taken during the excavation process, thus taking up more time and resources than other recording approaches.

When the participants were using the Quadrant Excavation method to excavate the grave simulation, the participants would divide the grave into four quadrants and excavate each separately. When excavating the first quadrant as each new context was identified it would be allocated a context number and a context form would start to be filled out. As stated earlier, with the Standard Context Recording system, the context form is the same as the one used in the Single Context Recording system, however, as with the Demirant Excavation method, participants would not complete an individual context form until all of the quadrants and the entirety of each context had been excavated. Again, similarly to the Demirant Excavation approach, some participants would record separate context forms for each of the four quadrants that were excavated, and would record the four quarters of a singular context using different context numbers, but would document them as being the 'same as' one another.

In addition to recording data using context forms some participants also used journals. Through using a journal the participants were able to keep track of what material evidence was found where and take note of theories relating to how the grave was constructed, that in turn, helped them to form their narrative of the grave's formation process at the end of the investigation.

Participants would also record a half section and long section drawing of the grave. These drawings would be completed after a participant had excavated two opposing quadrants and demonstrated how the grave was constructed across its length and across its width. This ensured that the dimensions of each of the contexts within the grave structure were recorded, and that even if such contexts sloped and did not reach the half section point, as was the case in this experiment, their presence and dimensions would still be captured in the long section drawing. Through recording the grave structure in this manner it meant that the participant was able to determine and demonstrate immediately after the completion of the investigation, how the grave was constructed visually in two diagrams, and subsequently, the participants were not required to spend additional time after the excavation reconstructing the formation sequence of the grave in accurate diagrammatical form. This recording approach also presented the findings in a way that interested parties, such as police officers in the case of a forensic investigation, could easily understand and interpret with very little explanation being required.

The Unit Level Recording system

Whilst excavating using the Arbitrary Level Excavation method the participants used the Unit Level Recording system to document their findings. Similarly to the Single Context Recording system and the Standard Context Recording system, the Unit Level Recording system relies on the use of pro-formas and plans to record data during an excavation. When using the Unit Level Recording system as each 10cm spit of soil was removed the participants would complete a unit level recording form, on which they would describe any deposits, fills, artefacts, or disturbances that they had come across during the excavation of that spit. They would also record any samples and photographs that were taken during the excavation of that particular spit on the unit level recording form.

The participants also drew plans of the excavation unit after each spit was removed, which in the case of this experiment was delineated by the grave walls by nine participants with only one participant choosing to establish an excavation unit outside the boundaries of the grave cut. On this plan of the excavation unit, the participants would draw the spit that was excavated, and record any artefacts, fills, or deposits that were present.

The participants also recorded their findings using journals. Within these journals they would describe what had been found, in terms of material evidence and the different contexts, and write down theories relating to how the grave was constructed in an informal manner. The participants then used this journal to help create their narratives at the end of the investigation.

Although this recording system is systematic in its approach to documenting findings, the fact that the excavation method that is associated with its use results in the destruction of individual contexts as it progresses, particularly those contexts which have any gradient in their deposition, means that the data that is recorded in this system has no use in terms of constructing or understanding the formation sequence of an archaeological feature, or in the case of this experiment, the grave. The recording system results, basically, in the creation of a list of artefacts and the spatial location in which they were found, with no contextual information surrounding it. This is acceptable if the individual conducting the investigation is wishing only to recover material evidence, but if they wish to understand the sequence of events that resulted in that piece of evidence ending up in that location this approach is unsuitable.

The Control participants recording system

The Control participants did not use a standardised approach to document their findings. Instead, the choice of how to record their findings was left to the discretion of individual participants. The majority of participants documented their findings using a journal. Within this journal they would draw sketches of the different contexts that were identified and would also write down the location of any material evidence that was found. As these participants had no archaeological training, their emphasis whilst recording was on the material evidence rather than the stratigraphic sequence, and so they would not draw accurate plans of the different contexts contained within the grave. Nevertheless, the participants did take photographs as their excavations progressed, although these photographs tended to be of material evidence items rather than of the contexts from which these items derived. This is unsurprising as these participants had no formal archaeological training and therefore couldn't be expected to know how the recording of grave features should be conducted.

In terms of deciding on which excavation method and which recording system is most suitable for use in forensic archaeological investigations, the methodology of this research project stated that it would be determined on the basis of: which excavation method was the most productive and consistent in terms of evidence recovery (including material evidence and stratigraphic contexts), and which recording system provided the most consistent and informative record of the evidence and deposition sequence present in the grave.

When considering these variables it is apparent that the Quadrant Excavation method proved to be the most productive and consistent in terms of evidence recovery as it identified the second highest amount of material evidence, had the lowest variation range in the rate of material evidence recovery, it found the second highest amount of material evidence in situ, it successfully identified the most contexts, it had the second highest overall accuracy score and the highest consistency rate for the overall accuracy score. In addition, the Standard Context Recording system that was applied whilst using the Quadrant Excavation method produced the most consistent and informative record of the deposition sequence of the grave simulation, as it accurately recorded all of the contexts and their dimensions in two diagrams in a clear and understandable manner, which, in forensic contexts, is advantageous as it means that lay jurors will be able to easily understand the formation sequence of the grave under investigation. Furthermore, as such recording can be done in the field, it requires less time than the other recording systems in terms of editing, and therefore saves the investigation both time and money, both of which are crucial variables to be taken into consideration during a forensic investigation. The Quadrant Excavation method is also the most applicable in forensic contexts as the method enables an archaeologist to excavate a single quarter of a suspect feature first. By excavating this quarter the archaeologist is then able to determine if the suspect feature is of forensic interest much quicker than by excavating the feature in its entirety as would be done using the Stratigraphic Excavation method, or in half as would be done using the Demirant Excavation method. Again, saving investigators more time in terms of investigative hours.

Having stated that the Quadrant Excavation method is the most suitable method for forensic archaeological investigations, the Demirant Excavation method was also highly productive in terms of evidence recovery, ranking in as equal first in terms of excavation methods. However, due to the aforementioned weaknesses of the Standard Context Recording system associated with the Demirant Excavation technique, and the increased tendency for people to fail to identify contexts whilst using this excavation method, it would suggest that this method should only be used in situations in which the Quadrant Excavation method is unable to be applied, for example, when archaeologists are investigating particularly small grave structures such as those of children, or when dealing with particularly loose soils, as the quadrants have a tendency to collapse in such situations.

The Stratigraphic Excavation method and Single Context Recording system was also reasonably successful in terms of evidence recovery, being the only technique in which a participant successfully located all possible evidence contained within the grave. However, this method was less consistent in terms of evidence recovery than the previous techniques and proved to have a greater tendency for contexts to be failed to be identified in comparison to the previous techniques. Furthermore, the Single Context Recording system that is used in conjunction with this technique was much more time consuming than the previous excavation methods and requires much more extensive post-excavation work in order for the data obtained during the archaeological investigation to be presented in court. Therefore, in terms of methodological preference for forensic investigations, this method should only be used in situations in which the two aforementioned techniques are unable to be applied.

Although the three aforementioned techniques could each be used in various circumstances during the course of a forensic archaeological investigation, as statistically, there was no significant difference between recovery rates for each of these methods. The one excavation method and recording system that should not, under any circumstances, be used is the Arbitrary Level Excavation method and the Unit Level Recording system. This method proved to have an extremely poor evidence recovery rate and achieved the lowest overall accuracy score of all of the approaches tested, coming behind the Control participants who had no archaeological training and did not follow any advocated archaeological approaches. Furthermore, as this method failed to accurately record any of the contexts contained within the grave, and destroyed the stratigraphic sequence within the grave as the excavation progressed, this technique must be deemed as highly unreliable and therefore should not be used during forensic investigations, as the data that it captures will be inaccurate and misleading, and therefore, potentially lead to a miscarriage of justice if utilised in forensic archaeological contexts.

The impact of archaeological excavation methods and recording systems on the formation of interpretation-based narratives

One of the objectives of this research project was to evaluate how excavation methods and recording systems influence the formation of interpretation-based narratives, and to determine which excavation and recording method provided the most consistent narrative of the grave's formation process. In order to obtain data to complete this objective, each experimental participant was instructed to compose a narrative of the grave's formation sequence after they had completed their excavation of the grave simulation.

In terms of the impact that excavation method and recording system selection had on the formation of the participants' narratives of the grave's formation sequence, the major impact was in relation to determining what was found and therefore discussed by the participants in their narratives. Those participants who used the Stratigraphic, Demirant and Quadrant Excavation methods recovered the greatest number of contexts and material evidence therefore their narratives were the most accurate. Whereas, the Control participants and the Arbitrary Level Excavation participants recovered the least number of contexts and material evidence and so their narratives were the least accurate in terms of describing the grave's formation sequence, as they had failed to recover the majority of the stages associated with the grave's creation. Therefore, in terms of which excavation method and recording system produced the most accurate and consistent narrative of the grave's formation sequence, it was the Quadrant Excavation method and the Standard Context Recording system, as this methodological approach proved to be the most productive and consistent when all criteria were taken into consideration.

Despite the fact that excavation method and recording system selection determined the accuracy of the narratives produced. It is interesting to note that excavation method and recording system selection had little impact on the way in which the participants structured their narratives. There was little consistency in the structure and content of the narratives between participants using the same excavation and recording methods or between participants in general. There were two definable approaches that the participants took in order to describe the formation sequence of the grave.

The first approach was to provide a description of the sequence of deposition of the contexts and material evidence in the chronological order in which they were deposited, and was the approach that was used by the majority of participants. By breaking the formation sequence down into the different stages of deposition, the narratives produced by these participants explained the formation of the grave step-by-step making the entire formation process easy to understand. Although, the level of detail provided by the participants using this approach varied. Only 20 of the participants discussed the dimensions of the grave. This is rather concerning as the dimensions of the grave is an essential part of the formation sequence. The participants who failed to describe the dimensions of the grave presumably thought that their recording sheets would provide such data if needed. In terms of discussing the contexts within the grave, the level of detail again varied, some participants would describe the dimensions and composition of the contexts whereas others would not. The reasoning behind this lack of detail is again, presumably, due to the fact that the participants thought that readers could refer to their recording sheets to obtain more data if required.

However, it was when the participants that were using this approach were discussing the material evidence that was found that their narratives varied the most. The majority of participants would merely list the material evidence that was found and the context or spit from which each piece was recovered, but others would attempt to analyse the material evidence and provide explanations and scenarios to account for its presence within the grave.

Archaeologist 03, Archaeologist 13, Control 01 and Control 02 stated that the individual within the grave was female due to the presence of a dress. Although one might assume that the dress was indicative of a female occupant, stating so in forensic contexts is dangerous as it is presumptive and may be proven incorrect at a later stage in the investigation.

Archaeologist 06, Archaeologist 13, Archaeologist 36 and Control 01 stated that the perpetrator or the victim was a smoker due to the presence of smoking paraphernalia. Such statements are valid as the lighter and cigarette papers both support this theory. Nevertheless, Control 01's and Archaeologist 06's statements that the presence of a disposable lighter, cigarette papers and a Primark dress is indicative of low status represents conjecture, as such statements cannot be supported by any of the other evidence contained within the grave.

Archaeologist 03 stated that the lighter, ID card and fake nail had been thrown into the grave, however, did not explain what evidence there was to support this. Moreover, Archaeologist 20 stated that the fake nail and lighter had been tossed into the grave, and again, failed to explain what evidence there was for this. Archaeologist 23 stated that a pair of stud earrings had been accidentally deposited but again failed to provide any supporting evidence to justify this statement. The fact that these three archaeologists did not provide any supporting evidence for making such statements means that their theories are unsubstantiated and in forensic contexts would be dismissed.

The most concerning narrative that discussed material evidence was that of Archaeologist 36 who stated that a small body had been placed into the grave. Presumably this statement was based on the fact that there was clothing along the base of the grave and the small dimensions of the grave. However, seeing as no human remains of any kind were placed in the grave this statement is invalid and highly concerning. In forensic contexts, such exaggerative statements could reduce the credibility of the archaeologist and be used by the defence to criticise the integrity and accuracy of the entire forensic archaeological investigation.

The second approach that was used by Archaeologist 10, Archaeologist 20, Archaeologist 31, and Control 03 was to describe how they had excavated and recorded the grave, and to discuss what they had found during this process. These narratives were much more informal in structure and tone than those participants who used the first approach to describe the grave's formation process. Due to the fact that these participants had attempted to discuss their excavation and recording techniques as well as the material evidence and contexts that they had found, their interpretations of the formation sequence of the grave were particularly difficult to discern, as these participants had provided so much extraneous data.

Unlike some of the participants who used the first approach to construct their narratives, the participants using this approach did not over interpret the material evidence present in the grave, rather, they remained constrained, listing what was found rather than theorising about its implications. As a result, these narratives lacked the conjecture that was present in some of the other approaches participants' narratives.

When one compares the two approaches against one another it is evident that the participants who used the first approach to construct their narratives tended to produce narratives that were much more structured and logical, enabling readers to understand the formation sequence of the grave clearly. This is because the participants who used the second approach overloaded their narratives with irrelevant information, such as the methodological approaches that they had used. Therefore, when a forensic archaeologist is writing a narrative of the formation sequence of a grave within a forensic archaeological report, it is recommended that the first approach be used, as this will produce a narrative that is easy for jurors and legal practitioners to understand. Moreover, any descriptions of the methods that were used by the forensic archaeologist should not be discussed within the narrative, but in a separate methodology section. It is also advised that if the forensic archaeologist wishes to discuss the material evidence within the grave that such discussions are limited to describing what was found and the location in which it was found, as any further interpretations are not within the remit of the forensic archaeologist's expertise.

Overall, it is apparent that excavation method and recording system selection does dictate what material evidence and contexts will be identified, and consequently, determines how accurate an archaeologist's narrative will be. Therefore, given that the Quadrant Excavation method and Standard Context Recording system was proven to be the most effective technique at recovering all types of evidence from the simulated grave this method should be used during forensic archaeological casework, as it will ensure a high evidence recovery rate, and in turn, a more accurate narrative of the grave's formation sequence. In terms of how narratives should be structured, narratives should provide a description of the sequence of deposition of the contexts and material evidence in the chronological order in which they were deposited. Such narratives should be interlinked with the recording forms and drawings that were produced during the excavation process. As the Quadrant Excavation method and Standard Context Recording system is the recommended approach for forensic archaeological casework, and results in the production of a long section and a half section drawing of the entire structure of the grave, it is suggested that archaeologists use these drawings as illustrative tools and interlink their narratives with them. This will allow lay jurors and legal practitioners to visualise what is being discussed in the narrative, and enable them to understand the findings of the forensic archaeological investigation more clearly.

The influence of time on archaeological investigations

Although time was not noted as a factor for consideration in the objectives of this research project, variance in the length of time that it took individual participants to excavate and record the grave simulation, and differences in the average time spent excavating and recording using each of the different excavation methods and recording systems was analysed. This data was then used to determine what, if any, impact time might have on the overall quality and quantity of evidence recovered from an archaeological investigation.

The reason why this analysis was deemed to be important was due to research conducted by Landry (2012) and Scherr (2009). These researchers used experimental grave simulations, such as the one used in this research project, to evaluate the impact that time constraints had on the overall quality and quantity of evidence recovered. Both Landry (2012) and Scherr (2009) found that there was a significant reduction in the overall quality of the excavation when less time was spent on excavating their grave simulations. The results of these studies affirms what is generally assumed in archaeological practice – the longer one spends excavating an archaeological feature, the greater number of finds will be identified and a greater understanding of the formation process of the feature will be obtained. Therefore, this research project sought to further test this assumption, with the hypothesis that there would be a linear relationship between the length of time spent excavating and overall score, with the greater amount of time that a participant spent excavating and recording the grave simulation resulting in a higher overall score. The results from the analysis of time are displayed in Figure 170, Figure 171 and Figure 172.

The hypothesis was found not to be supported by the data analysed during this research project. There appears to be minimal correlation between the length of time that a participant spent excavating and recording the grave simulation and their overall score, with participants who spent 2, 3 ½, 4, 5 ½ and 6 hours excavating and recording the grave achieving the same overall score, 91.67%. The participant who achieved the highest score of 100% spent 4 hours excavating and recording the grave, and the participant who attained the lowest score of 20.83% spent 3 hours excavating and recording the grave. Although, when one compares the results of the participants who spent the longest time excavating and recording the grave, 14 hours, who achieved overall scores of 70.83%, 75.00%, and 83.33%, against the results of the participants who spent the shortest time excavating and recording the grave, 2 hours, who achieved overall scores of 29.17%, 33.33%, 37.50%, 41.67%, 50.00%, and 91.67% it is possible to discern an improvement in the participants' overall scores with the greater length of time that was spent excavating and recording the grave simulation.

However, the lack of a significant linear relationship between the length of time that a participant spent excavating and recording the grave simulation and their overall score is most apparent in Figure 172, which displays the results of a linear regression analysis. The data provided an R^2 value of 0.10344. This indicates that there is a minimal to slight relationship between the length of time that a participant spent excavating and recording the grave simulation and their overall score. Moreover, this result indicates that the length of time that an archaeologist spends excavating a grave can only be used to predict an archaeologist's ability to achieve a higher rate of evidence recovery, or in the case of this experiment, overall score, in 10% of cases.

In terms of comparing the average length of time that it took for the participants, both the archaeologists and controls, to excavate and record the grave simulation using the different archaeological techniques tested in this research project, the results indicated some slight differences. The Stratigraphic Excavation participants took an average of approximately 5 ½ hours, the Demirant Excavation participants took an average of approximately 5 hours, the Quadrant Excavation participants took an average of approximately 4 ½ hours, the Arbitrary Level Excavation participants took an average of approximately 2 ½ hours, and the Control participants took an average of approximately 2 ½ hours to excavate and record the grave simulation. The differences in the rate at which the excavation and recording of the grave simulation was completed using these different techniques can be attributed to the methodological approaches themselves. As the Stratigraphic Excavation method, the Demirant Excavation method and the Quadrant Excavation method required the participants to spend greater lengths of time delineating the boundaries of individual contexts contained within the grave, and required more in depth recording, including, the completion of context sheets, plans, and section drawings. Whereas, the Arbitrary Level Excavation method required no such skill, as the fills contained within the grave were removed in set 10cm spits with no attention paid to the presence of multiple fills or their dimensions. Furthermore, the Unit Level Recording method required a basic plan of each spit, which would appear as a rectangle with a brief description of - the location of any material evidence items that were found, and the composition of the spit that was excavated. This argument can also be applied to the Control participants, as the majority of the participants used hand shovels rather than trowels to remove the fills of the grave, and only spent a minimal amount of time logging artefacts in their journals, creating sketches of what was found, and describing the different fills that they had identified, rather than completing comprehensive recording forms.

The fact that the various methods of excavation and their associated recording systems, particularly the Stratigraphic Excavation, Demirant Excavation, and Quadrant Excavation methods showed little variation in the length of time that it took for participants to excavate and record the grave simulation, suggests that the utilisation of any one of these methods during the course of a forensic archaeological investigation will not require a significant increase in the amount of time needed to complete the excavation. Therefore, when forensic archaeologists are deciding on which approach to use, time should not be used as an exclusionary factor during this decision making process.

Furthermore, as the length of time spent excavating and recording the grave simulation proved to have a minimal correlation with the overall score of participants, it can be stated that time scarcely effects the quality and quantity of evidence recovered during an archaeological investigation. Therefore, other variables, such as the excavation method used, recording system used, and the ability of certain participants to be more observant and careful than others when excavating are likely to be the factors that impact the overall quality of the archaeological investigation the most. This explanation parallels with the findings of the researchers Evis (2009) and Tuller and Đurić (2006) who found that these variables were the major contributors in determining the overall quality of an archaeological investigation, irrelevant of the amount of time that was spent excavating and recording.

The impact of archaeological experience on archaeological investigations

In any industry, there is a presumption that as an individual's level of experience increases, so too will their ability to conduct the job that they specialise in. This is no different in the field of archaeology. Most commercial archaeological organisations regard archaeological excavation experience as a key factor from which to determine the rank to which to assign a particular archaeologist within a company, with six months commercial experience being regarded as the most basic level required for the lowest fieldwork position. Likewise, as discussed earlier, although no set experience level requirements exist in order to participate in a forensic archaeological investigation, scholars such as Cheetham and Hanson (2009), Cox *et al.* (2005), Crist (2001), France *et al.* (1992), Hunter *et al.* (2001), Sigler-Eisenberg (1955), Spennemann and Franke (1995), United Nations (1991) and Wright *et al.* (2005) state that experienced field archaeologists should be employed for forensic archaeological work. Furthermore, Scott and Connor (2001) specifically state that such casework should only be conducted by archaeologists with a minimum of "3 years full time fieldwork experience" (Scott and Connor 2001: 104). This, they argue, will ensure that the participating archaeologist will be competent and fully capable of conducting a forensic archaeological investigation, and will be recognised by the courts as an expert, and be deemed competent by the courts to provide an expert opinion regarding the findings of a forensic archaeological investigation.

This presumption, that a higher level of archaeological experience will increase the quality and quantity of evidence recovered from an archaeological investigation, was tested during this research project. The results of which are displayed in Figure 173. The hypothesis was that there would be a linear relationship between the overall score of individual participants and their level of archaeological experience, with participants with higher levels of archaeological experience having the highest overall scores.

This hypothesis, however, was disproved. There appears to be no correlation between archaeological experience and overall score, with participants of 3, 4, 5, 7, 9, 12, 15, 18, 28 and 30 years of archaeological experience achieving the same score, 91.67%. The participant who achieved the highest overall score of 100% had 19 years of archaeological experience, and the participant who attained the lowest overall score of 20.83% had 6 years of archaeological experience. If the hypothesis was correct, the participants with only one year of archaeological experience should have attained the lowest overall scores, and the participants with the highest level of archaeological experience, thirty years, should have achieved the highest overall scores. However, some of the most experienced participants had overall scores that were lower than individuals with only one year of archaeological experience, for example, a participant with 4 years experience achieved 29.17%, a participant with 5 years experience achieved 37.50%, a participant with 12 years experience achieved 41.67%, a participant with 13 years experience achieved 41.67%, and a participant with 30 years experience achieved 33.33%, whereas the participants with only 1 year of experience achieved at worst 45.83% and at best 75.00%.

The lack of a linear relationship between years of archaeological experience and overall score is further emphasised in Figure 174, which displays the results of a linear regression analysis. The data from this analysis provided an R^2 value of 0.04298. This indicates that there is a minimal relationship between a participant's level of archaeological experience and their overall score. Furthermore, this data indicates that archaeological experience can only be used to predict an archaeologist's ability to achieve a higher rate of evidence recovery, or in the case of this experiment, overall score, in 4% of cases.

Consequently, these results contradict statements made by scholars that archaeological experience will ensure a higher quality of forensic archaeological investigation. One might explain the variation in the aforementioned results by stating that the archaeologists with higher levels of archaeological experience are likely to be in more senior roles and therefore do not conduct fieldwork on a regular basis, as their focus will be on managing archaeological projects and post-excavation work. Whereas the archaeologists with lower levels of archaeological experience are more likely to be conducting the practical fieldwork aspect of archaeological investigations. However, in the case of this research project, all participants conducted fieldwork on a regular basis, particularly those with the higher levels of experience. Therefore, this explanation is not valid.

Another explanation for the significant variation in overall scores might be complacency and over confidence. As archaeologists with higher levels of archaeological experience tend to consider themselves to have gained sufficient experience to excavate and record archaeological features as second nature, and have experience in excavating a variety of different archaeological features. Such archaeologists thus have a large reference collection from which to predict what one might expect to find in a particular type of archaeological feature, such as a grave. This predictive capability may bias experienced archaeologists, as they will use their previous experiences and knowledge of excavating archaeological features to predict what a particular feature, such a grave, may contain and how it may be constructed. Therefore, when excavating a feature, such as the grave used in this research project, the experienced archaeologists may well have assumed that the grave structure and its contents would mimic the types of graves that they had previously excavated, and led to a complacent attitude and approach to excavating the grave. That, in turn, may well explain why such archaeologists failed to identify certain contexts and material evidence items contained in the grave, as these did not adhere to their pre-determined expectations.

In comparison, archaeologists with lower levels of archaeological experience are less likely to be confident in their abilities to excavate and record archaeological features. Such insecurities result in a tendency for less experienced archaeologists to be more cautious and considerate when excavating and recording archaeological features, or in the case of this research project, the grave, resulting in these archaeologists achieving higher overall scores than one might have expected them to.

The findings of this research project clearly demonstrate that the employment of archaeologists on the basis of their archaeological experience is not a sufficiently reliable criterion upon which to judge an archaeologist's ability to participate in archaeological fieldwork, be it in a commercial, academic or forensic investigatory setting. In terms of a forensic context, as the consequences of such investigations can be so impacting on the victims' families, the accused individual(s), and individuals and societies effected by atrocities such as genocide, war crimes, crimes against humanity, and crimes of aggression, the selection process whereby archaeologists are chosen to participate in forensic investigations must be stringent, and ensure that the investigation is undertaken to the highest possible quality attainable.

In order to achieve such an objective, it is clear that merely relying on the number of years of archaeological experience that an archaeologist has will not guarantee a high quality of excavation. Therefore, in order to counter such problems, it would be advisable for potential forensic archaeological investigation team candidates to participate in a skills test, which would rely on the candidate excavating and recording a controlled simulated grave, such as the one used in this research project. Such competency tests are already used in the discipline of forensic anthropology with the American Board of Forensic Anthropology being responsible for examining and accrediting competent forensic anthropological practitioners (American Board of Forensic Anthropology 2013). Through reviewing the results of such a test, the selection committee would then be able to determine which individuals would produce the highest quality of archaeological investigation, and therefore be most suitable for forensic casework. Due to the fact that there appears to be no definitive link with archaeological experience and overall score, it would also be advisable that such tests were repeated, for example, every five years, to ensure that the highest archaeological standards are being maintained. The threshold for deeming a potential team candidate capable is debateable, as there is not currently a forensic archaeology competency test standard in place, however an overall score of

80% or higher would seem sufficient, as this is the standard used by the American Board of Forensic Anthropology in their competency tests (American Board of Forensic Anthropology 2013). If a potential candidate failed to meet such a standard, a re-take or focused training scheme would, in theory, raise the individual to the desired standard.

The establishment of error rates for archaeological excavation methods and recording systems

As discussed in 'Legal concerns: How international legislation and admissibility regulations impact forensic archaeological investigations', there are five requirements that must be considered by the court to determine if the expert testimony and the evidence retrieved by the forensic archaeologists during the course of an investigation can be deemed as reliable. These include: empirical testing, peer review, professional standards, widespread acceptance, and error rates.

The two major requirements that were yet to be satisfied by the archaeological community, prior to this study, were the lack of empirical testing and the establishment of error rates for archaeological excavation methods and recording systems.

The requirement for archaeological excavation methods and recording systems to be empirically tested has now been satisfied as a result of the excavation experiments conducted and discussed in this study, and the work of other scholars including: Pelling (2008) and Tuller and Đurić (2006) whose findings correlated with those discussed in this chapter. This body of work has demonstrated that there is variability in the suitability of different archaeological excavation methods and recording systems for forensic archaeological casework, and that the Quadrant, Demirant and Stratigraphic Excavation methods and their associated recording systems, on average, produce more accurate results than lay persons and archaeologists utilising the Arbitrary Level Excavation method and Unit Level Recording system, when tested using grave simulations of known properties.

In regards to the establishment of error rates for each archaeological excavation method and recording system, it is unfortunate that this particular requirement cannot be met. This is due to the fact that there is great potential for variability in how individual clandestine graves are constructed, and in turn, what evidence they may contain. Therefore, any error rates generated using simulated graves, such as the one used in this study, will not be arbitrarily applicable to clandestine grave excavations, as there are too many variables that may differ from the simulated grave used to establish these error rates.

Another factor that prevents error rates from being established for individual excavation methods and recording systems is the variability in the evidence recovery rates between individual archaeologists. This study demonstrated that recovery rates varied greatly, between archaeologists in general with archaeological experience proving to have little bearing on evidence recovery, and within individual methodological approaches. Thus, at present, as there are no skills tests to prove that an individual forensic archaeological practitioner is competent and that they produce consistent results, an error rate established using averages from simulated grave excavation experiments such as those used in this study, may not be applicable to a forensic archaeologist excavating a clandestine grave, as their recovery rates may in fact be less productive or more productive than the rates used to establish the error rate for the methodological approach that they used.

Therefore, in light of these two issues, the sub-field of forensic archaeology is unable to meet this requirement for admissibility. However, the results of the grave excavation experiments indicate how each of the archaeological excavation and recording methods perform against one another in a control setting. These results can be used by both court personnel and archaeological practitioners as a guide to determine which archaeological excavation methods and recording systems are the most productive and reliable, and thus suitable to use during a forensic investigation.

Chapter 7 Conclusion

It is evident from the results of this research that a variety of different excavation methods and recording systems are used in the United Kingdom, Ireland, Australasia, and North America to conduct archaeological investigations. To facilitate the use of these different approaches, archaeological organisations operating in these areas have developed archaeological manuals/guidelines through which they have attempted to standardise and improve how their employees conduct archaeological investigations, in order to ensure, that they maximise the accuracy of the recovery and recording processes and produce consistent archives from which they are able to create publications. Although these organisations share the same overall goals, the fact that 153 different archaeological manuals/guidelines are used by different archaeological organisations indicates that there is no singular standardised approach to conducting archaeological investigations. Consequently, this has resulted in multiple excavation methods and recording systems being advocated by different archaeological organisations, and, in turn, led to each of these different methodological approaches to be regarded as standardised approaches.

Despite different archaeological organisations advocating different "standardised" archaeological excavation and recording approaches, there are marked similarities in terms of the methodological approaches that are being advocated between archaeological organisations operating in the United Kingdom and Ireland, and archaeological organisations operating in Australasia and North America. In the United Kingdom and Ireland archaeological organisations define, record and excavate archaeological stratigraphy using similar, if not the same, approaches. Whereas, Australasian and North American archaeological organisations use different excavation methods and recording systems, and define, record, and excavate archaeological stratigraphy differently from British and Irish archaeological organisations, although, use similar approaches to each other. Therefore, there are similarities in the "standardised" approaches being advocated but not on an international scale.

These international methodological differences have been caused as a result of the different archaeological site types that these archaeological organisations have been operating in. Both British and Irish archaeological organisations conduct archaeological investigations on urban sites with complex archaeological stratigraphy, and rural sites with widely dispersed archaeological evidence with very few complex stratigraphic units present. As a result, the methods that they have developed and advocate in their archaeological manuals/guidelines have been designed to be used and/or adapted for use on both of these archaeological site types. Whereas, Australasian and North American archaeological organisations tend to conduct archaeological investigations on rural Native American or Aboriginal archaeological sites. Such sites contain widely dispersed archaeological evidence and deep stratigraphic deposits, and as a result, the archaeological organisations working on these sites have developed different methodological approaches to investigate such sites, and subsequently, published and advocated these approaches in their archaeological manuals/guidelines.

However, the problem with having multiple "standardised" archaeological excavation and recording approaches advocated within academic literature and archaeological manuals/guidelines, is that archaeologists can apply any one of these techniques to excavate and record the same archaeological feature and justify this by stating that this is a "standardised" archaeological approach. One such example of this was identified during the grave excavation experiment, whereby each archaeologist was given the same grave feature to excavate and record, and yet different groups of archaeologists chose to use different excavation methods and recording systems to investigate the grave, each justifying their chosen approach by stating that the approach that they had chosen was standard practice and used regularly during archaeological investigations. However, as little research had been conducted, prior to this research, to comparatively test these different "standardised" excavation methods and recording systems to determine the impact that these different methodological approaches have on data recovery and interpretation-based narratives, advocates of these different approaches had no data from which to state that the methodological approaches that they currently advocate do in fact result in greater accuracy in terms of the archaeological recovery and recording processes as their archaeological manuals/guidelines claim that they do, leaving the question of whether these different "standardised" methodological approaches are equally applicable debatable.

The problem of multiple "standardised" excavation methods and recording systems co-existing and being used during archaeological investigations to excavate the same type of archaeological feature is also present in forensic archaeology. This is due to the fact that forensic archaeological practitioners have adopted and adapted their methodological approaches from traditionalist archaeological practice. Therefore, forensic archaeological practitioners operating in the United Kingdom and Ireland will use different methodological approaches to their Australasian and North American counterparts, and again, justify their use of these different methodological approaches by stating that they are standard practice in the geographical area in which they are working.

The issue, however, is exacerbated when it is considered in a forensic context, as the primary aim of forensic archaeological investigations is the provision of evidence to legal proceedings. Consequently, when archaeologists are conducting forensic archaeological casework the methods utilised and the evidence retrieved as a result of the investigation are held accountable to the admissibility regulations and the legislative acts of the courts in the country in which the investigation is being conducted. Such legislative acts and admissibility regulations require that any techniques used during the course of a forensic investigation are required to have been – subjected to empirical testing, peer review, have known error rates, have standards controlling their operation, and be widely accepted amongst the academic community from which the methodology originates (Edmond 2010; Edwards 2009; Glancy and Bradford 2007; Hanzlick 2007; Klinker 2009; Pepper 2005; Robertson 2009; Robertson 2010; Selby 2010; The Law Commission 2009; The Law Commission 2011). Therefore, for the evidence that was recovered during the course of the forensic archaeological investigation to be accepted by the court, the forensic archaeologist must be able to demonstrate that the methodological approach that they used during the course of the investigation adhered to a widely accepted and tested forensic archaeological investigatory process (Hunter and Knupfer 1996: 37). However, as previously discussed, no such investigatory process has been established and no substantial empirical testing had been undertaken regarding archaeological excavation methods and recording systems prior to this research, therefore, despite multiple methods being attributed the status of a "standardised" technique, within stringent legal contexts, much of the casework relating to clandestine burials that was undertaken by forensic archaeologists, prior to the completion of this research, failed to meet the admissibility regulations and the legislative requirements of the international court systems.

In order to address these issues this research was undertaken with the aim of determining, which, if any of the various excavation methods and recording systems currently used in the United Kingdom, Ireland, Australasia and North America fulfil the criteria for legal acceptance and best meet the needs of forensic archaeology, focusing on the case of single and mass graves as they are the most common situation faced by forensic archaeologists.

The results of this research indicated that the four archaeological approaches that are currently used to excavate and record cut features, such as graves, are the Stratigraphic Excavation method and Single Context Recording system, the Demirant Excavation method and Standard Context Recording system, the Quadrant Excavation method and Standard Context Recording system, and the Arbitrary Level Excavation method and Unit Level Recording system. Each of these techniques were subsequently tested against one another using a grave simulation to determine - which of these excavation methods was the most productive and consistent in terms of evidence recovery (evidence included both material evidence and stratigraphic contexts), which of these recording systems provided the most consistent and informative record of the evidence and deposition sequence present in the grave, and which of these excavation methods and recording systems provided the most consistent interpretation-based narrative of the simulated grave's formation process.

The results gained from the grave simulation experiments revealed that the Demirant Excavation method was the most productive in terms of overall evidence recovery retrieving an average of 78.75% of all of the evidence present (Figure 168). However, both the Stratigraphic and Quadrant Excavation methods achieved similar evidence recovery rates, retrieving an average of 75.83% of the evidence present (Figure 168). In contrast, the Arbitrary Level Excavation method had an average evidence recovery rate of 39.58% (Figure 168). Due to the fact that the average evidence recovery rates were found to be very similar between the Demirant, Quadrant and Stratigraphic Excavation methods and that these differences were proven not to be statistically significant, each of these excavation methods were compared against each other in terms of the consistency of evidence recovery. The data obtained indicated that the Quadrant Excavation method was the most consistent with a variation rate of 33.34%, followed by the Demirant Excavation method with a variation rate of 41.67%, and finally, the Stratigraphic Excavation method with a variation rate of 50% (Figure 168). These results suggest that the Quadrant Excavation method is the most reliable excavation technique out of the four methodological approaches tested, and therefore, the most suitable excavation method to use in forensic archaeological casework.

In terms of which recording system produced the most consistent and informative record of the evidence and deposition sequence present within the simulated grave, each recording system provided a systematic approach to document both the evidence and deposition sequence present. However, the Standard Context Recording system associated with the Quadrant Excavation method was deemed to be the most informative and consistent recording system used. This was due to the fact that through using this recording system the participants produced both a long section and half section drawing of the simulated grave. This meant that the records that they produced illustrated each of the context's dimensions in their entirety across the length and width of the simulated grave, in a clear and understandable manner, which is advantageous in forensic contexts as it means that lay jurors would be able to easily understand the deposition sequence of the grave. Furthermore, as such recording can be done in the field, it requires less time in terms of editing after the investigation has been completed, saving the investigation both time and money, which are crucial variables to be taken into consideration for work that is conducted within a forensic context.

The Standard Context Recording system associated with the Demirant Excavation method also proved to be systematic and thorough. However, the problem with this recording system is that it relied on using a half section drawing to illustrate the deposition sequence of the simulated grave, and as some of the contexts sloped and failed to reach the half section point, these contexts did not appear in the half section drawing. Therefore, participants using this recording system were required to draw additional plans of the contexts that did not reach the half section point. Additionally, the half section drawing only displayed the deposition sequence of the simulated grave at one point, and therefore, failed to illustrate the dimensions of individual contexts across the entire simulated grave. This meant that the participants using this recording system were required to spend additional time, after the investigation had been completed, reconstructing the deposition sequence of the grave using the measurements of the contexts that were noted down on their context sheets, thus costing the investigation both more time and money.

The Single Context Recording system associated with the Stratigraphic Excavation method also produced a comprehensive set of records relating to the evidence and deposition sequence present within the simulated grave. However, the planning process associated with this technique was particularly time consuming, and required further post-investigation processing to superimpose each of the separate context plans that were produced in order to illustrate the deposition sequence of the simulated grave in its entirety. Consequently, costing the investigation both more time and money.

The Unit Level Recording system associated with the Arbitrary Level Excavation method proved to be an effective recording system to use to document the location and relative deposition sequence of material evidence. However, the Arbitrary Level Excavation method associated with this recording systems use resulted in participants digging through and destroying the dimensions of individual contexts, particularly those with any gradient in their deposition. Therefore, this recording system resulted in the production of a list of artefacts and their spatial location without any accurate contextual information. Consequently, this recording system provided an inaccurate reconstruction of the grave simulation's deposition sequence and must be deemed as uninformative and unsuitable for use in forensic archaeological casework.

In regards to which excavation method and recording system provided the most consistent interpretation-based narrative of the simulated grave's formation process, it was evident from the results of this research that the accuracy of the narratives that were produced was directly correlated with the excavation method and recording system that the participants chose to use. Therefore, those participants that chose to use the Stratigraphic, Demirant and Quadrant Excavation methods to excavate the grave simulation recovered the greatest amount of evidence, and subsequently, their narratives were the most accurate. Whereas, the participants who used the Arbitrary Level Excavation method recovered the least amount of evidence, and therefore, their narratives were the least accurate. Consequently, as the Quadrant Excavation method produced the most consistent results in terms of both the evidence that was recovered and the records that were produced, the participants using this approach also produced the most consistent interpretation-based narratives of the simulated grave's formation sequence.

An additional variable taken into consideration when comparatively evaluating each of the different excavation methods and recording systems was the length of time that it took participants to complete the grave excavation experiment using each of the different approaches. The results indicated that on average, the Stratigraphic Excavation method took approximately 5 ½ hours to complete, the Demirant Excavation method took approximately 5 hours to complete, the Quadrant Excavation method took approximately 4 ½ hours to complete, and the Arbitrary Level Excavation method took approximately 2 ½ hours to complete (Figure 170). The reason why greater lengths of time were spent excavating and recording using the Stratigraphic, Demirant and Quadrant Excavation techniques was because these methodological approaches required participants to spend time delineating the boundaries of the individual contexts contained within the grave simulation, and also required more in depth recording, including - the completion of context sheets, plans and section drawings. In comparison, the Arbitrary Level Excavation method required no such skill as the contexts contained within the simulated grave were removed in set 10cm spits with no attention paid to the presence of multiple contexts or such contexts' dimensions. Overall, due to the fact that the length of time that was spent excavating and recording using the Stratigraphic, Demirant and Quadrant Excavation methods varied very little, it suggests that the utilisation of any one of these methods during the course of a forensic archaeological investigation will not require a significant increase in the amount of time needed to complete the investigation. Therefore, when forensic archaeologists are deciding on which methodological approach to use, time should not be considered as an exclusionary factor during the decision making process.

The length of time that the participants spent excavating and recording the grave simulation was also analysed to determine if time had any impact upon the amount of evidence that was recovered. The results indicated that there was a minimal correlation between the length of time that the participants spent excavating and recording and the amount of evidence that they recovered. Therefore, a greater length of time spent excavating and recording a grave will not necessarily result in an improvement in the quality and quantity of evidence that is recovered during a forensic archaeological investigation. Rather, other factors, such as the methodological approaches that are used and the observation skills of the archaeologist will determine whether evidence will be successfully recovered during a forensic archaeological investigation.

An additional objective of this research was to evaluate the impact that archaeological experience had on the quality and quantity of evidence recovered during the grave simulation excavation experiment. The results indicated that there was no correlation between archaeological experience and evidence recovery rates, and that archaeological experience can only be used to predict an archaeologist's ability to achieve a higher rate of evidence recovery in 4% of cases (Figure 173; Figure 174). The reason why archaeological experience had little impact on evidence recovery is due to the fact that experienced archaeologists have obtained significant experience in excavating a variety of different archaeological features. They then use this experience to predict what an archaeological feature, such as a grave, might contain and how it may be constructed, and presume that the archaeological feature that they are excavating will match similar archaeological features that they have excavated in the past, resulting in a complacent attitude and approach to the investigatory process. Inexperienced archaeologists, however, are less confident in their abilities and are resultantly more cautious and considerate when conducting archaeological investigations, to ensure that they have not missed any evidence or misinterpreted what they have uncovered.

As a result of this research, each of the four different methodological approaches tested during the simulated grave excavation experiments now satisfy the legislative and admissibility requirements of the international courts. Although, on the basis of the findings of this research, it is apparent that the Quadrant Excavation method and Standard Context Recording system best meets the needs of forensic archaeology. However, the Demirant Excavation method and Standard Context Recording system was also highly productive in terms of evidence recovery, though due to the weaknesses of the recording system, it is suggested that this approach should only be used in situations in which the Quadrant Excavation method is unable to be applied, for example, when excavating particularly small clandestine burials or when excavating in loose soils, as the quadrants associated with the Quadrant Excavation method have a tendency to collapse under such conditions. Moreover, the Stratigraphic Excavation method and Single Context Recording system was also reasonably successful in regards to evidence recovery, but due to the fact that this approach produced inconsistent results and that the recording procedure was highly time consuming, it should only be used in situations in which the two aforementioned techniques are unable to be applied. The one approach that should not be used in forensic archaeological casework is the Arbitrary Level Excavation method and Unit Level Recording system. This is because this methodological approach had an extremely poor evidence recovery rate, and destroyed the deposition sequence present within the simulated grave. Moreover, through statistical testing, this methodological approach was found to recover significantly less evidence than the three other archaeological techniques tested and the Control participants. Consequently, as this methodological approach is highly unreliable and produces inaccurate data, any forensic archaeological reports that have been produced from forensic archaeological investigations that have used this approach, should be treated with extreme caution.

Given the extent to which the Arbitrary Level Excavation method and Unit Level Recording system has been shown to be used during archaeological fieldwork outside the United Kingdom and Ireland, the findings of this research project are particularly alarming. It suggests that archaeological data, whether it has been obtained during a research, commercial or forensic archaeological excavation, using this methodological approach, is incomplete, and any archaeologists that have relied on this data to formulate interpretations about a site are likely to have produced inaccurate or at least questionable interpretations, as they have used a data collection which has, on average, 60.42% of its data missing (Figure 168). In light of this finding, scholars must be cautious of any archaeological reports or interpretations that have relied on data collections obtained using the Arbitrary Level Excavation method and Unit Level Recording system, and must focus their attention on re-assessing this archaeological data in light of the findings of this research.

Another cause for concern amongst the archaeological community is that this research project demonstrated that archaeological data recovery rates varied greatly, between archaeologists in general with archaeological experience proving to have little bearing on data recovery rates, and within individual methodological approaches. This lack of consistency both within methods and between practitioners suggests that one cannot assume that an experienced team of archaeological practitioners will necessarily retrieve the maximum amount of archaeological data available, which is the ultimate aim of any archaeological investigation. It is therefore necessary, for a skills test to be established that will ensure that archaeological practitioners are consistently producing high quality archaeological investigations, and that the interpretations that are produced on the basis of these excavations are reliable. It is particularly important that skills tests are introduced into the sub-discipline of forensic archaeology, as the consequences of forensic investigations can be profound.

Chapter 8 Recommendations

Specific recommendations for forensic archaeological investigations:

In light of the findings of this research, when conducting forensic archaeological investigations the Quadrant Excavation method and Standard Context Recording system should be used. If this approach is unable to be utilised the Demirant Excavation method and Standard Context Recording system, or the Stratigraphic Excavation method and the Single Context Recording system should be used. Any deviation from these recommended approaches should be justified in the forensic archaeologist's report.

When using the Quadrant Excavation method or the Demirant Excavation method all recording forms should be completed as the excavation progresses, not only after a half or quadrant has been removed. This will prevent any material evidence from later being incorrectly associated with the wrong context.

When conducting forensic archaeological investigations, all spoil related to different contexts should be stored separately. All spoil should also be sieved, either at site or in controlled laboratory conditions. This will ensure that any material evidence items that were missed during the excavation process are recovered and can be reassociated with the context from which they originated.

General recommendations:

An internationally recognised forensic archaeology organisation/committee should be formed in order to share ideas, research, case studies, and methodological developments within the sub-field. The formation of such an organisation/committee will improve the standard and consistency of forensic archaeological practice on an international scale. Consequently, when forensic archaeologists of different nationalities are deployed to work on international projects, such as mass grave excavations, they will then all be familiar with the internationally advocated archaeological excavation and recording approaches. This will help to prevent disagreements between forensic archaeologists about which methods to use.

An internationally recognised forensic archaeology investigation protocol should be developed for single and mass graves. The development of such a protocol could be informed by the results of this research, and could be facilitated by the formation of an internationally recognised forensic archaeology organisation/committee.

A forensic archaeology competency test should be developed for archaeologists who wish to work in forensic contexts. Such a test would require archaeologists to excavate and record a controlled grave simulation, such as the one used in this research. In addition, as archaeological experience proved to have little bearing on evidence recovery rates, such competency tests should be repeated every five years in order to ensure that the highest archaeological standards are being maintained. The implementation of such competency tests would be facilitated by the formation of an internationally recognised forensic archaeology organisation/committee.

The applicability of the Quadrant, Demirant and Stratigraphic Excavation methods and their associated recording systems should be tested on mass graves. This will determine whether these approaches are as suited to mass grave investigations as they are to single inhumation investigations.

All archaeological sites that have been investigated using the Arbitrary Level Excavation method and Unit Level Recording system need to be re-examined and interpretations formed using data collections obtained through the use of this method need to be re-evaluated.

Glossary

A

Access Trench: a ditch dug around the boundaries of the feature of interest.

Admissible: acceptable as evidence in a court of law.

Arbitrary Level Excavation Method: a method whereby the site is divided into square or rectangular excavation units. The deposits contained within a unit are removed in arbitrarily defined levels, usually 5cm or 10cm in depth.

Archaeological Experience: knowledge and skill gained through participating in archaeological investigations.

Archaeological Feature: a non-portable artefact of past human activity.

Archaeological Qualification: the successful completion of a course or examination related to the discipline of archaeology.

Archaeological Site: a defined area of land that contains evidence of past human activity.

Archaeological Site Report: a document that contains the results of an archaeological investigation.

Archaeologist: an individual who uses established archaeological methods and theories to excavate, record, interpret and understand past human activity.

Archaeology: the study of the human past through the excavation, collection and recording of physical remains of human activity present at archaeological sites.

Artefact: any object that has been created or modified by a human being.

B

Baulk: an unexcavated wall of archaeological deposits left standing between two or more trenches.

Bias: an inclination or prejudice for or against a particular method, theory or result.

Body: the entire physical structure of a human being or animal.

Box Excavation Method: a method whereby an archaeological site is divided into square or rectangular trenches with baulks left standing between them.

C

Chronology: the organisation of events into the order of their occurrence.

Civil Court Proceedings: deal with non-criminal cases.

Context/Stratigraphic Unit: the smallest identifiable unit of stratification. A context can be classified as a positive record, which is created by the placement of material, or a negative record, which is created by the removal of material.

Contextual Information: is information that relates to the physical location and association of an artefact, ecofact, deposit, fill or feature to other such evidence forms within the archaeological site.

Credibility: the quality of being trusted and believable.

Criminal Court Proceedings: deal with criminal cases.

Cut: formed as a result of the removal of material.

D

Deposit/Layer/Fill: formed as a result of the placement of material.

Deposition: a process whereby material is moved from its point of origin and is placed elsewhere.

Demirant: division of an object into two halves.

Demirant Excavation Method: a method whereby an archaeological feature is divided into two halves, each of which are excavated separately.

E

Ecofacts: are non-artefactual organic or environmental materials that have been used by humans.

Evidence: information, documents, or objects that will be used to establish facts during court proceedings.

Excavation: a process whereby layers/deposits/fills are removed in order to recover the physical remains of the human past.

Excavation Manual/Guideline: a set of instructions that inform an archaeologist of how to excavate and record an archaeological site.

Expert Witness: an individual who by virtue of education, experience, training or skill is deemed to have gained a sufficient level of knowledge, in a particular field, to render an opinion about a fact in issue during legal proceedings.

F

Field Archaeology: is a sub-discipline of archaeology that focuses on archaeological fieldwork, including: excavation, recording, interpretation and publication of archaeological site reports.

Fill: a deposit that is contained within an archaeological feature.

Forensic: used in a court of law.

Forensic Archaeologist: an individual who has received training in archaeology, biological anthropology and crime scene investigation techniques, who utilises this training to assist with forensic and humanitarian investigations, and mass disaster recovery operations.

Forensic Archaeology: is a sub-discipline of archaeology that applies archaeological principles and methods to recover evidence for forensic and humanitarian investigations, and mass disaster recovery operations.

Forensic Platform: a thin rectangular sheet of rigid metal or plastic that is used to support human remains as they are removed from their burial environment.

Formation Process: a term used to describe the way in which archaeological sites and their associated layers/deposits/fills were created and altered by the activities of humans, animals, natural and environmental phenomena, up until the point of their excavation.

G

Geotaphonomic Features: are non-portable artefacts that have been formed as a result of the interment of buried evidence. Such features are classified as belonging to one of the following six categories: stratification, tool marks, bioturbation, sedimentation, compression/depression and internal compaction.

Gezer Method: a method whereby an archaeological site is divided into sixteen 3m square excavation units or 12m x 15m rectangular excavation units in order to form a trench, once one trench has been fully excavated another is started next to it.

Grave: a hole that has been purposefully dug into the ground in order to receive human remains.

Grave Block: is created by excavating around the circumference of a grave whilst leaving the grave structure intact. This results in the grave being left as a standing block of soil that can be undercut and transported away from the excavation site.

Grave Cut/Grave Boundary: is the border between the grave and the undisturbed soil surrounding it.

H

Human Remains: the remnants of a human being's entire physical structure after death.

I

Inadmissible: not acceptable as evidence in a court of law.

In Situ: when artefacts are recovered from the location in which they were originally deposited.

Interface: the point at which two deposits/fills/layers/features meet.

Interpretation: the point at which the results of an archaeological investigation are analysed in order to explain their meaning.

M

Mass Grave: a grave which contains the remains of more than one individual.

Material Culture: a term used to refer to objects that have been created by human beings.

Multi-Context Recording: a process in which multiple contexts are recorded simultaneously.

N

Narrative: a written or oral account of the sequence of events that resulted in the creation of an archaeological feature or site.

Negative Record: results from the removal of material and includes features such as: pits, graves and ditches.

O

Out Of Situ: when artefacts are not recovered from the location in which they were originally deposited.

P

Positive Record: results from the placement of material and includes such things as: structures, deposits/ layers and fills.

Pre-determined Strata Recording: a process that uses a pre-determined list of strata and feature codes in order to record any stratum or feature uncovered during an archaeological investigation. *Example: A pit feature is allocated the code P. If, during the process of excavation a pit is discovered it will be referred to as P.*

Q

Quadrant: division of an object into four parts.

Quadrant Excavation Method: a method whereby an archaeological feature is divided into four parts, each of which are excavated separately.

R

Recording System: is a set of established procedures designed to document the results of an archaeological investigation.

S

Schnitt Excavation Method: a method whereby the site is divided into a series of rectangular excavation units. These units are then excavated in a series of pre-defined spits, usually 5cm or 10cm in depth.

Single Context Recording: a process in which each context is exposed and planned in isolation.

Single Grave: a grave that contains the remains of one individual.

Skeletal Remains: the remnants of a human being's skeleton after death.

Soil Pedestal: a raised platform of soil that is left below an object as the excavation proceeds.

Stratigraphic Excavation Method: a method whereby contexts are defined and removed in isolation, in the reverse order of their deposition.

Stratigraphic Relationship: refers to the relationship that one context has with another within the stratigraphic sequence.

Stratigraphic Sequence: refers to the order in which contexts were created over the course of time.

Stratigraphy: is the overall study and validation of archaeological stratification; the process by which layers/deposits are continually laid down over time in a series of sequential layers one on top of another as a result of human activities and/or natural phenomena. This process of archaeological stratification is governed by the following principles:

Principle Of Superposition: in a series of layers, as originally created, the upper units of stratification are younger and the lower older, as each must have been deposited on, or created as a result of the removal of, a pre-existing mass of archaeological stratification (Harris 1989).

Principle Of Original Horizontality: any layer deposited in an unconsolidated form will tend towards a horizontal position; strata found with tilted surfaces were originally deposited that way, or lie in conformity with the contours of a pre-existing basin of deposition (Harris 1989).

Principle Of Original Continuity: an archaeological layer as originally laid down, will be bounded by a basin of deposition or will thin down to a feather edge. If any edge of the layer is exposed in vertical view, part of its original context must have been removed by excavation or erosion and its continuity must be sought or absence explained (Harris 1979).

Principle Of Stratigraphical Succession: any unit of archaeological stratification takes its place in the stratigraphic sequence of a site from its position between the lowest of all units which lie above it and the uppermost of all those units which lie below it and with which it has physical contact (Harris 1979).

Principle Of Intercutting: if a feature or a layer/deposit/fill is found to cut into, or across, another layer/deposit/fill it must be more recent (Darvill 2008).

Principle Of Incorporation: all artefactual and ecofactual material found to be contained within a layer/deposit/fill must be the same age or older than the formation of that layer (Darvill 2008).

Principle Of Correlation: relationships can be inferred between layers that exhibit the same characteristics, contain the same range of artefactual and ecofactual material, and occupy comparable stratigraphic positions within related stratigraphic sequences (Darvill 2008).

U

Undercutting: the deliberate removal of soil from underneath the feature of interest.

Unit Level Recording: a process in which each horizontal surface, revealed by the removal of a level or spit, is planned.

Bibliography

Adams, R. E. W. and Valdez, F. 1997. Excavation and Analysis of Human Remains. In T. R. Hester., K. L. Feder and H. J. Shafer (eds), *Field Methods in Archaeology*: 235-252. Seventh Edition. California, Mayfield Publishing.

American Academy of Forensic Sciences (AAFS). 2012. Upgrade to a Trainee Affiliate or Associate Member. Available from: http://www.aafs.org/upgrade-trainee-affiliate-or-assocaite-member [Accessed on: 03/03/2012 13:30].

American Board of Forensic Anthropology (ABFA). 2012. Policies and Procedures. Available from: http://www.theabfa.org/policies%20with%20button.html [Accessed on 02/09/2013 14:50].

Arnold, B. 1990. The past as propaganda: totalitarian archaeology in Nazi Germany. *Antiquity* 64 (244): 464-478.

Arnold, B. 1996. The Past as Propaganda: Totalitarian Archaeology in Nazi Germany. In R. W. Preucel and I. Hodder (eds), *Contemporary Archaeology in Theory A Reader*: 549-569. Oxford, Blackwell Publishing Ltd.

Atkinson, R. J. C. 1946. *Field Archaeology*. London, Methuen.

Australian Law Reform Commission. 2005. *Uniform Evidence Law Report*. Sydney, Australian Law Reform Commission.

Balme, J. and Paterson, A. 2006. *Archaeology in Practice A Student Guide to Archaeological Analysis*. Oxford, Blackwell Publishing Ltd.

Barker, P. 1993. *Techniques of Archaeological Excavation*. Third Edition. London, Routledge.

Barker, P., White, R., Pretty, K., Bird, H. and Corbishley, M. 1997. *Wroxeter, Shropshire: the baths basilica excavation 1955-90, English Heritage Archaeological Reports 8*. London, English Heritage.

Bar-Yosef, O. and Mazar, A. 1982. Israeli Archaeology. *World Archaeology* 13 (3): 310-325.

Bass, W. M. and Birkby, W. H. 1978. Exhumation: The Method Could Make the Difference. *FBI Law Enforcement Bulletin* 47 (7): 6-11.

Bedford, S., Regenvanu, R., Spriggs, M., Buckley, H. and Valentin. 2011. Vanuatu. In N. Márquez-Grant and L. Fibiger (eds), *The Routledge Handbook of Archaeological Human Remains and Legislation: An international guide to laws and practice in the excavation and treatment of archaeological human remains*: 657-670. London, Routledge.

Bevan, B. W. 1991. The Search for Graves. *Geophysics* 56: 1310-1319.

Blau, S. 2004. Forensic Archaeology in Australia: Current situations, future possibilities. *Australian Archaeology* 58: 11-14.

Blau, S. 2005. One Chance only: advocating the use of archaeology in search, location and recover at disaster scenes. *The Australian Journal of Emergency Management* 20 (1): 19-24.

Blau, S. and Skinner, M. 2005. The Use of Forensic Archaeology in the Investigation of Human Rights Abuse: Unearthing the Past in East Timor. *The International Journal of Human Rights* 9 (4): 449-463.

Blau, S. and Ubelaker, D. H. 2009. *Handbook of Forensic Anthropology and Archaeology*. California, Left Coast Press Inc.

Boas, G., 2001. Creating Laws of Evidence for International Criminal Law: The ICTY and the Principle of Flexibility. *Criminal Law Forum* 12 (1): 41-90.

Borić, I., Ljubković, J. and Sutlović, D. 2011. Discovering the 60 years old secret: identification of the World War II mass grave victims from the island of Daksa near Dubrovnik, Croatia. *Croatian Medical Journal* 52: 327-335.

Brooks, S. T. and Brooks, R. H. 1984. Problems of burial exhumation, historical and forensic aspects. In T. A. Rathbun and J. E. Buikstra (eds), *Human Identification: Case Studies in Forensic Anthropology*: 64-86. Illinois, Charles C Thomas.

Browman, L. and Givens, D. 1996. Stratigraphic Excavation: The First "New Archaeology". *American Anthropologist* 98 (1): 80-95.

Burns, R. K. R. 2006. *Forensic Anthropology Training Manual*. Second Edition. New Jersey, Pearson Prentice Hall.

Carver, M. 2009. *Archaeological Investigation*. Oxon, Routledge.

Carver, M. 2011. *Making Archaeology Happen Design versus Dogma.* California, Left Coast Press Inc.

Çatalhöyük. 2011. Excavations of a Neolithic Anatolian Höyük. Available from: http://www.catalhoyuk. com/ [Accessed on: 03/02/2011 10:22].

Cheetham, P. 2005. Forensic Geophysical Survey. In J. R. Hunter and M. Cox (eds), *Forensic Archaeology: Advances in Theory and Practice*: 62-95. Oxon, Routledge.

Cheetham, P. and Hanson, I. 2009. Excavation and Recovery in Forensic Archaeological Investigations. In S. Blau and D. H. Ubelaker (eds), *Handbook of Forensic Anthropology and Archaeology*: 141-149. California, Left Coast Press Inc.

Cheetham, P., Cox, M., Flavel, A., Hanson, I., Haynie, T., Oxlee, D. and Wessling, R. 2008. Search, location, excavation, and recovery. In M. Cox., A. Flavel., I. Hanson., J. Laver and R. Wessling (eds), *The Scientific Investigation of Mass Graves: Towards Protocols and Standard Operating Procedures*: 183-267. Cambridge, Cambridge University Press.

Chilcott, J. H. and Deetz, J. J. 1964. The Construction and Uses of a Laboratory Archaeological Site. *American Antiquity* 29 (3): 328-337.

Clarke, D. V. 1978. Excavation and Volunteers: A Cautionary Tale. *World Archaeology* 10 (1): 63-70.

Colley, S. 2004. University-based archaeology teaching and learning and professionalism in Australia. *World Archaeology* 36 (2): 189-202.

Collis, J. 2002. *Digging up the Past An Introduction to Archaeological Excavation*. Stroud, Sutton Publishing Limited.

Congram, D. R. 2008. A Clandestine Burial in Costa Rica: Prospection and Excavation. *Journal of Forensic Sciences* 53 (4): 793-796.

Connor, M. A. 2007. *Forensic Methods Excavation for the Archaeologist and Investigator.* New York, AltaMira Press.

Connor, M. A. and Scott, D. D. 2001. Paradigms and Perpetrators. *Historical Archaeology* 35 (1): 1-6.

Cordner, S. and McKelvie, H. 2002. Developing standards in international forensic work to identify missing persons. *International Committee of the Red Cross*: 867–884.

Cox, M. 2009. Forensic Anthropology and Archaeology: Past and Present – A United Kingdom Perspective. In S. Blau and D. H. Ubelaker (eds), *Handbook of Forensic Anthropology and Archaeology*: 29-41. California, Left Coast Press Inc.

Cox, M., Flavel, A., Hanson, I., Laver, J. and Wessling, R. 2008. *The Scientific Investigation of Mass Graves: Towards Protocols and Standard Operating Procedures.* Cambridge, Cambridge University Press.

Crabtree, K. 1990. Experimental Earthworks in the United Kingdom. In D. E. Robinson (ed.), *Experimentation and Reconstruction in Environmental Archaeology*: 225-236. Oxford, Oxbow Books.

Crist, T. A. J. 2001. Bad to the Bone?: Historical Archaeologists in the Practice of Forensic Science. *Historical Archaeology* 35 (1): 39-56.

Daniel, G. 1978. *150 Years of Archaeology*. London, Duckworth.

Darvill, T. 2000. Recording systems in archaeological excavation: introducing ARTHUR. In

T. Darvill., G. Afanas'ev and E. Wilkes (eds), *Anglo-Russian Archaeology Seminar: Recording Systems for Archaeological Projects*: 30-36. Bournemouth, Bournemouth University.

Darvill, T. 2008. *The Concise Oxford Dictionary of Archaeology*. Oxford, Oxford University Press.

Daubert V. Merrell Dow Pharmaceuticals, 509 U.S. 579, 133 S.Ct 2786, 125 L.Ed. 2d 249, (U.S. Jun 28, 1993) (No. 92-102).

Davenport, A. and Harrison, K. 2011. Swinging the blue lamp: The forensic archaeology of contemporary child and animal burial in the UK. *Mortality* 16 (2): 176-190.

Davenport, G. C. 2001a. *Where is it? Searching for Buried Bodies and Hidden Evidence.* Maryland, SportsWork.

Davenport, G. C. 2001b. Remote Sensing Applications in Forensic Investigations. *Historical Archaeology* 35 (1): 87-100.

Davenport, G. C., Lindemann, J. W., Griffin, T. J. and Borowski, J.E. 1988. Geotechnical Applications 3: Crime Scene Investigation Techniques. *Geophysics: The Leading Edge of Exploration* 7 (8): 64-66.

Desert Archaeology Inc. 2008. *Desert Archaeology Inc. Field Excavation Manual.* Arizona, Desert Archaeology Inc.

Dilley, R. 2005. Legal Matters. In J. R. Hunter and M. Cox (eds), *Forensic Archaeology: Advances in Theory and Practice*: 177-203. Oxon, Routledge.

Dirkmaat, D. C. and Adovasio, J. M. 1997. The Role of Archaeology in the Recovery and Interpretation of Human Remains from an Outdoor Forensic Setting. In W. D. Haglund and M. H. Sorg (eds), *Forensic Taphonomy: The Postmortem Fate of Human Remains*: 39-64. Boca Raton, CRC Press LLC.

Donlon, D. 2009. The Development and Current State of Forensic Anthropology: An Australian Perspective. In S. Blau and D. H. Ubelaker (eds), *Handbook of Forensic Anthropology and Archaeology*: 104-114. California, Left Coast Press Inc.

Donlon, D. and Littleton, J. 2011. Australia. In N. Márquez-Grant and L. Fibiger (eds), *The Routledge Handbook of Archaeological Human Remains and Legislation: An international guide to laws and practice in the excavation and treatment of archaeological human remains*:

633-645. London, Routledge.

Drewett, P. L. 1999. *Field Archaeology: An Introduction*. London, UCL Press.

Drewett, P. L. 2000a. Grid Excavation. In L. Ellis (ed.), *Archaeological Method and Theory An Encyclopaedia*: 269. London, Garland Publishing Inc.

Drewett, P. L. 2000b. Open Area Excavation. In L. Ellis (ed.), *Archaeological Method and Theory An Encyclopaedia*: 413-414. London, Garland Publishing Inc.

Drewett, P. L. 2000c. Quartering. In L. Ellis (ed.), *Archaeological Method and Theory An Encyclopaedia*: 502-503. London, Garland Publishing Inc.

Drewett, P. L. 2000d. Trench Excavation. In L. Ellis (ed.), *Archaeological Method and Theory An Encyclopaedia*: 635-636. London, Garland Publishing Inc.

Drewett, P. L. 2000e. Unit-Level Method. In L. Ellis (ed.), *Archaeological Method and Theory An Encyclopaedia*: 645. London, Garland Publishing Inc.

Droop, J. P. 1915. Archaeological Excavation. Cambridge, Cambridge University Press.

Dupras, T. L., Schultz, J. J., Wheeler, S. M. and Williams, L.J. 2006. *Forensic Recovery of Human Remains Archaeological Approaches*. Boca Raton, CRC Press.

Ebsworth, E. 2000. The Council for the Registration of Forensic Practitioners. *Science and Justice* 40 (2): 134-137.

Edmond, G. 2010. Impartiality, efficiency, or reliability? A critical response to expert evidence law and procedure in Australia. *Australian Journal of Forensic Sciences* 42 (2): 83-99.

Edwards, H. T. 2009. *Strengthening Forensic Science in the United States: A Path Forward Statement of the Honourable Harry T. Edwards Senior Circuit Judge and Chief Judge Emeritus United States Court of Appeals for the D.C. Circuit*. Washington D.C., United States Senate Committee on the judiciary.

Evis, L. H. 2009. *A comparative study into the evidential recovery potential of arbitrary and stratigraphic excavation and the affect that excavation experience has on the extent of evidence recovered.* Unpublished MSc Dissertation, Bournemouth University.

Ferllini, R. 2003. The development of human rights investigations since 1945. *Science and Justice* 43 (4): 219-224.

Fowler, P. J. 1977. *Approaches to Archaeology*. London, Adam and Charles Black Ltd.

Fowler, P. J. 1980. The Experimental Earthworks: A Summary of the Project's First Thirty Years. *Annual Report of the Council for British Archaeology* 39: 83-98.

Fox, C. 1959. *Life and Death in the Bronze Age*. London, Routledge and Kegan Paul.

France, D. L., Griffin, T. J., Swanburg, J. G., Lindemann, J. W., Davenport, G. C., Trammell, V., Armbrust, C. T., Kondratieff, B., Nelson, A., Castellano, K. and Hopkins, D. 1992. A Multidisciplinary Approach to the Detection of Clandestine Graves. *Journal of Forensic Sciences* 37 (6): 1445-1458.

Frye v. United States, 293F. 1013 (D.C. Cir. 1923).

Gamble, C. 2001. *Archaeology the Basics*. London, Routledge.

Gardner, R. M. 1997. *Bloodstain Pattern Analysis*. Third Edition. Boca Raton, CRC Press.

Glancy, G. D. and Bradford, J. M. W. 2007. The Admissibility of Expert Evidence in Canada. *Journal of the American Academy of Psychiatry and the Law* 35 (3): 350-356.

Gojanović, M. D. and Sutlović, D. 2007. Skeletal Remains from World War II Mass Grave: from Discovery to Identification. *Croatian Medical Journal* 48: 520-527.

Gould, R. A. 2007. *Disaster Archaeology*. Utah, The University of Utah Press.

Greene, K. and Moore, T. 2010. *Archaeology: An Introduction*. Fifth Edition. London, Routledge.

Haglund, W. D. 2001. Archaeology and Forensic Death Investigations. *Historical Archaeology* 35 (1): 26-34.

Haglund, W. D., Connor, M. A. and Scott, D. D. 2001. The Archaeology of Contemporary Mass Graves. *Historical Archaeology* 35 (1): 57-69.

Hampton, S. 2004. The role of the expert witness. *Journal of Wound Care* 13 (10): 435-436.

Hanson, I. 2004. The importance of stratigraphy in forensic investigation. *Forensic Geoscience: Principles, Techniques and Applications* 232: 39-47.

Hanzlick, R. 2007. *Death Investigation Systems and Procedures*. Boca Raton, CRC Press.

Harris, E. C. 1979. The Laws of Archaeological Stratigraphy. *World Archaeology* 11 (1): 111-117.

Harris, E. C. 1989. *Principles of Archaeological Stratigraphy*. Second Edition. London, Academic Press Limited.

Harris, E. C. 2002. Stratification, soil. In C. E. Orser (ed.), *Encyclopaedia of Historical Archaeology*: 598-599. London, Routledge.

Harris, E. C. 2006. Archaeology and the Ethics of Scientific Destruction. In S. N. Archer and K. M. Bartoy (eds), *Between Dirt and Discussion: Methods, Methodology, and Interpretation in Historical Archaeology*: 141-150. New York, Springer Science + Business Media.

Harris, E. C., Brown III, M. R. and Brown, G. J. 1993. *Practices of Archaeological Stratigraphy*. London, Academic Press.

Hester, T. R. 1997. Methods of excavation. In T. R. Hester., K. L. Feder and H. J. Shafer (eds), *Field Methods in Archaeology*: 69-112. Seventh Edition. California, Mayfield Publishing.

Hester, T. R., Feder, K. L. and Shafer, H. J. 1997. *Field Methods in Archaeology*. Seventh Edition. California, Mayfield Publishing.

Hochrein, M. J. 1997. Buried Crime Scene Evidence: The Application of Forensic Geotaphonomy in Forensic Archaeology. In P. G. Stimson and C. A. Mertz (eds), *Forensic Dentistry*: 83-100. Boca Raton, CRC Press.

Hochrein, M. J. 2002. An autopsy of the grave: Recognising, Collecting and Preserving Forensic Geotaphonomic Evidence. In D. W. Haglund and M. H. Sorg (eds), *Advances in Forensic Taphonomy Method, Theory and Archaeological Perspectives*: 45-70. Boca Raton, CRC Press LLC.

Hunter, J. R. 2009. Domestic Homicide Investigations in the United Kingdom. In S. Blau and D. H. Ubelaker (eds), *Handbook of Forensic Anthropology and Archaeology*: 363-373. California, Left Coast Press Inc.

Hunter, J. R. and Cox, M. 2005. *Forensic Archaeology: Advances in Theory and Practice*. Oxon, Routledge.

Hunter, J. R. and Dockrill, S. 1996. Recovering buried remains. In J. R. Hunter., C. A. Roberts and A. L. Martin (eds), *Studies in Crime: An Introduction to Forensic Archaeology*: 40-57. London, B.T. Batsford Ltd.

Hunter, J. R. and Knupfer, G. C. 1996. The police and judicial structure in Britain. In J. R. Hunter., C. A. Roberts and A. L. Martin (eds), *Studies in Crime: An Introduction to Forensic Archaeology*: 24-39. London, B.T. Batsford Ltd.

Hunter, J. R. and Martin, A. L. 1996. Locating buried remains. In J. R. Hunter., C. A. Roberts and A. L. Martin (eds), *Studies in Crime: An Introduction to Forensic Archaeology*: 86-100. London, B.T. Batsford Ltd.

Hunter, J. R., Brickley, M. B., Bourgeois, J., Bouts, W., Bourguignon, L., Hubrecht, F., De Winne, J., Van Haaster, H., Hakbijl, T., De Jong, H., Smits, L., Van Wijngaarden, L. H. and Luschen, M. 2001. Forensic Archaeology, forensic anthropology and Human Rights in Europe. *Science and Justice* 41 (3): 173-178.

Hunter, J. R., Heron, C., Janaway, R. C., Martin, A. L., Pollard, A. M. and Roberts, C.A. 1994. Forensic Archaeology in Britain. *Antiquity* 68 (261): 758-769.

Hunter, J. R., Roberts, C. A. and Martin, A. L. 1996. *Studies in Crime: An Introduction to Forensic Archaeology*. London, B.T. Batsford Ltd.

Hunter, J. R., Simpson, B. and Sturdy Colls, C. 2013. *Forensic Approaches to Buried Remains*. Chichester, John Wiley and Sons Ltd.

Hunter v. United States, 48F. Supp. 2d 1283, 1288 (D. Utah. 1998).

Hutt, S. 2006. The Acceptance of Archaeological Value as Evidence in Court. In S. Hutt., M. P. Forsyth and D. Tarler (eds), *Presenting Archaeology in Court Legal Strategies for Protecting Cultural Resources*: 143-152. New York, AltaMira Press.

ICTY. 2007. Trial Transcript: Vujadin Popovic. Available from: http://www.icty.org/x/cases/popovic/trans/en/070313ED.htm [Accessed on: 03/02/2012 12:33]

Isaacson, J., Hollinger, R. E., Gundrum, D. and Baird, J. 1999. A Controlled Archaeological Test Site Facility in Illinois: Training and Research in Archaeogeophysics. *Journal of Field Archaeology* 26 (2): 227-236.

Janaway, R. C. 1996. The decay of buried remains and their associated materials. In J. R. Hunter., C. A. Roberts and A. L. Martin (eds), *Studies in Crime: An Introduction to Forensic Archaeology*: 58-85. London, B.T. Batsford Ltd.

Janaway, R. C. 2002. Degradation of Clothing and other Dress Materials. Associated with Buried Bodies of Both Archaeological and Forensic Interest. In D. W. Haglund and M. H. Sorg (eds), *Advances in Forensic Taphonomy Method, Theory and Archaeological Perspectives*: 379-402. Boca Raton, CRC Press LLC.

Jankauskas, R., Barkus, A., Urbanavièius, V. and Garmus, A. 2005. Forensic archaeology in Lithuania: the Tuskulёnai mass grave. *Acta Medica Lituanica* 12 (1): 70-74.

Jessee, E. and Skinner, M. 2005. A typology of mass grave and mass grave-related sites. *Forensic Science International* 152 (1): 55-59.

Joukowsky, M. 1980. *A Complete Manual of Field Archaeology*. Englewood Cliffs, Prentice Hall Inc.

Juhl, K. 2005. *The Contribution by (Forensic) Archaeologists to Human Rights Investigations of Mass Graves*. Stavanger, Stavanger Museum of Archaeology.

Kershaw, A. 2001. Expressing a standard. *Science and Justice* 41 (3): 226-228.

Kidder, A. V. and Kidder, M. A. 1917. Notes on the pottery of Pecos. *American Anthropologist* 19 (3): 325-360.

Killam, E. W. 2004. *The Detection of Human Remains*. Illinois, C.C. Thomas.

Kittichaisaree, K. 2001. *International Criminal Law*. Oxford, Oxford University Press.

Kjolbye-Biddle, B. 1975. A Cathedral Cemetery: Problems in Excavation and Interpretation. *World Archaeology* 7 (1): 87-108.

Klinkner, M. 2008. Proving Genocide? Forensic Expertise and the ICTY. *Journal of International Criminal Justice* 6: 447-466.

Klinkner, M. 2009. Forensic science expertise for international criminal proceedings: an old problem, a new context and a pragmatic resolution. *The International Journal of Evidence and Proof* 13: 102-129.

Komar, D. and Buikstra, J. E. 2008. *Forensic Anthropology: Contemporary Theory and Practice*. New York, Oxford University Press.

Krogman, W. M. 1943. The role of the physical anthropologist in the identification of human remains. *FBI Law Enforcement Bulletin* 12 (4): 17-40.

Landry, C. 2012. *The Effects of Time Pressures in Forensic Archaeology: A Preliminary Investigation Into Establishing an Optimum Speed for Excavation*. Unpublished MSc Dissertation, Bournemouth University.

Larson, D. O., Vass, A. A. and Wise, M. 2011. Advanced Scientific Methods and Procedures in the Forensic Investigation of Clandestine Graves. *Journal of Contemporary Criminal Justice* 27 (2): 149-182.

Litherland, S., Márquez-Grant, N. and Roberts, J. 2012. Forensic archaeology. In J. Roberts and N. Márquez-Grant (eds), *Forensic Ecology Handbook From Crime Scene to Court*: 23-48. Chichester, John Wiley and Sons Ltd.

Lucas, G. 2001. *Critical Approaches to Fieldwork: Contemporary and Historical Archaeological Practice*. London, Routledge.

Lyell, C. 1830. *Principles of Geology*. Volume 1. London, Murray.

Lyell, C. 1832. *Principles of Geology*. Volume 2. London, Murray.

Lyell, C. 1833. *Principles of Geology*. Volume 3. London, Murray.

Lyman, R. L. and O'Brien, M. J. 1999. Americanist Stratigraphic Excavation and the Measurement of Culture Change. *Journal of Archaeological Method and Theory* 6 (1): 55-101.

McDonald, J. J. 1999. *Excavation at Berry Island, N.S.W.* Unpublished Report, New South Wales Police.

Menez, L. L. 2005. The place of a forensic archaeologist at a crime scene involving a buried body. *Forensic Science International* 152 (2-3): 311-315.

Morse, D., Crusoe, D. and Smith, H. G. 1976. Forensic archaeology. *Journal of Forensic Sciences* 21 (2): 323-332.

Morse, D., Dailey, R. C., Stoutamire, J. and Duncan, J. 1984. Forensic Archaeology. In T. A. Rathbun and J. E. Buikstra (eds), *Human Identification: Case Studies in Forensic Anthropology*: 53-63. Illinois, Charles C Thomas.

Morse, D., Duncan, J. and Stoutamire, J. 1983. *Handbook of Forensic Archaeology and Anthropology*. Tallahassee, Morse, D.

NAS. 2009. *National Academy of Sciences Report Strengthening Forensic Science in the United States: A Path Forward*. Washington D.C., The National Academies Press.

Nash, D. T. and Petraglia, M. D. 1987. Natural formation processes and the archaeological record: Present problems and future requisites. In D. T. Nash and M. D. Petraglia (eds), *Natural formation processes and the archaeological record BAR International Series 352:* 186-204. Oxford, British Archaeological Reports.

Nuzzolese, E. and Borrini, M. 2010. Forensic Approach to an Archaeological Casework of "Vampire" Skeletal Remains in Venice: Odontological and Anthropological Prospectus. *Journal of Forensic Sciences* 55 (6): 1634-1637.

Oakley, K. 2005. Forensic Archaeology and Anthropology an Australian Perspective. *Forensic Science, Medicine, and Pathology* 1 (3): 169-172.

Owsley, D. W. 2001. Why the forensic anthropologist needs the archaeologist. *Historical Archaeology* 35 (1): 35-38.

Pallis, S. A. 1956. *The Antiquity of Iraq*. Copenhagen, E. Munksgaard.

Pavel, C. 2010. *Describing and Interpreting the past European and American Approaches to the Written Record of the Excavation*. Bucharest, University of Bucharest.

Pelling, S. J. 2008. *A comparative study of evidential recovery in arbitrary and stratigraphic excavation*. Unpublished MSc Dissertation, Bournemouth University.

Pepper, I. K. 2005. *Crime Scene Investigation Methods and Procedures*. Maidenhead, Open University Press.

Phillips, P., Ford, J. A. and Griffin, J. B. 1951. *Archaeological Survey in the Lower Mississippi Alluvial Valley, 1940-1947. Papers of the Peabody Museum of American Archaeology and Ethnology*. Volume 25. Cambridge, Harvard University Press.

Pickering, R. B. and Bachman, D. C. 1997. *The Use of Forensic Anthropology*. Boca Raton, CRC Press.

Powell, J. F., Steele, D. G. and Collins, M. B. 1997. Excavation and Analysis of Human Remains. In T. R. Hester., K. L. Feder and H. J. Shafer (eds), *Field Methods in Archaeology*: 252-282. Seventh Edition. California, Mayfield Publishing.

Powers, N. and Sibun, L. 2011. *Standards and guidance for forensic archaeologists*. Reading, The Institute for Archaeologists.

Praetzellis, A. C. 1993. The Limits of Arbitrary Excavation. In E. C. Harris., M. R. Brown III and G. J. Brown (eds), *Practices of Archaeological Stratigraphy*: 68-86. London, Academic Press.

Pringle, H. 2006. Hitler's Willing Archaeologists. *Archaeology* 59 (2): 44-49.

Rainio, J., Karkola, K., Lalu, K., Ranta, H., Takamaa, K. and Penttilä. 2001. Forensic investigations in Kosovo: experiences of the European Union Forensic Expert Team. *Journal of Clinical Forensic Medicine* 8: 218-221.

Renfrew, C. 1973. *The Explanation of Culture Change: Models in Prehistory*. London, Duckworth.

Riley, T. J. and Freimuth, G. 1979. Field Systems and Frost Drainage in the Prehistoric Agriculture of the Upper Great Lakes. *American Antiquity* 44 (2): 271-285.

Roberts, N. 2009. *The possible application of archaeological theory in forensic archaeology*. Unpublished MSc Dissertation, Bournemouth University.

Roberts, P. and Willmore, C. 1993. *The Role of Forensic Science Evidence in Criminal Proceedings, Royal Commission on Criminal Justice Research Study Number 11*. London, HMSO.

Robertson, J. 2009. Forensic Evidence Goes on Trial. *Australian Journal of Forensic Sciences* 41 (1): 1-2.

Robertson, J. 2010. Professionalising Forensic Science. *Australian Journal of Forensic Sciences* 42 (3): 157-158.

Rodriguez, W. C. and Bass, W. M. 1985. Decomposition of Buried Bodies and Methods that May Aid in Their Location. *Journal of Forensic Sciences* 30 (3): 836-852.

Roskams, S. 2001. *Excavation*. Cambridge, Cambridge University Press.

Ruffell, A., Donnelly, C., Carver, N., Murphy, E., Murray, E. and McCambridge, J. 2009. Suspect burial excavation procedure: A cautionary tale. *Forensic Science International* 183: 11-16.

Ruwanpura, P. R., Perera, U. C. P., Wijayaweera, H. T. K. and Chandrasiri, N. 2006. Adaptation of archaeological techniques in forensic mass grave exhumation: the experience of 'Chemmani' excavation in northern Sri Lanka. *Ceylon Medical Journal* 51 (3): 98-102.

Saul, J. M. and Saul, F. P. 2002. Forensics, Archaeology, and Taphonomy: The Symbiotic Relationship. In D. W. Haglund and M. H. Sorg (eds), *Advances in Forensic Taphonomy Method, Theory and Archaeological Perspectives*: 71-98. Boca Raton, CRC Press LLC.

Schabas, W. A. 2006. *The UN International Criminal Tribunals: The Former Yugoslavia, Rwanda and Sierra Leone*. Cambridge, Cambridge University Press.

Scherr, J. 2009. *Speed of Excavation: A study on how time implications affect evidence recovered from individual graves*. Unpublished MSc Dissertation, Bournemouth University.

Schultz, J. J. and Dupras, T. L. 2008. The Contribution of Forensic Archaeology to Homicide Investigations. *Homicide Studies* 12 (4): 399-413.

Scott, D. D. and Connor, M. A. 2001. The Role and Future of Archaeology in Forensic Science. *Historical Archaeology* 35 (1): 101-104.

Selby, H. 2010. When sciences comes to court. *Australian Journal of Forensic Sciences* 42 (3): 159-167.

Sigler-Eisenberg, B. 1985. Expanding the Concept of Applied Archaeology. *American Antiquity* 50 (3): 650-655.

Skinner, M., Alempijevic, D. and Djuric-Srejic, M. 2003. Guidelines for International Forensic Bio-archaeology Monitors of Mass Grave Exhumations. *Forensic Science International* 134: 81-92.

Skinner, M. and Bowie, K. 2009. Forensic Anthropology: Canadian Content and Contributions. In S. Blau and D. H. Ubelaker (eds), *Handbook of Forensic Anthropology and Archaeology*: 87-103. California, Left Coast Press Inc.

Skinner, M. and Sterenberg, J. 2005. Turf wars: Authority and responsibility for the investigation of mass graves. *Forensic Science International* 151: 221-232.

Sklenář, K. 1983. *Archaeology in central Europe: the first 500 years*. New York, St. Martin's Press.

Smith, W. 1815. *A Delineation of the Strata of England and Wales with part of Scotland*. London, J. Cary.

Smith, W. 1816. *Strata Identified by Organised Fossils*. London, W. Arding.

Snow, C. C. 1982. Forensic Anthropology. *Annual review of Anthropology* 11: 97-131.

Sonderman, R. C. 2001. Looking for a Needle in a Haystack: Developing Closer Relationships between Law Enforcement Specialists and Archaeology. *Historical Archaeology* 35 (1): 70-78.

Spennemann, D. H. R. and Franke, B. 1995. Archaeological techniques for exhumations: a unique data source for crime scene investigations. *Forensic Science International* 74: 5-15.

Steno, N. 1669. *De Solido Intra Solidum Naturaliter Contento Dissertationis Prodromus*. Florence, Ex Typographia Sub Signo Stellae.

Stover, E. and Ryan, M. 2001. Breaking Bread with the Dead. *Historical Archaeology* 35 (1): 7-25.

Tayles, N. and Halcrow, J. 2011. New Zealand. In N. Márquez-Grant and L. Fibiger (eds), *The Routledge Handbook of Archaeological Human Remains and Legislation: An international guide to laws and practice in the excavation and treatment of archaeological human remains*: 647-655. London, Routledge.

The International Criminal Court. 2013. *Rules of Procedure and Evidence*. The Hague Netherlands, The International Criminal Court.

The Law Commission. 2009. *The Admissibility of Expert Evidence in Criminal Proceedings in England and Wales: A New Approach to the Determination of Evidentiary Reliability: Consultation Paper 190*. London, The Law Commission.

The Law Commission. 2011. *Expert Evidence in Criminal Proceedings in England and Wales*. London, The Law Commission.

Thompson, T. J. U. 2003. Supply and demand: the shifting expectations of forensic anthropology in the United Kingdom. *Science and Justice* 43 (4): 183-186.

Tuller, H. and Đurić, M. 2006. Keeping the pieces together: Comparison of mass grave excavation methodology. *Forensic Science International* 156: 192-200.

Turnbaugh, W., Jurmain, R., Nelson, H. and Kilgore, L. 2002. *Understanding Physical Anthropology and Archaeology*. Eighth Edition. California, Wadsworth Publishing.

Ubelaker, D. H. 1989. *Human Skeletal Remains: Excavation, Analysis, Interpretation*. Second Edition. Washington D.C., Taraxacum.

United Nations. 1991. Model protocol for disinterment and analysis of skeletal remains. In United Nations Office at Vienna Centre for Social Development and Humanitarian Affairs (ed.), *Manual of the Effective Prevention and Investigation of Extra-Legal, Arbitrary and Summary Executions*: 30-41. New York, United Nations.

Vanezis, P. 1999. Investigation of clandestine graves resulting from human rights abuses. *Journal of Clinical Forensic Medicine* 6: 238-242.

Wainwright, G. J. 1968. The excavation of a Durotrigian farmstead near Tollard Royal in Cranborne Chase, southern England. *Proceedings of the Prehistoric Society* 34: 102-147.

Wainwright, G. J., 1979. *Gussage All Saints – an Iron Age settlement in Dorset, Department of the Environmental Archaeological Reports 10*. London, HMSO.

Wheeler, R. E. M. 1954. *Archaeology from the Earth*. Oxford, Clarendon Press.

Willey, G. R. and Sabloff, J. A. 1980. *A History of American Archaeology*. Second Edition. San Francisco, W.H. Freeman.

Wolfe Steadman, D., Basler, W., Hochrein, M. J., Klein, D. F. and Goodin, J.C. 2009. Domestic Homicide Investigations: An Example from the United States. In S. Blau and D. H. Ubelaker (eds), *Handbook of Forensic Anthropology and Archaeology*: 351-362. California, Left Coast Press Inc.

Wright, R. 1995. Investigating War Crimes – The Archaeological Evidence. *The Sydney Papers* 7 (3): 39-44.

Wright, R., Hanson, I. and Sterenberg, J. 2005. The archaeology of mass graves. In J. R. Hunter and M. Cox (eds), *Forensic Archaeology: Advances in Theory and Practice*: 137-158. Oxon, Routledge.

Zahar, A. and Sluiter, G. 2008. *International Criminal Law: a critical introduction*. Oxford, Oxford University Press.

Legislation:

England and Wales: The Police and Criminal Evidence Act (1984); the Criminal Justice Act (2003); the Criminal Evidence (Experts) Act (2011); the Criminal Procedure Rules (2011).

Scotland: The Criminal Procedure (Scotland) Act (1995); the Criminal Justice (Scotland) Act (2003).

Northern Ireland: The Criminal Evidence Act (Northern Ireland) (1923); the Police and Criminal Evidence Act (1984); the Criminal Justice (Evidence) (Northern Ireland) Order (2004).

Ireland: The Criminal Justice (Forensic Evidence) Act (1990); the Rules of the Superior Courts (Evidence) (2007).

Australia: The Criminal Procedure Ordinance (1993); the Evidence Act (1995); the Evidence Amendment Act (2008); the Practice Note CM7 Expert Witnesses in Proceedings in the Federal Court of Australia (2009).

New Zealand: The Evidence Act (2006); the Practice Notes – Expert Witness – Code of Conduct (2011); the Criminal Procedure Act (2011).

Canada: The Canada Evidence Act (1985); the Criminal Code (1985); the Federal Court Rules – SOR/98-106.

United States of America: The Frye Standard (1923); the Daubert Standard (1993); Federal Rules of Evidence (2011).

Appendix A: List of contributors

List of contributors to the Central Theoretical Archaeology Group Conference Session, Birmingham, December 2011:

Timothy Darvill, Bournemouth University; Paul Cheetham, Bournemouth University; Ian Hanson, Bournemouth University; Catalin Pavel, Kennesaw University; Frank Meddens, Pre-Construct Archaeology; Gary Brown, Pre-Construct Archaeology; Thomas Whitley, Brockington and Associates Inc.; Kevin Wooldridge, Bergen Museum; Robert Hedge, Foundations Archaeology; Rebecca Hunt, Birmingham University; Steven Roskams, University of York; Martin Carver, University of York; Chiz Harward, Cotswold Archaeology; Karl Harrison, Cranfield University.

List of contributors to the Society for American Archaeology Conference Session, Memphis, April 2012:

Timothy Darvill, Bournemouth University; Robert Heckman, Statistical Research Inc.; Richard Ciolek-Torrello, Statistical Research Inc.; Michael Heilen, Statistical Research Inc.; Edward Harris, Bermuda Maritime Museum; Russell Quick, Cultural Resource Analysts Inc.; Randall Cooper, Cultural Resource Analysts Inc.; Paul Bundy, Cultural Resource Analysts Inc.; Andrew Bradbury, Cultural Resource Analysts Inc.; Jason Anderson, Cultural Resource Analysts Inc.; Scott Palumbo, College of Lake County; Virginia Petch, Northern Lights Heritage Services Inc.; Donald Thieme, Valdosta State University; Jane Whitehead, Valdosta State University; Kit Wesler, Murray State University; Douglas Wilson, Northwest Cultural Resources Institute; Elizabeth Horton, Northwest Cultural Resources Institute; Gerald Schroedl, University of Tennessee, Knoxville; Stephen Yerka, University of Tennessee, Knoxville; Nicholas Herrmann, University of Tennessee, Knoxville; Jeffery Shanks, Southeast Archaeological Centre; Craig Dengel, Southeast Archaeological Centre; Michael Russo, Southeast Archaeological Centre; Katherine Turner-Pearson, URS Corporation.

List of contributors to the Institute for Archaeologists Conference Session, Oxford, April 2012:

Birgitta Hoffmann, University of Liverpool; Reuben Thorpe, University College London; Chiz Harward, Cotswold Archaeology; Timothy Darvill, Bournemouth University; Phil Mills, Freelance Ceramic Specialist, Leicester; Beth Werrett, Wiltshire Conservation Service; Mary Neale, Berkshire Archaeology.

List of excavation manual/guideline contributors:

United Kingdom: Archaeological Project Services (APS); Archaeology South East; Birmingham Archaeology; CFA Archaeology Ltd.; Cornwall Archaeological Unit (CAU); Cotswold Archaeology; Dyfed Archaeological Trust Ltd.; John Moore Heritage Services; L-P Archaeology; Museum of London Archaeology; Nexus Heritage; Oxford Archaeology; Phoenix Consulting Archaeology Ltd.; Pre-Construct Archaeology Ltd.; SHARP Archaeology; SLR Consulting; Surrey County Archaeology Unit (SCAU); Thames Valley Archaeological Services (TVAS); The Environmental Dimension Partnership (EDP); The University of Winchester; University College London (Institute of Archaeology); University of Leicester; University of Leicester Archaeological Services; Warwickshire Museum Field Services; Waterman Group; Wessex Archaeology; Worcestershire Archaeology Service.

Ireland: Achill Field School; Eachtra Archaeological Projects Group; Gahan and Long Ltd.; Headland Archaeology Ltd.; Judith Carroll and Company Ltd.; Margaret Gowen and Co. Ltd.; Northern Ireland Environmental Agency (NIEA); The Archaeology Company; University College Dublin (UCD).

Australasia: Archaeological and Heritage Management Solutions Pty Ltd.; Archaeology at Tardis Cultural Heritage Advisors; CFG Heritage; Clough and Associates Ltd.; Comber Consultants Pty Ltd.; Department of Planning and Community Development, Victoria; Dortch and Cuthbert Pty Ltd.; *Freelance* - Oliver Brown, Field Archaeologist; Noelene Cole, Field Archaeologist; Colin Pardoe, Bioanthropology and archaeology specialist; Elizabeth White, Consultant Archaeologist; Hunter Geophysics; James Cook University; Jo McDonald Cultural Heritage Management Pty Ltd.; Scarp Archaeology; Susan McIntyre-Tamwoy Heritage Consultants.

North America: AECOM; AF Consultants; Alaska Department of Transportation; Alutiiq Museum and Archaeological Repository; AMEC Earth and Environmental Inc.; ANCHOR QEA LLC.; Archaeological Consultants of the Carolinas Inc.; Applied Earthworks Inc.; Archaeological Damage Investigation and Assessment (ADIA); Archaeological Legacy Institute; Arizona State University; Aspen CRM Solutions; Bennett Management Services LLC.; Big Bend Ranch State Park; BLM-Fairbanks District Office, Alaska; BonTerra Consulting; Boonshoft Museum of Discovery; Boston Landmarks Commission; Brian F. Smith and Associates Inc.; Brockington and Associates; Brunson Cultural Resource Services LLC.; Bryn Mawr College; Buckhorn Archaeological Services; Buffalo State College; Bureau of Land Management; Bureau of Ocean Energy Management; Bureau of Reclamation; CAIRN Underwater Unit; California Department of Transportation; California State College; California State Parks; Calvert County Planning and Zoning, Maryland; Centuries Research Inc.; Chattahoochee-Oconee National Forests; Chicora Foundation Inc.; CH2M HILL; Coastal Carolina Research; Coastal Heritage Society; College Lake County; College of Southern Nevada; Colorado Historical Society; Commonwealth Cultural Resources Group Inc.; Connecticut College; Cornerstone Environmental Consulting LLC.; Cox/McLain Environmental Consulting Inc.; Crow Canyon Archaeological Centre; Cultural Resource Analysts Inc.; Desert Archaeology Inc.; Earth Search Inc.; ECORP Consulting Inc.; ECS Mid-Atlantic LLC.; Environmental Services Inc.; Far Western Anthropological Research Group Inc.; Federal Preservation Institute; Florida Atlantic University; Florida History LLC.; Francis Heritage LLC.; *Freelance* - Anne Beaubien, Archaeologist; Dayle Cheever, Archaeological Consultant; Scott Crull, Archaeologist; Elizabeth Davidson, Archaeologist; Erica Degelmann, Archaeologist; William Johnson, Archaeologist; David Maxwell, Archaeologist; Antoni Paris, Archaeologist; Angel Rodriguez, Archaeologist; Elisabeth Sheldon, Archaeologist; Suzanne Stewart, Archaeologist; Fudan Museum Foundation and Museum of Asian Art; Gateway Archaeology; Georgia Department of Transportation; Hartgen Archaeological Associates Inc.; Haywood Archaeological Services; HDR Inc.; Historical Research Associates Inc.; Historic Properties Consultants; History and Museums Division of California State Parks; Human Systems Research Inc.; Illinois Department of Transportation; Indiana University – Purdue University; Interpreting Time's Past LLC.; John Milner Associates Inc.; Knudson Associates; Logan Simpson Design; Louisiana Department of Justice; Louisiana State University; Maryland Historical Trust; Mercyhurst College; Metcalf Archaeological Consultants Inc.; Montana Heritage Commission; Montana Historical Society; Murray State University; Museum of New Mexico; National Park Service Archaeology Program; New Hampshire Department of Transportation; New South Associates Inc.; Northern Illinois University; Northern Lights Heritage Services Inc.; Northwest Cultural Resources Institute; Ohio Historic Preservation Office; Oregon State Historic Preservation Office; Oregon State Parks and Recreation Department; Orloff. G. Miller Consulting; Pacific Northwest Resource Consultants; Paciulli, Simmons and Associates Ltd.; Panamerican Consultants Inc.; Past Forward Inc.; Planning and Buildings Department in St. Augustine, Florida; Presidio Archaeology Lab; Redwood National Park; Robert. M. Lee. Trust; Santa Barbara County Planning and Development; SouthArc Inc.; South Carolina Institute of Archaeology and Anthropology (SCIAA); Southeast Archaeological Centre; Southeastern Archaeological Research Inc. (SEARCH); Stantec; Stratum Unlimited LLC.; Summit Envirosolutions; SWCA Environmental Consultants; Territory Heritage Resource Consulting; Tesla Offshore LLC.; Tetra Tech EC Inc.; Texas Parks and Wildlife; The Department of Architectural and Archaeological Research; The LAMAR Institute; The Missouri Department of Transportation; The

University of Oklahoma; The University of Texas, Austin; Thunderbird Archaeology; Transportation and Land Management Agency – Planning Department, County of Riverside; TRC Solutions Inc.; Tribal Historic Preservation Office; Universidad Autónoma del Estado de Morelos; Université Laval; University of Georgia; University of Hawaii Joint POW/MIA Accounting Command; University of Louisiana at Lafayette; University of Oregon; University of Pennsylvania; University of South Florida; University of Tennessee, Knoxville; University of Wisconsin-Baraboo; University of Wyoming; URS Corporation; US Army Corps of Engineers; US Army Garrison Joint Base Lewis-McChord; US Army Garrison Picatinny Arsenal, New Jersey; US Forest Service; US Government; Valdosta State University; Walton Enterprises; Wilbur Smith Associates; William Self Associates Inc..

List of organisations who participated in the experimental research:

Birkbeck College; Bournemouth University; Clwyd-Powys Archaeological Trust; Context One Archaeological Services Ltd.; North Cornwall Heritage; SHARP Archaeology; Trust for Thanet Archaeology; University Centre Peterborough; University College London; University of Cambridge; University of Nottingham; University of York.

Grave illustrations by:

Seán Goddard, Artist and Archaeological Illustrator.

Appendix B: Archaeological manual/guideline analytical criteria

Manual usage: Organisation has its own manual; Organisation has not got its own manual; The organisation are in the process of updating their manual; The organisation uses another organisation's manual.

Manual creation year: 1960-1969; 1970-1979; 1980-1989; 1990-1999; 2000-2009; 2010-2019.

Objectives of the manual: Recording and excavation protocols are developed on a project-by-project basis; The manual states which excavation methods are to be used; The manual suggests which excavation methods are to be used; The manual does not discuss excavation methods; The manual states which recording systems are to be used; The manual suggests which recording systems are to be used; The manual does not discuss recording systems; To instruct the archaeologist in which excavation method should be used; To instruct the archaeologist in which recording systems should be used; To instruct the archaeologist in how recording sheets are to be completed; To instruct archaeologists as to what should be recorded; To instruct archaeologists as to when recording should be conducted; To provide archaeologists with a default approach to excavation; To provide archaeologists with a default approach to recording; To instruct archaeologists in the excavation methods to be used on small-scale sites; To instruct archaeologists in the excavation methods to be used on large-scale sites; To instruct archaeologists in the excavation methods to be used on rural sites; To instruct archaeologists in the excavation methods to be used on urban sites; To instruct archaeologists in the recording systems to be used on small-scale sites; To instruct archaeologists in the recording systems to be used on large-scale sites; To instruct archaeologists in the recording systems to be used on rural sites; To instruct archaeologists in the recording systems to be used on urban sites; To ensure that the most cost effective methods of excavation are utilised; To ensure that the most efficient excavation methods are utilised; To ensure that the most cost effective recording systems are utilised; To ensure that the most efficient recording systems are utilised; To ensure that archaeological investigations maximise the accuracy of the recovery process; To ensure that the archaeological investigations maximise the accuracy of the recording process; To ensure consistency in approach to archaeological investigations; To ensure systematic approaches are used during archaeological investigations; To ensure standardised approaches are used during archaeological investigations; To ensure that objectivity is maintained throughout an archaeological investigation; To ensure that a complete a record as possible is obtained; To ensure that records produced are able to be re-interpreted by interested parties; To enable the creation of an ordered and systematic archive; To enable archaeologists to accurately describe the archaeological site; To enable the archaeologist to accurately interpret the archaeological site; To provide a structure to recording from which the organisation is able to publish its findings; To ensure that archaeological investigations adhere to ethical standards; To ensure that a high quality archaeological investigation is undertaken; To improve the current state of archaeological practice; To instruct interested parties in how the process of excavation should be conducted; To instruct interested parties in how recording should be conducted; To demonstrate to interested parties that the excavation methods being used are those that are widely accepted; To demonstrate to interested parties that the recording systems being used are those that are widely accepted.

Sector: Organisation works in the commercial archaeology sector; Organisation works in the research archaeology sector; Organisation works in the government archaeology sector.

Definition of stratigraphy: The manual does not discuss stratigraphy; A stratigraphic unit is any discrete unit or feature that has potentially recognisable archaeological integrity; Stratigraphic sequences are formed by a process of deposition and/or removal; Stratigraphic sequences are comprised of positive stratigraphic units and features; Stratigraphic sequences comprise of stratigraphic units which are either positive or negative; Urban sites are most likely to have complex stratified archaeological sequences present; Rural sites are least likely to have complex stratified archaeological sequences present; All archaeological sites are composed of stratified sequences; Chronology in an archaeological context refers to the relative date of activity between one stratigraphic unit and another; Within a stratified sequence the chronologically earliest stratigraphic unit will always be found to be sealed or cut by a chronologically later stratigraphic unit; Each stratigraphic feature will have at least one stratigraphic relationship – later than/earlier than/ same as; Each stratigraphic unit will have multiple physical relationships with other stratigraphic units.

Purpose of recording stratigraphy: Primary route to understanding the activity represented at the archaeological site; Primary route to understanding the relative chronology of the archaeological site; Primary route to understanding the development of the archaeological site; Physical relationships enable the archaeologist to identify possible sources of contamination in recorded features; Physical relationships are useful for identifying the function of individual features; The recording of stratigraphy provides the raw data upon which higher levels of interpretation can be constructed.

Who records stratigraphy: Individual archaeologists will be responsible for identifying stratigraphic features present at the site; Senior archaeologists will be responsible for identifying stratigraphic features present at the site; Geoarchaeological specialists will be responsible for identifying stratigraphic features present at the site.

How to record stratigraphy: When recording stratigraphic sequences one must only record the stratigraphic relationships present; When recording stratigraphic sequences one must only record the physical relationships present; When recording stratigraphic sequences one must record both the physical and stratigraphic relationships present; All stratigraphic units present at the archaeological site must be considered equally when one is discerning the stratigraphic sequence; Only positive stratigraphic units (fills, deposits, layers etc.) are classified as stratigraphic units; Negative stratigraphic units (cuts etc.) will be allocated the status of a feature rather than a distinguishable stratigraphic unit and is considered in this manner when compiling the stratigraphic sequence; Only negative stratigraphic units will be considered when determining the stratigraphic sequence; Stratigraphic sequences should be described from the earliest stratigraphic unit to the latest; Each stratigraphic unit recognised will be examined individually with separate paperwork; Each feature identified will be examined individually with separate paperwork; Each stratigraphic unit identified will be given a unique identification number; A pre-determined feature 'type' designation system will be used; Unique identification numbers will be used for particular features identified; If the stratigraphic sequence is inaccurately recorded the archaeological site will not be able to be reconstructed during post-excavation; The stratigraphic account is likely to be modified in light of other information derived from other sources; Stratigraphic units should be initially grouped by intervention; Phases/context groups should be explained; Floating units must be explained; Any uncertainties/ ambiguities must be stated; Site phasing/concordance of stratigraphic units must be described/justified; When excavation proceeds using arbitrary levels, the stratigraphic accumulation of the site is determined post-facto; During post-excavation spot dates/other dating information can be added to allow for the phase activity of the site to be determined.

Stratigraphic matrices: The organisation uses the stratigraphic matrix system to record any identified stratigraphic units; The organisation creates a 'running stratigraphic matrix' to represent all stratigraphic units present at the archaeological site; The organisation will use spot dates from analysed artefacts and/ or ecofacts to verify the stratigraphic matrix; The manual does not discuss phasing; Once the stratigraphic matrix for the whole site has been completed it will be divided into phases; Once the stratigraphic matrix for the whole site has been completed and phased it will be divided into periods; The periodisation phase of analysing the stratigraphic matrix will indicate periods of deposition; The periodisation phase of analysing the stratigraphic matrix will indicate both periods of deposition and non-deposition; The manual does not discuss the periodisation phase of analysing the stratigraphic matrix.

Stratigraphic units: Negative stratigraphic units are recognised and recorded as features rather than individual stratigraphic units; Positive stratigraphic units are identified and recorded as individual stratigraphic units; Stratigraphic units identified in the field will be allocated a primary class (cut/deposit); Stratigraphic units will be allocated a secondary class (which gives an indication of the context's function); A positive or negative stratigraphic unit represents a single action in the past; A stratigraphic unit is a record of a single action in the past; A stratigraphic unit can be either a positive or a negative feature; Each identified stratigraphic unit is recorded on an individual sheet; A positive stratigraphic unit will be defined by its type (fill/layer/deposit/structure); A feature will first be defined by using a pre-determined feature-type designation system; A stratigraphic unit will first be defined by its category (fill/layer/cut/structure etc.); An individual stratigraphic unit is the smallest unit into which the archaeological record is divided, this will correspond to the minimum separately identified stratigraphic units; A stratigraphic unit will represent the minimum detectable event of human or natural activity located during archaeological excavation; Each positive stratigraphic unit identified will be given equal weight in terms of recognition and description; Each stratigraphic unit (positive or negative) is given equal weight in terms of completeness of recognition and description; Interpretive comments are encouraged; If a stratigraphic unit is sealed by another a suffix is added to the stratigraphic unit's unique identification number; If the stratigraphic unit being excavated is potentially composed of mixed deposits a suffix will be appended to the stratigraphic unit's unique identification number; If artefacts or samples are recovered from a disturbed stratigraphic unit a suffix is appended to the stratigraphic unit's unique identification number; A suffix is appended to the end of stratigraphic units contained within a whole or reconstructable vessel.

Definition of deposits: A positive stratigraphic unit can be a fill - a deposit of material contained within a feature; A positive stratigraphic unit can be a layer - a horizontal deposit of material that is not contained within the edges of a feature; A positive stratigraphic unit can be a structure - a body of material that has been deliberately placed; A structure is defined as a feature rather than an identifiable stratigraphic unit; Positive stratigraphic unit groups will be created in which layers of fills are grouped together (fills of one pit etc.); Positive stratigraphic units present at a site are identified, defined and recorded according to a pre-determined stratigraphic unit 'type' designation system; Each positive stratigraphic unit will be allocated an individual number; Each positive stratigraphic unit identified is recorded individually on separate paperwork.

Definition of cuts: A cut is the result of an activity which has involved the removal of material to form a negative feature; Each identified cut is considered individually and is recorded on a separate sheet; Cuts are allocated the status of a feature rather than a singular stratigraphic unit; Cuts are allocated the status of a sub-feature and are recorded in a sub-feature log if they are associated with a larger feature (posthole of a roundhouse etc.); Feature groups can be formed by collating the cut numbers for a particular feature, the creation of a 'cut group' will automatically link the fills from these cuts into the feature group; Cuts are not recorded as an individual stratigraphic unit.

Section drawings: Section drawings are only used if site circumstances prevent single context planning being used; Running section drawings may be recorded across the site; Section drawings are used to record the long sections of any of the trenches excavated at the site; Section drawings are only used to record the long sections of trenches that reveal evidence of features or significant stratigraphic deposits; Profile drawings are used to record one or more wall profiles of any test excavation unit excavated at the site; Section drawings are used to record any features that have been sectioned during the course of the excavation; Once the section has been recorded it will be photographed; Section drawings will be given a unique identification number; Section drawings will contain the site code and site name; Section drawings will contain the date that the section was recorded; Section drawings will contain a scale; Section drawings will contain the cardinal points; Section drawings will contain datum points; Section drawings will indicate who drew the section; Section drawings will contain the elevations for the section drawn; Section drawings will contain a key; Section drawings will illustrate the line level; Section drawings will illustrate the presence of any artefacts present in the section; Section drawings will illustrate the presence of any disturbances in the section (animal activity, utilities etc.); Section drawings will contain the grid co-ordinates/location of where the section was drawn; Section drawings will contain a brief matrix indicating the relationships between each of the stratigraphic units recorded; Section drawings will illustrate both negative and positive stratigraphic units and will contain annotations that indicate the stratigraphic unit's unique identification number; Section drawings will illustrate each positive stratigraphic unit identified and the edge of its associated feature, it will also contain annotations that indicate each positive stratigraphic unit's unique identification number; Section drawings will include a separate code which will indicate the feature number/type; Section drawings will contain additional annotations describing the stratum's munsell colour, texture, structure, consistence and cementation; Stratigraphic boundaries drawn will be annotated with descriptions of their distinctiveness – abrupt/clear/gradual/diffuse; Stratigraphic boundaries drawn will be annotated with descriptions of their topography – smooth/wavy/irregular/broken; Long section drawings will indicate each sub-feature identified; Long section drawings will indicate each feature identified.

Plans: Each feature identified is recorded in plan; Each positive stratigraphic unit identified is recorded in plan; The standard context recording method is usually used; The single context planning method is used unless site conditions prevent its application; The single context planning method is only used (normally urban sites); The arbitrary unit planning method is usually used; The multicontext planning method is usually used (normally rural sites); A new plan will be drawn before a stratigraphic unit is excavated; A new plan will be drawn when a stratigraphic unit has been partially excavated (cut and top fill); A new plan will be drawn when a stratigraphic unit has been fully excavated (usually cuts); A new plan will be drawn prior to an arbitrary level being excavated; A new plan will be drawn before a positive stratigraphic unit is excavated; Once the plan has been recorded the stratigraphic unit will be photographed; Plans will be given a unique identification number; Plans will contain the site code and site name; Plans will contain the date that the plan was recorded; Plans will contain a scale; Plans will contain the cardinal points; Plans will contain datum points; Plans will indicate who drew it; Plans will contain the elevations for the area drawn; Plans will contain a key; Plans will indicate the location of where samples were retrieved from; Plans will contain a brief description of what the plan shows; Plans will illustrate the presence of any artefacts; Plans will illustrate the presence of any disturbances (animal activity, utilities etc.); Plans will contain the grid co-ordinates/location of where the plan was drawn; Plans will contain a brief matrix indicating the relationships between each of the stratigraphic units recorded; Plans will illustrate each context identified and will contain annotations that indicate each stratigraphic unit's unique identification number; Plans will illustrate each strata identified and will contain annotations that indicate each strata's unique strata identification number; Plans will include a separate code which will indicate the feature number/type; A site plan will be created at the end of the excavation indicating the location of any excavation activity; A site plan will be created at the end of the excavation indicating the location of any recording activity.

Process: Separate teams are responsible for the excavation and recording; Individual archaeologists are responsible for excavating and recording any stratigraphic unit which they themselves have excavated; Vertical control is ensured through the use of a fixed datum point; The site is divided into pre-determined unit types (trench(es), stripping unit(s), feature unit(s), test exploration unit(s), collection unit etc.); The site will be divided into three units – operational unit/sub-operational unit/lots; The site will be divided into sections according to significant archaeological features; A site grid will be established prior to excavation work; Each excavation square is divided into quadrants and excavated individually; Usually, if a large number of archaeological features are present a representative sample will be excavated; Usually, if a small number of archaeological features are present all features will be excavated; All stratigraphic units identified will be excavated; When features are identified they are excavated in isolation; Artefacts found in situ will be recorded by the level in which they were found, grid reference and mapped if arbitrary levels are used; Artefacts found in situ will be recorded by the stratigraphic unit in which they were found, grid reference and will be mapped; Artefacts found out of situ will be recorded by the level in which they were found and grid reference if arbitrary levels are used; Artefacts found out of situ will be recorded by the stratigraphic unit from which they were recovered and grid reference; Non-structural features will be bisected; Cut features will be sectioned and excavated using the Demirant or Quadrant excavation methods; If a feature is sectioned the feature number will have a sub-number added (feature 1.1 etc.); For large features a representative slot will be excavated through the fills of the feature; For curvilinear features several representative slots will be excavated along the length of the feature; For complex or large structures baulks may be maintained throughout the site as a means of recording the stratigraphy; Shovel test pits are excavated via shovel in areas which lack substantial surface evidence; Test excavation units are excavated individually; Structures will first be sampled using test excavation unit(s) or evaluated by excavating a quadrant within the boundaries of the feature; Feature excavation units – if a test excavation unit identifies archaeological features they will be re-designated as a feature excavation unit; Once the archaeological investigation area is uncovered any stratigraphic unit is excavated and recorded in isolation; Excavation will proceed using fixed arbitrary levels; Excavation will proceed using identifiable stratigraphic levels; Excavation will proceed in arbitrary levels until the boundary of the next stratigraphic unit has been identified; Excavation will proceed using fixed arbitrary levels if stratigraphic levels are unable to be identified; Manual rapid recovery (MRR) is a method used for non-feature deposits; Controlled manual excavation (CME) is a method used for precision excavation.

Recording forms: Provenience card; Operation card; Sub-operation card; Log card; Milling station record; Linear feature record; Primary record; Archaeological site record; District archaeological site record; Rock art record; Unit log; A site notebook is maintained by individual archaeologists; Trench log; Stripping log; Feature unit log; Context log; Photographic log; Section log; Profile log; Artefact log; Plan log; Test unit summary form; Sub-feature inventory form; Pit house excavation form; Room excavation summary form; Feature profile form; Unit level excavation form; Pit feature summary form; Surface room summary form; Inhumation excavation form; Cremation excavation form; Sample collection form; Excavation unit summary form; Singular stratigraphic unit recording form; Hearth recording form; Structure recording form; Other feature recording form; Timber recording form; Masonry recording form; Skeleton recording form; Coffin recording form.

Appendix C: Interview questions

Section 1: <u>Participating archaeologist's profile</u>

1- Please select the highest academic qualification you have obtained in the field of archaeology/ anthropology:

Diploma; Higher National Diploma (HND); Bachelor of Arts/Science (BA/BSc); Postgraduate Diploma (PGDip); Master of Arts/Science (MA/MSc); Master of Philosophy (MPhil); Doctor of Philosophy (PhD); Doctor of Science (DSc).

2- In which country did you receive your archaeological training?

3- Please select the archaeological sector(s) in which you currently work. You may select more than one category.

Academic sector; Research sector; Commercial/Cultural resource management sector; National government sector; Regional government sector; Local government sector.

4- Please select the job category in which you are currently working. You may select more than one category.

Academic archaeologist; Research archaeologist; Field archaeologist; Supervising field archaeologist; Senior field archaeologist; Archaeological site director; Archaeological company director; Other (please specify).

5- How many years have you been working within the archaeological industry?

6- In which country do you currently work in the field of archaeology?

7- When conducting archaeological fieldwork, do you, or the organisation with which you are affiliated, follow a set of established archaeological guidelines?

8- Are you, or the organisation with which you are affiliated, required to report the findings of an archaeological investigation to a governing body?

9- Do you, or the organisation with which you are affiliated, conduct archaeological fieldwork on any of the following site types? You may select more than one category.

Urban sites; Rural sites; Pre-contact/Prehistoric sites; Post-contact/Historic sites; Cemetery sites; Underwater sites; Other (please specify).

Section 2: <u>Archaeological excavation methods</u>

10- Do the excavation methods you use vary according to the type of archaeological site you are working on?

11- Do you, or the organisation with which you are affiliated, have an excavation manual?

12- When excavating an archaeological site, do you follow the excavation procedures outlined in your organisation's excavation manual, or do you excavate according to your own methodological preferences?

13- Please rate each of the following factors by the extent to which they influence your selection of an excavation method. 1= Most influence 5= Least influence

	Influence				
Factors	1	2	3	4	5
Literary sources					
Previous archaeological training					
Requirements of the local governing body					
Research aims and objectives					
Field experience					
Communication with other archaeologists					
Site type					
The recording method that will be used					

14a- When excavating a negative feature, which of the following four methods would you choose to use?

Stratigraphic Excavation method; Demirant Excavation method; Quadrant Excavation method; Arbitrary Level Excavation method.

14b- Please provide a summary of how the excavation method you selected is used during the excavation of a negative feature.

14c- Please explain the reason(s) why you chose this excavation method.

Section 3: <u>Archaeological recording techniques</u>

15- Do the recording techniques you use vary according to the type of archaeological site you are working on?

16- Do you, or the organisation with which you are affiliated, have an archaeological recording manual?

17- When recording an archaeological site, do you follow the recording procedures outlined in your organisation's recording manual, or do you record according to your own methodological preferences?

18- Do you, or the organisation with which you are affiliated, use pro-formas when recording archaeological data?

19a- When recording the excavation of a negative feature, which of the following recording techniques would you choose to use? You may select more than one category.

Plans; Sections; Context sheets; Excavation unit forms; Unit level form; Photographs; Sketches; Journal; Other (please specify).

19b- Please explain the reason(s) why you chose these recording techniques.

Appendix D: Grave excavation experiment locations

Experiment Locations

Slaughterbridge, Cornwall.

Langport, Somerset.

Bournemouth, Dorset.

West Dean, West Sussex.

Birchington, Kent.

Raglan, Monmouthshire.

Caerwys, Flintshire.

Nottingham, Nottinghamshire.

Sedgeford, Norfolk.

York, Yorkshire.